The Business of Sustainable Forestry

CASE STUDIES

A PROJECT OF THE SUSTAINABLE FORESTRY WORKING GROUP
THE JOHN D. AND CATHERINE T. MACARTHUR FOUNDATION
CHICAGO

The John D. and Catherine T. MacArthur Foundation
140 South Dearborn Street
Suite 1100
Chicago, Illinois 60603-5285

ISBN 1-55963-615-7

Library of Congress Catalog Card Number: 98-065553

Printed on recycled, acid-free paper

North Shore Printers, Inc.
10 9 8 7 6 5 4 3 2 1

TABLE OF CONTENTS

The Sustainable Forestry Working Group

Individuals from the following institutions participated in the preparation of this publication.

Environmental Advantage, Inc.

Forest Stewardship Council

The John D. and Catherine T. MacArthur Foundation

Management Institute for Environment and Business

Mater Engineering, Ltd.

Oregon State University
Colleges of Business and Forestry

Pennsylvania State University
School of Forest Resources

University of California at Berkeley
College of Natural Resources

University of Michigan
Corporate Environmental Management Program

Weyerhaeuser Company

The World Bank
Environment Department

World Resources Institute

The Business of Sustainable Forestry

PROJECT DIRECTOR:

Michael B. Jenkins
John D. and Catherine T. MacArthur Foundation
Program on Global Security and Sustainability

PROJECT COORDINATION AND SUPPORT:

Greg Lanier
John D. and Catherine T. MacArthur Foundation

PROJECT CONTRIBUTORS:

Matthew Arnold
World Resources Institute,
Management Institute for Environment and Business

Tasso Rezende de Azevedo
Instituto de Manejo e Certificação Florestal e Agrícola

John Begley
Weyerhaeuser Company

Bruce Cabarle
World Resources Institute,
Management Institute for Environment and Business

David S. Cassells
The World Bank, Environment Department

Rachel Crossley
Environmental Advantage, Inc.

Robert Day
World Resources Institute,
Management Institute for Environment and Business

Betty J. Diener
University of Massachusetts, Boston

Richard A. Fletcher
Oregon State University Extension Service

Jamison Ervin
Forest Stewardship Council

Eric Hansen
Oregon State University,
Department of Forest Products

Stuart Hart
University of Michigan,
Corporate Environmental Management Program

Tony Lent
Environmental Advantage, Inc.

Stephen B. Jones
Auburn University

Isak Kruglianskas
Universidade de São Paulo,
Faculdade de Economia, Administração e Contabilidade

Keville Larson
Larson and McGowin, Inc.

Stephen Lawton
Oregon State University, College of Business

Catherine M. Mater
Mater Engineering, Ltd.

Scott M. Mater
Mater Engineering, Ltd.

James McAlexander
Oregon State University, College of Business

Bill McCalpin
John D. and Catherine T. MacArthur Foundation

Mark Miller
Two Trees Forestry

Mark Milstein
University of Michigan,
Corporate Environmental Management Program

Larry A. Nielsen
Pennsylvania State University,
School of Forest Resources

Diana Propper de Callejon
Environmental Advantage, Inc.

John Punches
Oregon State University,
Department of Forest Products

Richard Recker
Oregon State University,
Sustainable Forestry Partnership

A. Scott Reed
Oregon State University, College of Forestry

Jeff Romm
University of California at Berkeley,
College of Natural Resources

Nigel Sizer
World Resource Institute, Biological Resources and
Institutions/Forestry Frontiers Initiative

Michael Skelly
Environmental Advantage, Inc.

Thomas Vandervoort
Vandervoort Public Affairs & Communications

Court Washburn
Hancock Timber Resource Group

Michael P. Washburn
Pennsylvania State University,
School of Forest Resources

Mark Webb
Mark Webb & Co.

Charles A. Webster
Environmental Advantage, Inc.

Peter Zollinger
FUNDES/AVINA Group

COMMUNICATIONS AND SUPPORT:

Ted Hearne
Ted Hearne Associates

Erika Fishman
Ted Hearne Associates

Tess Hartnack
Hartnack Design
Case Study Design

Ray Boyer
John D. and Catherine T. MacArthur Foundation

Emily T. Smith
Environmental Journalist
Case Studies Editor

The Pursuit of Innovation

CASE STUDY

PREPARED BY:
JEFF ROMM

A Case Study from "The Business of Sustainable Forestry"
A Project of The Sustainable Forestry Working Group

The Sustainable Forestry Working Group

Individuals from the following institutions participated in the preparation of this report.

Environmental Advantage, Inc.

Forest Stewardship Council

The John D. and Catherine T. MacArthur Foundation

Management Institute for Environment and Business

Mater Engineering, Ltd.

Oregon State University
Colleges of Business and Forestry

Pennsylvania State University
School of Forest Resources

University of California at Berkeley
College of Natural Resources

University of Michigan
Corporate Environmental Management Program

Weyerhaeuser Company

The World Bank
Environment Department

World Resources Institute

CCC 1-55963-616-5/98/page 1-1 through page 1-9

The Pursuit of Innovation

CASE STUDY

PREPARED BY:
JEFF ROMM

A Case Study from "The Business of Sustainable Forestry"
A Project of The Sustainable Forestry Working Group

Contents

The Pursuit of Innovation

Introduction

Sustainable forestry is a term used so often—and in such seemingly disparate contexts—that its meaning may be confusing or obscure. In fact, sustainable forestry describes a complex social process for advancing the long-term social, economic, and ecological viability of forests in a world dominated by short-term need and normal instability. It involves pursuing the means to use forests without diminishing them, to avoid and replace any losses that may diminish forests, and to ease the economic, environmental, and social pressures that cause forests to deteriorate. The essence of sustainable forestry is innovation and investment in the capacities to take such actions. The practice of sustainable forestry stimulates experiments in policy, organization, and enterprise that promise to expand these capacities. In the dynamic context of the world's forests today, all of the institutions involved in sustainable forestry, whether wood products companies, government agencies, environmental groups, or the public are engaged in adaptive learning to accelerate innovation, increase investments in forests, and develop institutions that favor such innovation and investment.

Business enterprises are making important contributions to the process of sustainable forestry. They are doing so as innovators, as investors, as advocates, and as leaders in institutional reforms that strengthen motives and capacities to sustain forest systems. This paper characterizes the process in which sustainable forestry enterprises are engaged, and the means by which their experience can be used to accelerate the expansion of such initiatives.

New Context for Forests and Forestry

Sustainable forestry has emerged as a widespread aspiration because the contexts and conditions of forests are changing at an unprecedented rate and in ways never before possible. Long viewed as hinterlands valued primarily for meeting the extractive needs of societies, or as preserves of wildness, forests are now mainstream concerns in the United States and throughout most of the world. Increasingly, forests are recognized as pervasive and crucial features of the social landscape that supply fundamental human needs for wood, paper, water, food, jobs, medicines, minerals, and energy; form watersheds, agricultural systems, and reservoirs of genes, species, and ecosystems; and regulate climate. In the process they distribute resources and services among groups, communities, and nations. In this new context, people have come to view forests as critically scarce systems within the bounds of direct human interest rather than as abundant resources beyond those bounds.

The new perceptions of forest scarcity have a biophysical as well as a social basis. Throughout most of the world the frontier is gone. Few places remain inaccessible or protected from the impulses of global modernization. Population and economic growth explain this fact, in part, while images from space satellites, transmitted through televisions in most villages and farms, graphically demonstrate these impacts. But worldwide democratization of political and cultural access has also given local and indigenous communities voice to stake their claims in forests. And the globalization of forest issues, politics, and organizations has diversified and strengthened once-weak or unorganized national and local interests.

Scientific advances have augmented the perceptions of forest scarcity. The ecological sciences are now sophisticated enough to measure and explain the causes and consequences of forest change at the household and farm, at the regional and global scales. They reveal forests as systems that function interdependently as water, climate, and biological infrastructure, as complexes of social rights and responsibilities, and as enduring expressions of the dynamics of human and natural history in which people have always shaped forest conditions everywhere. No longer are forests understood to function simply as large blocks of wood or wilderness "out there:" they are now understood to be patterns of trees that permeate every scale and scene of social life.

The time-honored approaches to forest management, which developed in simpler times, assumed that forests were isolated from broader social forces, and had singular, unambiguous purposes. These approaches are insufficient in an era when forests are recognized as forceful features of the social fabric. Older approaches to forestry cannot cope with the modern intensity, variation, and complexity of human expectations. The new sense of forest scarcity makes the "cut and run to the next hill or to the next country" mentality toward forests as marginal as the opposing "stop-everything" attitude. It replaces these views with the urge to manage forest systems as valuable, diverse, and vulnerable assets.

Today's pervasive conflicts about forest uses are symptoms of the disparities among people about what they want and what forests are actually able to do. For the same reasons, forests have become the focus of innovations that promise to achieve regimes of action consistent with the scale and complexity of forests' new mainstream role. Sustainable forestry, the pursuit of the means to govern, finance, and manage forests as central features of social life, signals that forestry has come of age.

THE SUSTAINABLE FOREST

Forestry is a regime of actions, but the forest is a concrete biophysical system. The sustainable forest is an aggregation of trees that people preserve in a dynamic social and natural environment for the ecological qualities, services, and yields that they want. It requires active effort to prevent unwanted changes that human activities and natural phenomena would otherwise produce. Preservation requires human investment to control forest uses and the social forces that determine them, to conserve and replace necessary ecological capacities, and to develop the organizations, policies, and technologies that forest replacement, maintenance, and enhancement are likely to require. The sustainable forest strikes a dynamic balance among economic, environmental, and social forces that people control to prevent the loss of whatever forest state they prefer.

This definition, while it explains the central role of sufficient investment to ensure forest sustainability, does not explain the ubiquitous conflicts that exist among people who fully agree on the need to sustain forests. Conflicts over forest use arise because people have multiple, different interests in forests. No one forest structure and composition can satisfy all those interests and their infinite combinations. The meaning of "sustainable forest" is inseparable from the interests, capacities, values and contexts of those who define it. A short-rotation monocultural plantation or a wild preserve can be sustainable forests through the eyes and actions of those who champion them. They are not sustainable from the perspectives of those with different values.

Scale Matters

Nor can all the diverse interests in forests be attained on the same spatial scales. Some interests reflect global perspectives, as is the case with stabilizing climate, preserving biodiversity, and sustaining the international wood products industry. Most interests, however, fit along a continuum of possible scales from the farm to a public forest to a region of nation-states. When it comes to sustaining forests, scale matters. A forest that is sustainable at one scale, such as a national forest, will not be so at another. An intensive forest plantation, for example, may relieve pressures to harvest wood from natural forests, thereby increasing their sustainability, even if the plantation itself is unsustainable from the perspective of natural diversity. Restrictions on the harvesting of wood from public forests may shunt consumption pressures to privately held lands, which may reduce the sustainability of the forest region as a whole. The paradox of scale in forest management often scuttles constructive debate on sustainability.

Moreover, the forms and scales of desired forests change readily with shifts in market forces and credit institutions, demographics, social and political trends, developments in science and technology, communications networks, and natural catastrophes. For any specific interest, the sustainable forest in one period will differ from that in the next. If all interests were taken together in their own terms, the sustainable forest would be a swarm of interacting scales, structures, and compositions that change for different reasons over time.

Although sustainable forests are as different as the people who define them, they share common needs. The foremost of these is protection from and adaptation to changes that would otherwise disrupt the desired pattern, whatever that may be. The sustainable forest is a concept of stability in an unstable world. It is a longer-term vision in the midst of short-term forces that shape forests differently than people prefer. Fire, hunger, disease, financial crisis, market booms and busts, discoveries, conflicts, and political necessities, are the normal instabilities that change any forest structure, composition, and spatial pattern if allowed to do so. The sustainable forest is protected from such forces, by excluding or modifying them, and is able to adapt to those forces that penetrate it in a manner that maintains its desired attributes.

Even though most people may agree about the need to sustain forests, few agree about which kind of forest should be sustained, or about what the rates, forms, and sources of forest innovation and investment should be. Ironically, such disagreements too often erode the climate for innovation and investment from which all visions of the forest would gain, and a "tragedy of the commons" arises in the realm of the social capacity to resolve forest interests.

SUSTAINABLE FORESTRY

Forestry is the social process through which people organize the effort to perpetuate a forest's desired attributes. It consists of the regime of actions that shapes the forest's attributes for specific purposes. It is supposed to maintain the desired balances within the forest and between the forest and the external conditions that shape it. Its actions regulate the structure and processes within the forest by strategies of harvest, conservation, enhancement, and exclusion. Its actions also protect the forest from external disruption through systems of ownership, cooperation and coordination, fire protection, and market organization, to name a few. These actions are designed to use public policy, education, and science, for example, to reduce the pressure from sources of disruption. They are accountable to the social purposes that apply in the particular case.

The essential difference between "forestry" and "sustainable forestry" lies in the complexity and scale of relations between people and trees that are now recognized to define forests and to motivate the actions that affect them. Regimes of action that were suitable in an era of abundant frontiers are insufficient in a world in which the population has quadrupled in this century, most people have been freed from the more obvious forms of political and economic tyranny, economies and states are urbanized and integrated on a global scale, and the social and scientific basis for valuing and acting upon forest conditions has been revolutionized. Forests are increasingly becoming diverse overlays of different systems of social interests that interact uniquely in any one place. The actions that predominate are those that affect the pressures in that context and the opportunities that happen to converge in that place.

Pursuit of Innovation

As yet, sustainable forestry does not provide a regime of actions to respond to these emerging circumstances and shifts of priority. Rather, it is the pursuit of innovation in science and technology, the organization of forest ownership and jurisdiction, the development of business, financial, and market institutions, and the formation of political interests and modes of popular engagement that conventional forestry did not address. As a social process, sustainable forestry is characterized by the expansion of ideas, instruments, and applications that diversify experience and the social capacities to learn from it. Sustainable forestry challenges us to appreciate change, to respect differences, to encourage innovative endeavor, and to grow means to capture and use the lessons it will yield.

OPPORTUNITIES

Sustainable forestry has already spawned a variety of innovations that are demonstrating practical worth. These innovations are taking place in several key arenas that open up new opportunities and pose new challenges.

From Silviculture toward Ecoculture

Silviculture, an elegant field of applied ecology, has historically focused on cultivating timber products. In sustainable forestry, however, it encompasses a much wider range of goods and services that forests can provide, and addresses ways to manipulate the entire forest ecosystem to diversify, finance, organize, and sustain its aggregate productive capacities. Foods, spices, medicines, ornamental plants, feeds

and fuels, as well as "waste" tree species and wood are developing as marketable sources of income. Environmental services, such as water yields and water quality protection, recreational opportunity, and wildness and open space, increasingly represent values that can yield economic returns through new forms of exchange and asset valuation. The values of nontimber products from a forest are now competing with those of timber products in an increasingly wide range of conditions. In these circumstances, the purposeful manipulation of forest ecosystems requires a much more diversified scientific program than conventional silviculture has provided. Researchers and practitioners are just beginning to respond to the challenges of "ecoculture."

From Volume toward Quality

Previously, products of the forest typically were considered in terms of volume rather than quality. Market and industrial structures have favored the production of bulk wood products that satisfy a minimum quality and economies of scale associated with highly capital-intensive technologies rather than ways to value and exploit differences in the quality of products. As these strategies have displaced workers, it is not surprising that quality differentiation is related to efforts in sustainable forestry to create knowledge-and labor-intensive job opportunities in forest ecosystem management.

This is occurring through, for example, the development of specialty products and markets, the restoration of stream systems, the regeneration of forests, the reduction of hazards to forest health, and the development of systems of finance and insurance that suit these activities. Although household enterprises are the pervasive ones in the world's forests, the potential organization, scales, and technologies to produce, process, market, and finance a diversified and quality-oriented forest economy are only beginning to be explored.

From Stands toward Landscapes

Until recently, forests were treated primarily as collections of timber stands—discrete homogeneous units of timber and timber potential. Today, they are increasingly considered patterned aggregations of trees in landscapes of functionally interdependent units. The distinction between treating forests as collections of separate timber stands or as unified landscapes of interdependent forest elements takes on qualities of a religious war in some arenas, in large part because of the strong value content embedded in any such classification. The concept of landscape itself contains similar sources of difference because people want those particular landscapes that express the interactions they think are important. The tension between stand and landscape perspectives, and among perspectives of landscapes, has provoked innovations that are accelerating the transition to a landscape view of forests and enriching the concepts and techniques involved. Ecosystem management, watershed agreements, agroforestry, community forestry, riparian forests, urban forestry, carbon forests, and bioregions, merely characterize a broad field of trials that have come to dominate public understanding of what the forest is and can be.

These innovations increase the economic value of all the forest's attributes by establishing quantifiable comparative relations among them. The transition depends upon how well these relative weights among attributes, meaning their values, can be transformed into financial returns that will sustain reinvestment in the forest. The growth of knowledge, organizations that dilute institutional barriers, and proven techniques are critical factors in the transition. Advances in information science, environmental survey, simulation and monitoring, and geographic information systems have brought these possibilities within view, but not yet into normal practice.

From Ownerships toward Councils and Communities

Forest management has typically been carried out within the lines, authorities, and interests of specific forest ownerships and jurisdictions. Sustainable forestry diversifies the concepts, patterns, and scales of forests to such a degree that no unilateral form of control can encompass them. Indeed, many conflicts in forestry persist because of the absence of mediating processes where formal controls and de facto interests have grown increasingly inconsistent with one another. The response of sustainable forestry is to diversify the sources of authority. Including local communities in forest management is particularly significant in disputes that have previously been waged, and stalemated, primarily among national interest groups. Although local communities depend upon the forest and possess human resources critical for its viability, they have long been disenfranchised by the characteristic patterns of absentee control of forests, both public and private. The legitimization of their interests is helping to build a "third force" means to resolve conflicts and to increase the resiliency of dispute resolution generally.

Engaging communities in forest management is a central force in sustainable forestry. It usually involves ratifying local entitlements to share in forest governance and creating motives for local investments in forest stewardship. Experimentation in types of partnerships, cooperative management, community management, enterprise and job creation, and the like are increasing. The rise of multijurisdictional approaches to watershed management, and of community-centered systems of forestry are but two examples of these new approaches. They are part of a broad search for the mixes of entitlement that will mesh the interests and capacities necessary to achieve forests that, even from diverse points of view, are sustainably productive. To sustain such processes, knowledge of forest ecosystems must be sufficiently predictive, and the benefits to communities sufficiently large and equitable, to justify the effort involved.

From the Forest as Product toward the Forest as Capital

From an economic perspective, sustainable forestry is intended to enhance forest assets, to essentially increase nature's capacities for the growth of economic value. But few economic systems yet have the capacity to capitalize the full benefits that forests provide, to demonstrate the true capital gains that investments in forests yield, to distribute the returns to those who make the effort to create them, and to thereby motivate the levels and combinations of investment that forest benefits justify.

Despite the obvious economic contributions of forests—in water storage and flow, to energy supply and conservation, as infrastructure that is exceedingly expensive to replace, to local subsistence and enterprise, as a source of foreign exchange through sale of materials and tourism, in housing construction and landscape organization of settlements—a small fraction of these contributions is assessed to forest value. And a much smaller fraction is invested to sustain them. Those who benefit from those contributions typically have little obligation and no mechanism to invest in the source of the benefits. The problem of assessment is even more severe for less obvious contributions—wildlife, climate, open space, biodiversity—or potential future contributions, such as specialty foods, medicines, and woods.

Sustainable forestry is stimulating efforts to close the loop between those who benefit from the forest and the forest sources of those benefits. Efforts to incorporate natural assets in national capital accounts are strengthening the capacity to treat forest investment as an aspect of national economic strategy. Institutional reforms are breaking down tenurial and jurisdictional barriers to exchanges between sources and recipients of forest benefits, thus increasing opportunities for return flows of investment from forest beneficiaries. The concepts underlying such efforts are emerging in public discourse to complement and enrich longstanding debates about preferred forest conditions.

From Current Income toward Natural Capital and Green Finance

Forests are exceedingly vulnerable to the force of financial markets. Landowners typically accrue large debts to purchase timberlands and develop processing facilities, then convert timber stocks in accordance with the pressures to repay their loans. Governments exploit forests to transform "free" natural capital into money rather then run up debts to expand their infrastructure or industrial base. Throughout the world, forest composition and structures are historic expressions of financial forces rather than ecological and silvicultural judgment. A critical aspect of sustainable forestry is the pursuit of means to regulate relations between the dynamics of forest ecosystems and of the financial markets that dictate patterns and rates of forest exploitation.

Innovations are occurring on a number of fronts. New institutions are translating ecological and financial values into comparable terms that permit monetary payment for retained ecological attributes. Some institutions, such as ecosystem management organizations, create arenas of exchange that force ecological, social and financial interests to come to terms with one another. Others, such as differentiated markets for forest qualities and attributes, strengthen financial incentives for forest preservation or enhancement. Still others, including "green" brokerage and mutual funds, debt-for-equity swaps, and forest trusts, are reducing the costs of holding and buying forest assets for long-term conservation interests. As knowledge of long-term forest values improves, enterprises oriented toward asset growth, such as insurance companies and pension funds, are increasing their shares of forest ownership, a trend that is countering the tendencies of financial markets to value forest assets primarily for current income. Such changes are expanding the investment in natural capital.

From Blind Consumption toward Consumer Awareness

Forest products embody "gifts of nature" this generation did not create and as yet does not replace. Forest nutrients, moisture, and ecological structures and functions consumed in the production of forest goods are treated as "free." In a competitive economy their unpaid costs are excluded from the prices people pay for the goods that forests produce. Consumers are rarely aware of the "costs to nature" of the depletion of this production capacity, which their purchases embody. Sustainable forestry includes pursuing the means to close this loop.

The increase in forest certification efforts in the past several years represents an effort to inform consumers about the conditions of the specific forest operations from which the products they buy are derived. These efforts distinguish products in the market by the sustainability of their source operations. Some of these efforts are made by neutral bodies that certify forestry operations as sustainable only if they meet stringent standards. Certification entitles producers to label their products as certified, which distinguishes them from others that come from operations that are not validated as replacing the gifts from nature that they use. Certification enables consumers to vote for forest management in the marketplace. In the past several years a number of wholesale and retail outlets have begun to carry certified products and advertise the differences in content among forest goods that may otherwise appear to be identical.

CHALLENGES

Sustainable forestry is proving to be a catalyst for innovations in forest management. But it is a pursuit, not a prescription. Many unresolved issues will demand new ideas in the future. Several are indicative of the challenges that remain in transforming sustainable forestry from concept to reality.

Unified Concept vs. Reality of Conflicting Practices

Although unified within the process of sustainable forestry, the emerging visions of the sustainable forest will prove to be competitive or inconsistent with one another when put to practical test in the same place. Economic, community, watershed, and biodiversity visions of the sustainable forest, for example, are unlikely to shape the same sets of choices and actions, or to yield the same ecosystem structures and functions. Community visions include sustainable and growing forest employment opportunities while biodiversity visions tend to limit or exclude human activity. The process used to resolve these diverse views will determine how effective sustainable forestry will be as a practical pursuit. As yet, these processes have not faced a serious test.

One-Scale Visions vs. Many-Scale Realities

Sustainable timbersheds, watersheds, community forests, habitat ranges, biodiversity regions and metropolitan forests exemplify the differences of scale at which different sustainable forests function. A forest that is sustainable when viewed at one scale is not necessarily sustainable when viewed at another. The diversification of forest concepts involves equivalent diversification of scales at which forests are managed. Relations among actions at these various scales become fundamental to the sustainability of the parts and the whole.

Because relations among scales are poorly recognized and therefore often counterintuitive, they tend to seed conflict rather than constructive strategy. An intensively managed monoculture plantation, for example, can be sustainable on a small scale, but unsustainable for biodiversity at a regional scale. Yet it may satisfy sufficient economic need at the smaller scale to reduce production pressures on larger-scale forests that then can be managed more easily for biodiversity and other natural attributes. A diverse, regional mosaic of intensive, small-scale monocultures can be more sustainable than a natural forest at the same scale, even if many of the monocultures are not individually managed in a sustainable way. Moreover, a forest that seems sustainable on any scale may not be sustainable on the global scale of financial forces that will determine its long-term viability.

Forest resilience and sustainability on a global scale require a reasonable complementarity among actions at different forest scales. Achieving that, however, requires a readiness among different interests to withdraw from absolute forest prescriptions that may suit one scale or circumstance, but that adversely affect the whole if applied generally. Better explanation of the relationships among scales should facilitate the process by demonstrating the potential complementarities at current edges of conflict.

General Government Authorities vs. Flexible Forest Institutions

The pervasive spontaneous experimentation fueled by sustainable forestry is rapidly transforming forestry from a rigidly controlled to a remarkably inventive field of action. If the past is a guide, such creative flexibility will eventually run into the more general and less resilient authorities of formal governments al all levels. Governments have broader concerns than just forest conditions and have strong needs to apply uniform rules. These tendencies tend to confine institutional innovations within limits that are relatively insensitive to the requirements of the problem and the motives for its resolution. Confinement may be avoidable to the extent that the configurations of governmental policy begin to integrate forest conditions as mainstream rather than specialized and marginal public concerns. In other words, the framework that governs all public actions would change in ways that strengthen the context of sustainable forestry.

SUSTAINABLE FORESTRY AS ADAPTIVE LEARNING

Forestry has existed as a marginal, specialized, and narrowly vested interest segment of the environment, economy, and society for too long and at great cost. Sustainable management of timber could be accomplished within conventional ownerships and jurisdictions. Sustainable management of forest ecosystems cannot. The knowledge, techniques, and institutions it requires cannot exist without general public interest that has expanded the systems and scales of the forest, its governance and its finance. The public increasingly weighs forest choices against the impact those choices may have, and it is increasingly aware of how the broader forces of society affect the forest. This recognition started the processes of innovation, investment, and institutional reform that constitute sustainable forestry.

Are these processes necessarily spontaneous pursuits, or can method accelerate their progress? The answer is "yes" to both questions. In complex and uncertain situations, nothing substitutes for creativity, initiative, and enterprise. Imaginative action has social virtues that reflex caution can never claim. But method can help to show why different endeavors are relatively more and less successful, how they are likely to benefit from various changes in approach and context, and what changes are likely to be feasible and effective.

Methodical approaches to the business of sustainable forestry might begin with fundamental requirements for long-term viability. Sustainable forestry businesses must be sufficiently profitable to sustain the necessary levels of investment, sufficiently suitable ecologically to avoid depletion of nature, sufficiently responsive socially to avoid human harm and conflict, and sufficiently dynamic to learn rapidly from experience over time. As the operational conditions under which any sustainable forestry enterprise exists are relatively more and less forgiving within the envelope of these thresholds of sufficiency, a viable strategy positions sustainable forestry activities to maintain an effective balance within their contextual envelope, utilizing its opportunities and managing its risks. Since operational conditions can be characterized, different strategies can be assessed or defined for different ranges of conditions, including those that do not offer reasonable prospects.

Assessing Strategies

Such contextual conditions might be assessed along three planes. The first, the horizontal, includes relations over space that affect the advantages and disadvantages—financial, ecological, social—of conducting activities in a particular place. An isolated enterprise, for instance, may face labor scarcities, but be relatively free of social pressures. Or it may be in a natural environment that is relatively fragile or resilient, and may enjoy broad managerial discretion, but limited scope, for its use. A metropolitan business will face a different configuration of issues, and will have better opportunities for cooperation with others who confront similar disadvantages.

The second plane, the vertical, includes the relationships upon which a business depends. Sustainable forestry enterprises may be particularly affected by: 1) the quality, resilience, and security of supply and sales chains; 2) the strength, suitability, and cost of access to finance; and 3) links to responsive sources of science and technology. The type, scale, and history of an enterprise would affect the relative influence that these conditions have on its viability and, thus, the strategy that it may need to adopt to compensate for the risks it confronts.

The third plane of context, policy, includes explicit influences such as interest and inflation rates, trade and investment mechanisms, taxation, regulation, subsidies, and the distribution of functions among central, provincial, and local governments. It also includes implicit expectations regarding how the enterprise operates. These expectations are difficult to characterize—their significance may not be apparent until problems arise—but they can be sufficiently powerful to dominate long-term considerations. They include cultural and political notions of corporate responsibility and the plausible streams of future liabilities and opportunities that they entail.

Conclusion

Sustainable forestry is a process of adaptive learning that depends upon spontaneous innovation, investment, and institutional reform. It needs enterprise to progress, organizations that reward it, and organizations that embrace creativity, critique, and change as sources of success. It is a process of learning and adapting for entrepreneurial organizations that are attuned to their circumstances and committed to the long-term. The business cases in this project present such organizations. They provide examples of innovation and leadership, and collectively demonstrate the value of methodical efforts to capture and explain the experience and to improve the approaches and conditions that shape it. The sustainability of forests depends, in large measure, upon how well these activities serve one another over time.

Sustainable Forestry within an Industry Context

CASE STUDY

PREPARED BY:
DIANA PROPPER DE CALLEJON
TONY LENT
MICHAEL SKELLY
CHARLES A. WEBSTER

A Case Study from "The Business of Sustainable Forestry"
A Project of The Sustainable Forestry Working Group

The Sustainable Forestry Working Group

Individuals from the following institutions participated in the preparation of this report.

Environmental Advantage, Inc.

Forest Stewardship Council

The John D. and Catherine T. MacArthur Foundation

Management Institute for Environment and Business

Mater Engineering, Ltd.

Oregon State University
Colleges of Business and Forestry

Pennsylvania State University
School of Forest Resources

University of California at Berkeley
College of Natural Resources

University of Michigan
Corporate Environmental Management Program

Weyerhaeuser Company

The World Bank
Environment Department

World Resources Institute

CCC 1-55963-617-3/98/page 2-1 through page 2-39

Sustainable Forestry
within an Industry
Context

CASE STUDY

PREPARED BY:
DIANA PROPPER DE CALLEJON
TONY LENT
MICHAEL SKELLY
CHARLES A. WEBSTER

A Case Study from "The Business of Sustainable Forestry"
A Project of The Sustainable Forestry Working Group

Contents

(continued)

Contents

Sustainable Forestry within an Industry Context

Executive Summary

The forest products industry ranks as one of the world's most important industries—for the global economy and the environment. It represents close to 3% of the world's gross economic output. The forests upon which it depends are among the most critical ecosystems for the health of the planet and for human well-being. The size of the industry, its links to the rest of the world economy, and the importance of its resource base for environmental services make it the target of intense public scrutiny and government regulation. Understanding sustainable forestry requires understanding the evolving dynamics of the forest products industry—an evolution that is increasingly making the cost of wood a smaller fraction of the final value of a forest product.

Two frameworks are used here as prisms through which to view the industry. The first section describes how the major business and environmental trends sweeping the industry are transforming Sustainable Forest Management (SFM) into a major industry force. It then outlines the most critical nonenvironmental drivers that make or break all businesses within the industry, and explains how they will influence sustainability issues. The second section describes how all these forces play out within each of the three major industry segments: paper, solid wood, and engineered wood products, and maps out in which parts of the industry sustainable forestry is already a major issue, where it is not, and why.

This approach makes sense given the history of SFM. Most sustainable forestry businesses have started from the forest, then tried to move forward to the market. An analysis that assesses the industry and links market conditions back to sustainable forestry supply capabilities reveals where sustainable forestry is well integrated, where it may not have much current opportunity, and where opportunity for closer end-market integration remains untapped.

The forces transforming the industry include: tightening supplies, a shift in production regions, global-ization, increased raw material efficiency, intensified product consistency, and heightened government regulation. Just as these forces are affected by environmental pressures, they also have environmental impacts of their own.

As population growth and burgeoning economies spur the consumption of forest products, wood supplies are tightening worldwide. While no crisis is imminent, the industry is turning to new regions, especially South America and South Asia, as a source for wood. It is also gradually shifting from a supply based largely on natural forests to one that depends on plantations, many located in the southern hemisphere. Just when environmental restrictions are curtailing wood production in many northern countries, heightened demand elsewhere is causing the industry to expand into delicate ecosystems in the Southern Hemisphere. Meanwhile, the industry is becoming increasingly globalized, with raw materials sourced throughout the world to create products for equally diverse markets.

Shifts in producing regions and globalization are creating new opportunities for value-added industries in the southern hemisphere. Primary and secondary processing industries will follow wood supplies for financial reasons, as timber producing nations try to capture a larger share of the production from forest products. These changes will draw significant investment to the Southern Hemisphere.

Globalization brings improvements in communications, shipping, and distribution that facilitate the transfer of knowledge about state-of-the-art forest management techniques. These same developments make the emergence of an international trade in certified forest products possible. As capital travels to formerly untapped forest reserves, for example those in eastern Russia, the forces unleashed by globalization will exert even greater pressures on forests worldwide in the next twenty years.

Evermore efficient raw material use and increasing product standardization are also contributing to the industry's transformation. Over the past several decades, the industry has created many technological silver bullets that enable it to create more product from less wood.

The industry-wide drive for standardization and consistency is moving down the value chain from final consumer products through to the forest. Instead of emphasizing efforts to use individual species such as oak and cherry, resources are now allocated to figure out how to make a vanilla feedstock such as rubber wood look and perform like oak or cherry. Eventually, this trend will lead to more investment in processing assets that can guarantee consistency, and a movement toward either tree plantations or homogenization during primary and secondary processing.

Environmental forces have flexed their political and market muscles, placing the forest products industry under intensifying public scrutiny and government regulation of its environmental performance. New regulations and market initiatives are curtailing access to government controlled forest resources, and influencing the management of private forests. While a number of international agreements designed to improve forest practices might eventually affect the industry, few now have the teeth to do so.

In the past five years "certification" has emerged as a nongovernmental initiative that may further transform the way the industry manages its forests. Certified forest products are defining the market for wood products grown in an environmentally sound fashion. While the full impact of certification is still unknown, if it focuses the concerns of consumers and purchasers on the quality of the forest from which a product is harvested, and if certification is widely adopted, it could dramatically improve forest management and change markets.

How the business and environmental forces affect the paper, panels, and sawnwood segments of the industry will determine, in large measure, the future of sustainable forest products. The paper industry, with its massive capital investments, huge pollution abatement costs, extreme business cycles, and susceptibility to buyer power, has long been beleaguered. The paper industry's recent shift to greater use of recycled paper demonstrates both its vulnerability to outside pressures and its ability to adapt rapidly to a new business environment.

Panels and engineered wood products may be a model for the future. Products in this segment, capitalizing on rapid-fire technological advances, are among the fastest growing in the industry. From an environmental perspective, these products' ability to use a variety of woods now makes them more attractive than plywood, the once dominant panel product. On the other hand, certified panel products will be much tougher to bring to market because it is so difficult to ensure that all the woods used in them come from sustainably managed forests.

Sawnwood products draw most of the attention from the certification community. The sawnwood segment is more fragmented, less capital intensive and adds relatively less value to its products than paper or panels. Sawnwood companies in temperate regions that produce hardwood will have opportunities to sell to markets opened up by a new resistance to tropical hardwoods. The forest management practices of softwood producers, however, are under heavy scrutiny, and they will find fewer opportunities to leverage superior forest management. Although tropical countries are under enormous international pressure to improve their forest management practices, most of the internal and Pacific Rim markets they serve, so far, remain relatively uninterested in the environmental qualities of forest products. Niche opportunities, though, are available in Europe to tropical producers that can produce certified forest products.

In the future, the successful forest products company will understand and embrace the forces that are transforming the industry. Environmental trends are at the leading edge of these changes, and will be instrumental in determining the industry's winners and losers. Companies that understand the role of the environment will profit by doing so: Those that underestimate the force of environmental issues will do so at their peril.

Introduction

The forest products industry represents the best and the worst of today's global economy. It produces a diverse assortment of goods to match an ever-expanding array of wants and needs. Through technological advances, these products are created ever-more efficiently and reach consumers around the world through increasingly specialized and efficient marketing and distribution channels. The industry harnesses capital, technological, and human resources on a scale matched by few other human endeavors. It makes written communications cheaply available to citizens throughout the world, helps provide housing for hundreds of millions, and contributes in countless other ways to human well-being.

An Industry Beset by Crises

On the other hand, the forest products industry is beset by crises. The natural resource base it depends on is being rapidly consumed by the world's hunger for more forest products. Forests are being depleted by millions of the world's poor who use fuelwood for cooking and heating. Population and economic development are boosting the numbers of consumers of forest products worldwide, so that the consumption of roundwood soared from 1.5 billion cubic meters in 1950 to 3.65 billion cubic meters in 1995, according to a study by Pira International. Fortunately, the forest products industry has developed technologies that enable it to create more products using less wood. Yet, many of these processes generate harmful pollutants. To prevent them from doing environmental damage,

the industry is forced to invest huge amounts of capital. Economic development, which has spurred such dramatic growth in the forest products industry, has also led people to value the forest in new ways. Formerly prized mainly for their timber production, forests are now coveted for the biodiversity they foster and shelter, for the scenic beauty they provide, and for the water resources they harbor and protect. These new claimants to forest values blame the destruction of the forests squarely on the industry. At the very time when the industry is faced with dwindling resources, powerful environmental forces are claiming those resources. Ironically, an industry that once prided itself upon its use of a renewable resource, is now under siege from environmentalists, groups it might once have considered its natural allies.

Nor has the forest products industry escaped the monumental currents that are buffeting other industries. Globalization, shifts in production locations to optimize costs, worldwide capital mobilization, rapid technological advances, and a constant pressure to innovate are all reshaping the forest products industry. Environmental forces will influence all of the restructuring underway. These forces are accelerating technology development, creating new market niches, shifting wood availability, and driving new capital investments. In short, environmental forces are propelling an industry already in a hurry to keep pace with a changing world.

Focus on Intersection of Structural Change and Environmental Forces

The intersection between structural change in the forest products industry and the environmental forces that are at once the causes and effects of those changes is the focal point of this study. It will describe the broad economic and market forces that are acting on the industry, then analyze how environmental trends are affecting the principal product segments. Risks and opportunities in the nascent "sustainable" and certified forest products

A Word on Forest Management

Forest management has changed dramatically over time and across countries.

- Initially, societies considered forests nearly unlimited resources. They were often even viewed as obstacles to the development of more profitable activities, such as agriculture. Wood prices were determined chiefly by extraction and transportation costs.

- Around the mid- 20th century, an "industrial paradigm" of forest management held sway. Under this system natural forests were managed to yield as much timber as possible. Pests, undergrowth, and fires were tightly controlled.

- Today, two new models are becoming influential. In the developed world, forest management is slowly adopting an approach that tries to maximize a range of values, including timber production, wildlife habitat, biodiversity, and scenic beauty. Under ecosystem management, most prevalent in temperate Northern Hemisphere countries, forests are managed as ecosystems with wood being just one of many desirable products from the system. Wood yields are only optimized to the extent that other "goods" produced by the forest are not diminished. The U.S. Forest Service has officially adopted this system as have several other industry players. It may well represent the future for the management of "natural" forests.

- In other temperate areas and throughout the Southern Hemisphere, however, the dominant forestry thinking holds that single-species plantations are the most efficient way to produce wood fiber. Genetics, maximum yield management, and rigorous integration with downstream processing facilities are the cornerstones.

industry created by the intersection of industry evolution and environmental forces over the next twenty years will be identified and analyzed.

The approach is designed to help sustainable forestry enterprises understand the changing industry dynamics and identify market opportunities. Most sustainable forestry efforts have begun at the forest by improving management practices. Only later have they analyzed the market dynamics of the products grown in a well-managed forest.

Unfortunately, there are a number of "stranded" sustainable forest enterprises—companies with excellent forest management practices but with products out of line with the market. Most sustainable forestry products are solid wood products, but they represent only a fraction of the overall industry. If sustainability is to enter the mainstream, and if certified forest products companies are to successfully continue to operate in niche markets, they and new entrants in the field need to fine-tune their strategies to accommodate the evolving industry, adapt to shifts in the niche markets, and react to the inevitable threats and opportunities these changes create. In the dynamic and competitive forest products industry, merely being a certified product entrepreneur is no longer sufficient.

By the same token, understanding the forces that are reshaping the industry is critical to preserving the health of forests. Fuelwood consumption, expansion of agriculture, and other issues are important factors in determining the long-term health of the world's forests. But the forest products industry remains the single most influential economic activity that will determine how forests are managed, whether SFM is widely implemented, and whether the health of the world's forests will be preserved.

The Forest and the Industry

The forest products industry is one of the most economically important in the world. Accounting for nearly three percent of worldwide GDP, the industry is the mainstay of key economic sectors, such as construction, publishing, and furniture on the output side, and chemical and machinery on the input side. The economies of numerous developed and developing countries, Finland and Indonesia respectively, for instance, depend heavily on the forest products industry. The forests of some developing countries hold potential not only to meet their own fiber needs, but also to earn foreign exchange, and to develop a platform for further economic growth. For the foreseeable future the forest products industry will remain the "anchor industry" in the forest because its monetized value is so much larger than that of any other industry, such as pharmaceuticals.

Forest products can be divided into two basic categories, softwoods, which come from coniferous forests, and hardwoods, which are harvested from nonconiferous trees. Both types of trees feed the principal forest products industries of pulp and paper, panels, and sawnwood.

A Snapshot of Global Fiber Supply and Demand - 1995

	Millions of Cubic Meters	Percent
Fuelwood Demand	1,971	54%
Industrial Roundwood	1,680	46%
Lumber and Sawnwood Products	949	26%
Plywood, Panels, Indus. Wood Products	329	9%
Pulp and Paper Products	402	11%
Total Demand	**3,651**	**100%**

Source: Robert W. Hagler, 1995

Table 1

Table 1 shows the use of fuelwood for cooking and heating, which accounts for the majority of the world roundwood harvest. The second leading category is lumber and sawnwood products, followed by pulp and paper products. In addition to the 402 million cubic meters of roundwood destined directly for the pulp and paper industries, an additional 205 million cubic meters of residuals, or scrap, flow from other wood processing industries into the pulp and paper industry. Panels of all types consume 9% of industrial roundwood.

Major Drivers of Industry Transformation

A number of forces are placing consistent, long-term pressure on the forests products industry. Each affects various segments of the industry differently, and will present significant opportunities and threats to the development of sustainable forestry business opportunities. This section addresses seven major trends: supply/demand, plantations development, globalization of trade, standardization of products, efficiency increases, demand for environmental sustainability, and increasing government regulations.

FOREST SUPPLIES TIGHTENING

The forest products industry continuously gains greater efficiencies in fiber production and fiber processing. Even so, world fiber supplies are likely to become tighter over the next twenty years. Population growth and greater economic development caused worldwide consumption of industrial roundwood to rise at a rate of approximately 1.3% per year between 1983-1993, according to industry supply expert Robert W. Hagler. This relatively modest growth rate, however, cloaks wide regional variations. Most developed countries had little or no growth in their consumption of forest products; but countries of the Pacific Rim have experienced huge spurts in demand. Furthermore, the growth in demand for individual sectors of the industry generally outstripped increases in roundwood consumption. The ability of the industry to more efficiently convert roundwood into forest products explains the difference.

Supply Side Pressures Reducing Forest Availability

A number of supply variables, most closely related to environmental pressures, are also constraining the availability of wood fiber. Some regions' forests, including those in West Africa and some countries in Latin America, have almost reached commercial extinction. But other factors are also important:

- Withdrawal of some forests from production for environmental reasons to create national parks or to protect certain species. Thousands of acres, for instance, were taken out of production in the Pacific Northwest to protect the spotted owl.

- Historical overcutting of forests, such as that taking place in Indonesia.

- Lack of investments to increase productivity, including required reforestation, which was the case until recently in British Columbia.

- Lack of infrastructure to cost-effectively harvest and transport timber, which has occurred in the Russian Far East.

No world wood crisis is imminent. But as wood supplies gradually tighten over the next 20 years, local, and even regional, supply shortfalls may become common. For primary and secondary processing facilities, which are generally located close to forests, supply issues will become critical when area forests are exhausted or are taken out of production and sources of wood supplies shift to other areas. Logistics and transportation have improved dramatically in recent years, which will help enable processors to buy wood from a wide variety of sources. Capital is more mobile, as well, so processing assets will continue to follow wood supplies.

These shifts will define many of the opportunities in the medium and long term.

Industry Will Respond to Tighter Supplies

Few industry analysts are willing to commit themselves to hard predictions about future supply and demand for wood. Were it simply a question of projecting inelastic demand and a steady amount of supply, assumptions would be more readily available.

Historically, when faced with supply challenges, the forest products industry breeds new trees, develops new technologies, substitutes products, takes advantage of formerly "weed" or lesser known species, and in short overcomes fiber-related costs and supply problems. On the other hand, recent experience has demonstrated that as economies develop and individual incomes rise, most societies come to value their forest resources for services other than wood production. Newly enriched middle classes in some parts of the world may demand the curtailing of harvests. Moreover, analysts disagree about the future of logging in Russia, rates of afforestation in South America and Southeast Asia, and demand responses to price increases for forest products, which are critical variables in the supply and demand for wood. For these reasons, any projections of future supply need to be reviewed with an appropriate dose of skepticism. Although a full analysis of worldwide fiber supply and demand issues is beyond the scope of this discussion, a variety of views on world fiber supplies are available in analysis done by Sedjo, Reed, FAO, Jaakko Pöyry, and the U.N. Commission on Sustainable Development, among others.

Supply and Demand for Softwood

Softwood, or coniferous forests, is found largely in the temperate regions of the Northern Hemisphere. Most softwood is produced in Canada, Scandinavia, the United States, and Russia, and consumed mainly for construction lumber and long-fiber paper such as newsprint. Softwood production is typically double that of hardwood. Hardwood, or deciduous forests, grow throughout the tropics and the world's temperate regions, and are traditionally used for solid wood, plywood, and paper.

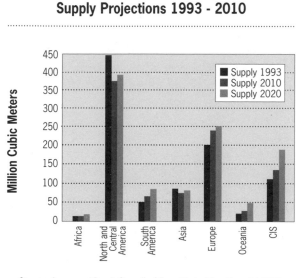

Softwood Industrial Roundwood Supply Projections 1993 - 2010

Source: Apsey and Reed, Council of Forest Industries (Canada) 1995

Graph 1

Total softwood production is estimated to increase from 939 million cubic meters to a total of 1,085 million cubic meters, less than 15% total growth, over a 25-year period. Demand for softwood is concentrated in the Northern Hemisphere. If one makes the rather crude estimate of maximum growth in demand of 1.5% per year, a gap of some 315 million cubic meters would develop by the year 2020. The system clearly has little slack to absorb consumption increases, and the forest products industry will have to respond to fill the gap. The responses will help determine the opportunities for sustainable forestry in the future.

Environmental Forces Will Restrict Softwood Supplies

Several regional developments, many of them environmental, will hinder the expansion of softwood supplies over the next twenty years:

- Russia holds half of the world's coniferous forests, yet production is stagnant and well below cutting potential. Although production is projected to increase slowly, significant infrastructure problems, an uncertain investment and political climate, and corruption are expected to continue to hamper development of these huge resources. In 1994, Weyerhaeuser, for instance, canceled plans to invest in a joint venture in Siberia with Koppensky Kombinant, a Russian timber and forest products group, citing corruption and political instability. According to the industry magazine *Wood Technology,* Weyerhaeuser also cited opposition from U.S. environmentalists as a reason for its pullout.

- Continued environmental restrictions on logging in government-owned coniferous forests are expected to limit North American production. The southern United States will increase production marginally, but not enough to offset the drop in sales from U.S. public lands.

- Production will drop in Canada as provincial governments implement increasingly strict timber harvest controls in response to environmental concerns.

- Softwood production in South America and Oceania will increase as extensive plantations come on line over the next twenty-five years. Given the short rotation cycle for these plantations (as little as fifteen years), over the long term, supplies from these areas will be relatively more elastic than elsewhere in the world.

- Europe's harvests are expected to remain below the annual incremental growth, but production will increase nonetheless. Europe will continue to import much of the fiber it needs.

Hardwood Supplies Tightening

All trends point to a period of much tighter hardwood supplies, despite large stocks of hardwood in tropical and temperate regions. As indicated in Graph 2, total production of hardwood industrial roundwood is expected to grow from 565 million cubic meters in 1993 to 599 million in 2020, and most production will continue to be concentrated in the United States. Clearly, the availability of hardwood fiber will have a major effect on the forest products industry over the long term. Most tropical forests are hardwood, and the issues of rain forest conservation and the availability of tropical hardwoods are closely linked.

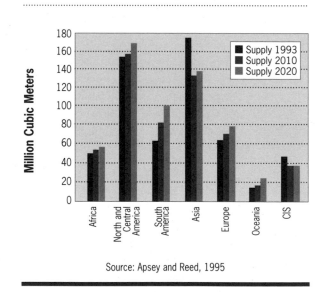

Hardwood Industrial Roundwood Supply Projections 1993 - 2020

Source: Apsey and Reed, 1995

Graph 2

The following developments will be instrumental in limiting the growth of hardwood supplies:

- Malaysia and Indonesia, Asia's top hardwood producers, rely almost exclusively upon their natural forests. Both governments are expected to slow down the liquidation of those forests over the next decade. Production out of Indonesia, Asia's second largest producer, is expected to drop significantly as natural forests are further harvested. Both countries are attempting to develop hardwood plantations that should produce harvests over the next twenty years, although the quality of statistics about the total area of these plantations is open to question. But neither country has planted plantations in most of the previously logged areas. Those areas will continue to supply local paper mills with mixed second growth, which produces low-quality pulp.

- North America has large hardwood reserves, most of them privately held, so production of U.S. hardwoods is less dependent than softwoods on government lands. Despite this insulation from direct government control, increased regulations of wetlands, biodiversity guidelines, and other government-mandated management practices will prevent hefty increases in supply from the private lands that produce most U.S. hardwood.

- Tree farms in South America and Asia, particularly intensely managed eucalyptus plantations in Brazil and acacia in Indonesia, will account for much of the growth in hardwood supplies. There is some debate as to the quality of these plantations, however. Reed and Apsey's estimates acknowledge the generally poor quality of the information on these plantations and heavily discount total planted areas and their expected performance.

- The Amazon basin, most analysts agree, will not be tapped for significant hardwood production or trade.

Under these circumstances, meaningful new sources of hardwood are unlikely to become available in the near future, which will place far greater demands on existing resources.

Asia Pacific Supply Situation Will Turn Negative in Near Future

Changing timber supply and demand will alter the competitive dynamics of forest products manufacturers around the world. Asia Pacific's growing population, rapid economic growth, and depletion of natural forests will compound the looming fiber deficit in the region. Japan continues to be an enormous consumer of forest products, and is by far the heaviest importer from other parts of Asia. According to a major study by Jaakko Pöyry commissioned by the American Forest and Paper Association, over the next twenty years Asia Pacific will reverse its status from being a major exporter of wood to importing much of the wood it needs, with Japan absorbing the bulk of the imports (see Graph 3). Major growth opportunities for wood producing countries will develop in this region. Despite the large volumes of plantation fiber expected from the southern Pacific, these plantations will only supplement fiber growth needs, and trade flows from North and South America are likely to continue to shift towards the Pacific Rim.

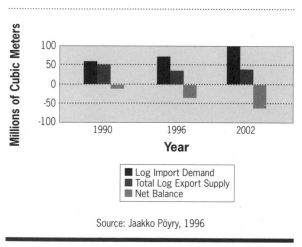

Pacific Rim Roundwood Balance

Source: Jaakko Pöyry, 1996

Graph 3

Environment and Sustainability Implications of Tightening Fiber Situation

Tighter worldwide fiber supplies have implications for increased or diminished industry sustainability, including:

- More of the world's wood fiber will be produced on plantations, often using exotic species. The environmental desirability and risks of such exotic plantations will continue to be a contentious issue.

- The quality of the world's fiber will continue to deteriorate. Old growth forests produce the best fiber. Evolution has designed such fiber to support large, heavy, dynamic, and resistant structures (trees) over very long stretches of time. The replacement of high-quality, old growth fiber by lower-quality plantation and secondary growth fiber has dictated many of the industry trends discussed in this study, such as the rise of engineered panels.

- As per capita incomes rise, people typically read more newspapers, purchase more packaged goods, and live in bigger houses. Ironically, they then begin to value more the nonwood products produced by forests. Competing claims on natural forests' resources will increase as fiber supplies diminish and other claimants to forest resources emerge. Over the very long term, the value of nonwood "products and services" may even determine how forest economics are structured in some areas, but otherwise the forest products industry's priorities will dominate.

- The forest products industry will be forced to become even more efficient and creative in its use of fiber resources, from the forest up the value chain to final distribution.

- In certain areas, such as Indonesia and Malaysia, the combination of still significant natural forest reserves, relatively effective governments, important biodiversity issues, and a mature national industry may eventually force all parties to balance wood production with other values. If such a balance is achieved, new opportunities may be created for companies that lead such developments.

SHIFT IN PRODUCTION TO PLANTATIONS AND SOUTHERN HEMISPHERE

As indicated in Graphs 1 and 2, plantations in the Southern Hemisphere will become a major supplier of the world's future wood needs. Today less than 10% of the world's industrial roundwood is produced on plantations, but this number is growing at a double-digit rate. Indeed, plantations are the only wood supply growing this quickly. Several conditions are fueling the rise of plantations and the Southern Hemisphere as a center of fiber production.

- Growing conditions, labor, and land costs in the Southern Hemisphere are more favorable, making it less expensive to grow fiber in this region.

- Plantations are predictable, reliable, malleable, and flexible. They produce relatively standardized fiber, which makes them extremely attractive to the pulp and paper industry, whose expensive equipment runs better on the predictable wood of plantations. They offer faster financial returns than natural forest management and are less subject to the risk of environmental regulation or intervention by environmentalists.

- The sophistication of modern communications, transportation, and capital movement make the international expansion of fiber production much more feasible today than in the past.

- A worldwide trend towards plantation forestry, highly-developed genetics, and intensive planting, and harvesting make the southern hemisphere more attractive to the forest products industry. A great deal of degraded land is available, which is appropriate for plantation forestry. And as Graph 4 indicates, much less land is required for such farms in the Southern Hemisphere than in the Northern because trees grow faster and yields per hectare are higher.

- Over the long term, capital markets will be more comfortable with investments in plantations because they can be made to behave in a more predictable fashion and are less subject to government regulations.

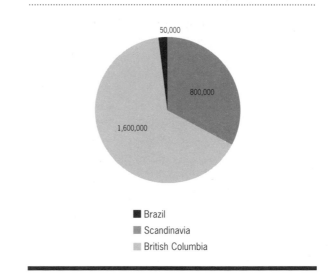

Forest Area Required for 500,000 ton/yr. Pulpmill (Hectares)

- 50,000
- 800,000
- 1,600,000

■ Brazil
■ Scandinavia
■ British Columbia

Graph 4

The areas receiving the largest infusions of investment capital for plantations are Chile, Brazil, and New Zealand. All these countries have had governments that at one point placed a great deal of importance on the development of a forest products industry. New Zealand's government-led promotion of the industry dates to the 1930s, and the Chilean and Brazilian governments have pushed plantation forestry since the 1960s.

Cost Advantage of Fiber Production in Southern Hemisphere

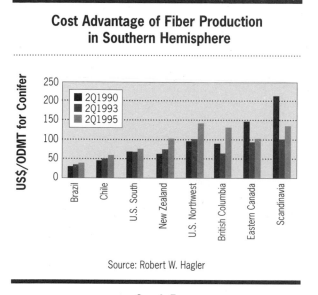

Source: Robert W. Hagler

Graph 5

Shifting Trade Flows, Plantation Fiber, and Southern Hemisphere Production: Implications for the Industry

The shift towards plantation forestry in the Southern Hemisphere has a number of implications for the forest products industry in general:

- The world may be moving towards a dual supply situation, in which a limited area of natural forest is preserved or minimally harvested while most of the world's fiber is produced on plantations or other "fiber farms." Some industry analysts, such as David Price of the U.K., suggest that all of the world's pulp needs could be met from a massive tree farm no larger than 2% of the land area of Brazil.

- As wood supplies shift, primary and secondary processing industries will follow the new sources, for financial reasons and because producer nations will exert pressure and offer incentives to do so (see box). The influx of capital investment and capital equipment will create a variety of new business opportunities.

- Producer nations may exercise more influence over market dynamics. Recently Indonesia and Malaysia have moved from merely banning the export of unprocessed logs to trying to set plywood prices. In late 1995, Malaysia and Indonesia agreed on minimum plywood prices for the Chinese, South Korean, and Japanese markets. Malaysia hopes to ban the export of sawn timber by the year 2000, in part to husband its forest resources, but also to capture value added locally.

Chile's Rio Iata - Processing Follows Fiber Supply

Over the past thirty years, Chile has successfully developed radiata pine plantations. Now, it is trying to leverage those plantations by building processing industries. The leading industry magazine, *Wood Technology*, reported that the Rio Iata group has added a sawmill and particleboard and veneer mills to its plantation operations. Rio Iata's radiata pine plantations are expected to supply 70% of the group's raw material needs in a few years. Since the company has moved into processing, half of the sawmill's output is remanufactured on site, the rest is sold to Japan. The company exports all of its finger-joint stock, moldings, edge-glued panels, and other value-added products.

Rio Iata has also developed a particleboard/veneer project to maximize the value of its plantation radiata pine in the face of worldwide shortages of wood products. The company is now "concentrating on improving the quality of their products and developing new overseas markets," according to *Wood Technology*.

Environmental Opportunities Created by Shifting Sources of Wood Supplies

The dramatic shift in the origins of wood supplies will create new, environmentally driven opportunities and threats:

- New producer countries may lack the institutional ability to respond to the pressures generated. *Tropical Timbers,* not a publication one would expect to be alarmist on environmental issues, recently opined, when "log exports from Gabon go from 75,000 to 700,000 cubic meters, it is less a matter for congratulation and more a reason for concern at the potential impact on...the country in the long term. Ghana was another country caught unawares by a massive increase in buying from Asia Pacific. Only by drastic measures and something of a confrontation between Government and parts of the industry has it managed to put things back on an even keel, principally by calling a halt to log exports...The situation will not get easier because demand for wood is rising and much of the increase is in markets where *environmental arguments carry little or no weight* (emphasis added)."

- The forest products industry in developed countries often argues that the net effect of environmental restrictions on wood production in their home countries accelerates the industry's migration into other countries and increases pressure on natural forests elsewhere. While undoubtedly self-serving, the argument that a window created by a drop in exports of U.S. product to Japan is quickly filled by highly unsustainable Asian producers is compelling.

- Natural forest management may attract less and less interest among major industry players. They may consider plantation forestry a more attractive investment on financial and technical grounds, to say nothing of the reduced exposure to environmental risks. This represents a significant threat to the development and expansion of a sustainable natural forest industry.

- Although plantations require less land for wood production than a natural forest, plantations also have a diminished capacity to supply the other valuable, though often nonmonetized services, of habitat, water supply, and recreation than do natural forests.

- Most economists disparage export bans and other such efforts to increase local value added industries because they are economically inefficient. But these measures have strong political appeal in most countries and are likely to increase over the short term. Export bans are also much easier to administer than forest management regulations, and in countries with underdeveloped administrative infrastructures they can be quite appealing.

- Long term, opportunities for companies that manage natural forests in an environmentally beneficial way may surface. Access to superior sites, greater cooperation from local governments, and better access to financial resources are among the possible rewards.

- Forward-looking countries with an ability to enforce regulations may be able to position themselves to take advantage of opportunities in the forest products industry. But they will need to prepare adequately—by investing in infrastructure, marketing, appropriate controls, and positioning themselves well with buyers of natural forests and investors in plantation forestry.

CHANGING TRADE FLOWS AND GLOBALIZATION OF THE INDUSTRY

International trade flows of forest products have accelerated in recent years, and will continue to do so over the next twenty years. The combination of high economic growth rates in certain geographic areas, gradual exhaustion of traditional supply sources, shifting sources of fiber supply, and the relaxation of trade barriers under new trade agreements such as the General Agreement on Trades and Tariffs (GATT), the North American Free Trade Agreement (NAFTA), and other regional accords will further stimulate international trade of forest products.

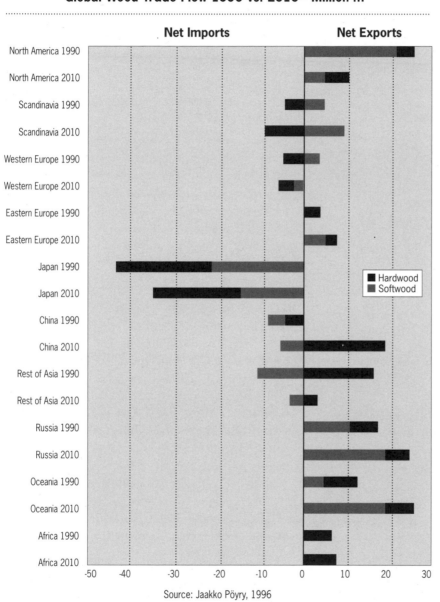

Global Wood Trade Flow 1990 vs. 2010 - Million m^3

Source: Jaakko Pöyry, 1996

Graph 6

products, even though imports will drop slightly over the next decade and a half. Greater shipments will flow from Africa to Asia, as well as from South America to Europe.

Industry Increasingly Global

Rising international trade flows, dramatic shifts in sources of supply and centers of consumption, and a movement towards plantation forestry have all contributed to the increased globalization of the industry over the last twenty years. Today, it is not uncommon for an international paper company to sell its equity in Tokyo, grow its fiber in Brazil, purchase papermaking equipment in Europe, and sell its products worldwide. Similarly, companies that use sawnwood often source around the world to service equally dispersed markets. Over the next twenty years the globalization of the industry will accelerate.

As globalization continues and the industry optimizes production costs across borders, freight and logistics will play an increasingly important role in the search for international efficiencies. Leading companies have already developed sophisticated systems to handle these logistics, which helps lower costs and will open up opportunities for competitive advantage.

As Graph 6 shows, North American and Asian exports are expected to drop considerably over the next fifteen years, with rising exports from Russia and Oceania making up for most of the decline in volume. Japan will continue to dominate trade of forest

Implications and Opportunities of Heightened Trade Flows and Globalization

- As mentioned earlier, most of the growth in the forest products industry lies either in the non-developed world or other areas of the Pacific Rim. None of these markets has demonstrated any great interest in evaluating the environmental performance of the companies from which they purchase.

- As companies roam the globe seeking opportunities, pressures to harvest in more remote and fragile areas, or marginally profitable areas, will increase, as for example in Guyana, Suriname, areas of the Amazon, and in the Golden Triangle of Southeast Asia. Many of these regions lack the regulatory infrastructure to deal with these well-organized companies, and the potential for considerable long-term environmental damage is very real.

- If supply pressures force an increase in hardwood prices, enhanced margins might allow for more careful and more costly forest management in countries with sophisticated regulatory systems.

- Worldwide information systems, enhanced telecommunications, and much more efficient distribution mechanisms created by the globalization of the industry, help make niche markets possible. Much of the "certified" wood products industry to date has developed out of these niche markets, and further globalization will create similar opportunities.

- With information on markets and management techniques readily disseminated globally, new trends like certification and sustainability quickly reach the entire industry. Large trade flows will help spread sustainability ideas into other markets, and backwards into fiber supply.

- Improvements in information transmission, shipping, and distribution should make it more feasible to introduce species previously not used commercially, particularly from tropical areas. These developments should bode well for tropical producers that harvest sustainably and try to maximize the value of natural forest species, instead of simply harvesting the species with greatest commercial value.

STANDARDIZATION AND ENGINEERING CONSISTENCY

Like other industries, the forest products industry is striving for greater standardization and consistency in its production. Just as consumers want consistent quality in the food they purchase and the cars they drive, they are also looking for paper and wood products with no surprises. This trend promises to have wide impact on forest management and the way the industry invests in forests, plants, and equipment.

The paper industry, for example, is increasingly using genetically engineered plantation species that yield fibers which will behave predictably inside a paper mill. New technologies allow Medium Density Fiberboard (MDF) to look like heavily dyed mahogany, even though the MDF's raw material may be a fast-growing secondary growth species. Oriented Strand Board (OSB) and other engineered wood products are successful because their performance qualities do not vary. Furniture designers are quite intolerant of blemishes, knots, and other natural wood features. However, when wood does contain such imperfections, designers like it to look alike in all identical products.

Environmental Implications of Standardization and Engineering Consistency

- In a world where consistency is king, with other considerations being equal, products that directly incorporate solid wood will be undervalued relative to their engineered peers. The former will always look less perfect than the latter.

- With downstream processing creating "wood" products out of a wide range of species, less and less of the final value of forest products will be determined by the forest itself. As a result, processing will attract disproportionately more investment than forests in the future.

- The paper industry owns and manages huge tracts of forestland around the world. In the drive to lower costs and produce a more standardized, often specialized product, the paper industry will favor pulp made from genetically consistent trees. This trend is likely to accelerate the evolution towards a biologically less diverse source of fiber for the paper industry.

- In the past, the forest determined the quality of final output: Old and well managed was better than new and unmanaged. The emergence of OSB, MDF, and plantation-cloned eucalyptus means that the quality of the forest will dictate the quality of products in fewer and fewer segments of the industry. The separation of forest quality from final product quality will make generating the largest amount of raw fiber the priority instead of creating as natural a forest as possible.

The drive towards standardization and consistency is relentless and inevitable in all except small niche markets. Under the circumstances sustainable producers need to seek out opportunities that go beyond the traditional production of sawnwood products.

INCREASING RESOURCE EFFICIENCY

Over the last several decades technological advances have enabled the industry to produce more forest products from the same amount of wood. This technological revolution helps explain why growth in industrial roundwood consumption has only risen by about 1.3% per year over the past decade, while consumption of many categories of forest products such as paper and panels has grown more rapidly. In the paper industry, alkaline pulping technologies and improved papermaking machinery have allowed the industry to post impressive efficiency gains. Similarly, the panel

industry has flourished on its ability to use scrap materials that would otherwise be used as boiler fuel.

Environmental Implications of Improved Resource Efficiency

Industry managers quietly worry that the opportunities for dramatic efficiency improvements may be, for the most part, exhausted. Should this be the case, rising demand for final products will translate into higher demand for industrial roundwood. Such a scenario would exacerbate the already tight supplies predicted above. In any event, the industry's emphasis on efficiency could mean that in the future sustainable forestry may attract less attention from the industry.

As resource efficiencies become more important and natural forests scarcer, more efficient systems and technologies will probably take firmer root in the Southern Hemisphere's industry. Today harvesting and milling in these regions is enormously wasteful. At present these operations convert just 25-30% of the wood used to product, compared to rates of 45-50% in more developed regions. These figures do not include the salvage of residuals, such as bark and sawdust, which has not started in most of the Southern Hemisphere.

INDUSTRY UNDER PRESSURE FOR CERTIFICATION OF ENVIRONMENTAL QUALITY OF FOREST PRODUCTS

The development of forest management certification constitutes an "early stage" driver of change in the forest products industry. The extent or speed of its impact is difficult to predict, and is likely to vary greatly among different countries and regions.

Industry Involvement in Certification

The concept of certification emerged from the nongovernmental sector as a voluntary, nonregulatory, market-based mechanism to improve forest management on the ground. Gradually, industry, mainly through its trade associations, has become engaged in the debate over certification programs, with associations in Europe, Canada, and the United States leading the way. Few major companies, however, publicly endorse certification. Nevertheless, some experts expect certification to reshape the forest products trade over the next ten years. Proponents predict that forest product certification will become as widely accepted and essential for market access as the Underwriters' Laboratory (UL) seal of approval on electrical appliances. On the other side, trade associations, such as the U.S. International Hardwood Products Association, dispute both the need for and value of certification. They contend that certification will only apply to a niche market of particularly environmentally conscious consumers.

Despite the diversity of opinion, some in the industry have accepted that market pressures for sustainability of one kind or another are here to stay. They consider participation the best strategy to help shape certification to meet their needs. That reasoning led Sweden and Finland to initiate third-party certification programs in 1997.

Forest Management Certification Programs

Industry and environmentalists have different ideas about certification programs, however. Conservationists advocate independent auditing of the management practices of specific forest sites by third parties as represented by the Forest Stewardship Council (FSC) program. Industry prefers schemes that it develops, which take a "continual improvement" approach, and are not subject to independent scrutiny. Such schemes include the U.S. American Forest and Paper Association Sustainable Forestry Initiative (SFI), and the U.K. timber trade's "Woodmark" system. Other producing countries are developing their own systems. There is also significant industry support for development of an international environmental management systems (EMS) standard, modeled on and promoted within the ISO system, a global standards body.

Industry generally resists the certification programs advocated by nongovernmental organizations (NGO) for several reasons. Many in industry contend that forests are already managed sustainably, and that certification is, therefore, unnecessary. They also maintain that it is too expensive, especially for small private landowners; that it sets unnecessarily high standards; and that there is no significant market demand for certified product. Nor does the industry want to submit itself to an assessment by outsiders, or make information on its performance publicly available.

Demand for Certified Forest Products

Certified forest products have given a name and a process to a market demand for wood produced in an environmentally sound fashion. At present demand for certified forest products exists mainly in western European and, to a lesser extent, North American markets. Buyers groups, made up of companies that want to buy certified products, are stimulating corporate demand. The creation of such groups is helping to educate major corporate and institutional forest products buyers, as well as end consumers about forest management and the role they can play through their buying decisions. Although final consumer demand is weak, buyers groups may kick-start greater demand for certified forest products.

Drivers of Certification

Clearly, producers in countries that supply the environmentally sensitive markets of Europe—such as Gabon, Cote D'Ivoire, Cameroon, Indonesia, Malaysia, Brazil, Sweden, Finland, and Norway—have the greatest incentive to pursue certification. Yet some companies, wood chips' producer Ston Forestal in Costa Rica, for one, appear to have pursued certification to demonstrate their environmental commitment to the government and the public.

Environmental Implications and Opportunities Created by Certification

- If large players in any major producing country "break from the pack" and seek FSC-type certification, the move is likely to induce competitors to follow suit to protect their market share. First movers stand the greatest chance of securing market share, and of potentially regaining any share previously lost to timber boycotts. Under this scenario, widespread changes in on-the-ground forest management practices could contribute significantly to the certification movement.

- In developing countries, companies that become certified could set a new standard for forest management at home. Such demonstrations of state-of-the-art forest management could help to redefine accepted practices over the long run and ensure access to concessions and other government incentives.

- If certification becomes a de facto requirement for access to European or other environmentally-conscious markets, as some experts predict, it could have a strong impact on improving forest management worldwide.

INCREASING GOVERNMENT REGULATIONS OF THE INDUSTRY'S FOREST MANAGEMENT PRACTICES

Worldwide, national, provincial, and local forestry management regulations are increasingly defining the industry's access to raw material, the terms of international trade opportunities, and forest management practices. Forestry regulations are fairly well established in developed countries. In developing countries regulation ranges from practically nonexistent to significantly structuring trade and opportunity. Every country in the world is now engaged in a serious debate over how forest assets should be managed, and more regulations are sure to come. New rules generally increase harvesting costs, either by forbidding certain types of cuts, or prohibiting logging of certain areas, such as water-sheds and habitats. They are also reducing the overall harvest of wood from natural forests.

No international agreements exist strong enough to materially affect forest management practices yet, but a number of developments may make such accords more plausible in the future:

- The near-certain failure of the International Tropical Timber Organization's attempt to achieve forest management sustainability by the year 2000 is quite certain to be the catalyst for some sort of international agreement on forest practice.

- The United Nations Commission for Sustainable Development's examination of forest practices will put modest pressure on countries to improve their management.

- The ability to use satellites to monitor forest management practices at a reasonable cost is arming NGOs and international bodies with much better information, which they will use to lobby for better forest practices.

- If governments take action on global warming, the forest products industry will be a target. Their actions would quite likely include greater restrictions on harvesting natural forests, which would diminish wood supplies in the short term; while incentives for reforestation to create "carbon sinks" would increase supplies long term.

The Pulp and Paper Industry

The following three sections analyze pulp and paper, panels, and sawnwood industries, identifying opportunities for and obstacles to the development of a more sustainable forest products industry. Even though sustainable forestry takes place in the forest, its success ultimately will depend on product end-use. These three industry segments represent the vast majority of forest products produced by the industry and account for most industrial round-wood consumption.

INDUSTRY STRUCTURE

Totaling more than $390 billion in annual sales, the global pulp and paper industry produces thousands of wood-based products ranging from newsprint and printing papers to tissue, cardboard, and other packaging products. In the United States, the world's leader in per capita consumption of paper products, the pulp and paper industry generates annual sales of more than $90 billion, and accounts for 2% of GDP. In Canada, according to the *1996 North American Pulp and Paper Factbook,* the paper industry represents about 6% of GDP. In emerging Asia, pulp and paper products generate roughly 3% of regional GDP. In total, the global pulp and paper industries account for the consumption of just over 600 million cubic meters of industrial roundwood, including trees harvested for pulp and scrap from other forest products industries. The principal business dynamics at work in this important segment, and the impact of environmental forces on growth and development, are discussed below.

Fragmented Supply but Stable Demand

Global demand for paper products is expected to rise between 2% and 3% annually over the next five years, driven principally by economic and population growth, according to the American Forest and Paper Association. The United States leads in per capita con-

sumption at 330 kg per year, followed by Finland (266 kg) and Japan (231 kg). Developing nations, such as China and India, lag well behind with per capita consumption of less than 30 kg each year. Any economic growth in these two countries will, of course, cause a considerable uptick in demand.

In recent years, the fast growing economies in Asia, Latin America, and even Eastern Europe, have registered the highest growth in paper consumption - a trend that is expected to continue. In China, for example, paper use has grown more 10% per year since 1980 (rising 16% in 1994 alone), even though per capita consumption levels remain a fraction of those in industrialized companies. Growth in demand for forest products in Asia is similar to the frenzied growth of the energy infrastructure in that region and is attracting massive amounts of capital and corporate investment from around the world.

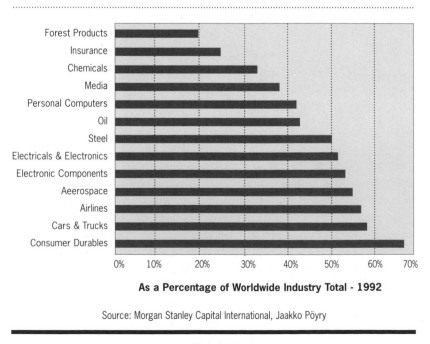

Industry Concentration—Global Sales of Top Five Companies

As a Percentage of Worldwide Industry Total - 1992

Source: Morgan Stanley Capital International, Jaakko Pöyry

Graph 7

The long-term impact of the electronics revolution and the "paperless" office is the unknown variable in paper demand. Newspaper readership in developed countries is already on the decline as readers turn to electronic media for their news and entertainment. The internal business market consumes 70% of all printing and writing paper. Over the next decade, as corporations more fully automate their payment, invoicing and "paperwork" systems, the business sector will use less and less paper. However, most observers predict that the rising appetites for paper in developing countries will compensate for any decline in the internal business market.

Demand for paper products may be relatively stable over the next decade, but the industry's supply dynamics are considerably more volatile. Ownership of both supplies and manufacturing facilities is fragmented in most world markets, although a gradual trend toward consolidation appears to be accelerating. Between 1980 and 1990 the number of paper and board mills in key markets plunged: 21% in the United States; 25% in Europe; and 23% in Japan. Further consolidation shook the U.S. market in 1995 and 1996, most notably the high-profile merger of Kimberly-Clark and Scott Paper that created an $11 billion combined company, and International Paper's $3.5 billion acquisition of Federal Paper Co. But as the Graph 7 indicates, the forest products industry still has a long way to go before consolidation reaches the levels found in most other global industries.

Growing Capital Intensity

One of the industry's most distinguishing characteristics is its capital intensity. In the United States, capital expenditures of pulp and paper manufacturers average 10-11% of sales, or roughly double that of other highly capital-intensive industries like chemicals, primary metals and manufacturing. Moreover, the industry's capital intensity is rising. In the United States, from 1980 to 1990 the percentage of all mills producing more than 450,000 tons per year (tpy) of pulp rose from 40% to almost 60%, while the number of mills producing more

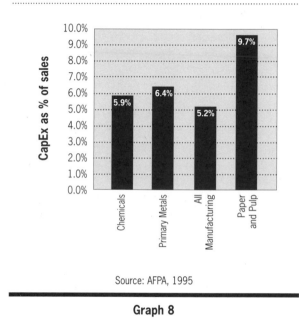

Capital Intensity of Pulp

Source: AFPA, 1995

Graph 8

than 500,000 tpy doubled, according to the *1992 Pulp and Paper Factbook*. As might be expected, the costs for mills of ever greater size has grown proportionately. Today, a new pulp mill installed to match the industry standard capacity of 500,000-700,000 tons costs roughly $1 billion.

The huge scale of production required to maintain a competitive cost structure is the chief reason for the pulp industry's high capital requirements. With few exceptions, pulp and paper products are difficult to differentiate, leaving manufacturers little ground but price on which to compete. Distribution channels and customer relationships remain important to business success. However, price drives most purchasing decisions, and buyers retain great power over producers. This intense price sensitivity requires producers to continually strive for lower production costs by using new technology and seeking greater efficiencies in growing, harvesting, and processing trees.

Capital intensity is exacerbated because a handful of equipment makers supply the industry. The few manufacturers of papermaking machines generally receive their orders in lumpy increments during peaks in the business cycle, and they counter by keeping prices high.

Variable Profitability

The industry's fragmentation, capital intensity, commodity nature, and tendency to overspend on plant and equipment contribute to a cycle of steep swings in pulp pricing with proportionate variations in corporate profits. Over the past thirty years, the U.S. industry's net profit margin has averaged roughly 4.5%. But margins fluctuate sharply, sometimes doubling and tripling in the space of several years, only to collapse again in the same period of time. Similarly, the industry's return on net worth, which has averaged roughly 9% since 1960, has suffered dramatic gyrations. This unimpressive ROE profile—for most companies, frequently beneath their average cost of capital —has been adversely affected by excessive debt, which many companies used aggressively in the late 1980s.

According to Morgan Stanley, long-term returns in the paper industry underperformed other global financial securities by more than 35% from 1970 to 1996. As the industry's dividend yields tend to be modest (a 2.7% average in 1995), investors have concentrated on short-term pricing swings that correlate with industry profits. The valuations of major pulp and paper companies reflect these dynamics, which is why stocks tend to trade at a discount to their markets. In the United States in 1995, according to McIntosh-Baring, pulp and paper product stocks traded at nearly a 50% discount on a price-earnings basis compared to their peers.

ENVIRONMENTAL FORCES AND PULP AND PAPER

In the midst of these challenging dynamics, environmental forces have hit the paper industry with rising intensity over the past twenty years. Initially, they centered on pollution, and pollution abatement during the 1970s, then moved on to recycling in the 1980s. During the 1990s—and for the foreseeable future—forest management has emerged as the pressure point. Other sources of paper fibers may become an environmental flash point at some time, but so far no major actors have taken up the issue.

Emissions Pressures

The industry's wastes are already heavily regulated. The pulp and paper industry uses a variety of chemicals and, in the process of pulping and bleaching wood fiber, generates gaseous and liquid wastes. In the United States, the cost of complying with environmental regulations to control these waste products has been high, requiring total expenditures by the industry of more than 10% of capital expen-

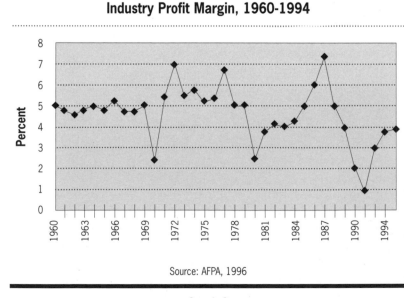

Industry Profit Margin, 1960-1994

Source: AFPA, 1996

Graph 9

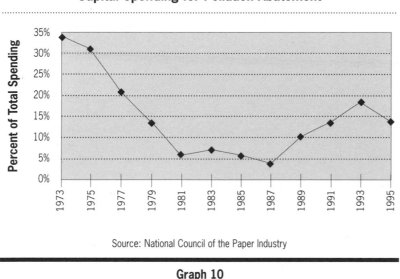

Capital Spending for Pollution Abatement

Source: National Council of the Paper Industry

Graph 10

dito in most years. As Graph 10 shows, however, the financial burden of pollution control has varied considerably over the past two decades, ranging from almost 35% of annual capital expenditures in the early 1970s to approximately 14% in 1995. As currently proposed in the U.S. Environmental Protection Agency's Cluster Rules combining air and water regulations, the cost of compliance is still high. Industry sources estimate that regulatory compliance with those measures is expected to cost the industry between $10 billion and $13 billion by the end of 1997. Technology is now so "clean" that no major equipment manufacturers even produce the "dirty" equipment of a decade ago. It is almost impossible to build a new paper mill without purchasing new generation clean equipment.

Recycled Fiber

Environmental pressures have caused the dramatic growth of recycled fiber as a substitute for virgin fiber in paper (see Graph 11). In the United States, concerns about landfill capacity initially triggered increased demand for recovered fiber in the late 1980s. Government mandates then called for an increase in recycled fiber content in paper supplies.

After initial opposition, U.S. paper producers discovered economic advantages to the use of recovered paper, including the lower fixed capital costs of recycled mills and the lower cost of recovered paper relative to virgin. The combination of environmental pressure and economic advantage pushed demand for recovered paper to rise more than twice as quickly as that for virgin fiber over the last ten years. U.S. paper mills increased their use of recovered papers by 94% between 1985 and 1995, from less than 17 million tons to more than 32 million tons. Recycled fiber, which accounted for less than one-quarter of fiber input into the U.S. paper industry as late as 1986, now represents roughly 35% of total production, according to the U.S. Forest Service. Moreover, recycled fiber capacity grew over 100% in the United States from 1985 to 1995, and the American Forest and Paper Association reports that it will continue to increase.

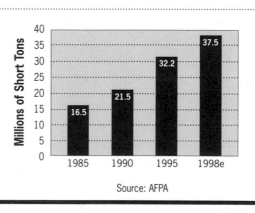

Growth in U.S. Recovered Paper Use

Source: AFPA

Graph 11

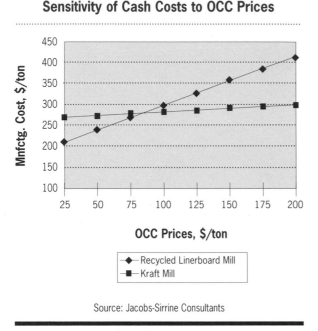

Sensitivity of Cash Costs to OCC Prices

Source: Jacobs-Sirrine Consultants

Graph 12

While demand for recovered fiber is expected to remain strong, volatile recycled paper prices as well as limits to potential recovery rates, are expected to moderate the growth in industry capacity.

Benchmark recycled paper prices, including old corrugated cardboard (OCC) and old newspapers (ONP), remained attractively low through the early 1990s, but rose sharply in 1994, greatly reducing the price advantage over virgin pulp.

Although OCC and ONP prices have fallen since their mid-1995 peaks (again providing cost savings to virgin materials), for new mills, recovered pulp costs less only if prices remain below $75 per ton.

Forest Management Practices and Pulp and Paper

The industry has adapted to pollution control and recycling at a time when the rise in the size and costs of modern mills makes a steady, reliable fiber supply far more important. To assure those supplies, pulp producers have increasingly adopted plantation forestry. Many industry leaders have directed their investments towards the faster tree growing areas of the Southern Hemisphere.

Although the source of new fiber supplies is clearly evolving towards exotic plantations in the Southern Hemisphere, the industry still actually obtains more pulpwood from original old growth forest than it does from such plantations—16% vs. 11% in 1993.

The pulp and paper industry is vulnerable to buyer pressures over forest management practices, particularly in Europe. Several large producers have lost important contracts over their poor practices, and European producers are now required to monitor their new forestry and mill operations in eastern Europe to ensure that the sustainability issues important to their European customers are carefully controlled.

Nonwood Fiber

Over the past five years, the industry has renewed its attention to nonwood fiber sources that come from annual crops such as kenaf, hemp, and wheat straw. The potential for nonwood fiber production has periodically generated enthusiasm, but worldwide, fiber from nonwood sources made up a modest 10.6% of the worldwide pulping capacity in 1995. China, and to a lesser extent India, are the major producers. Nonwood fibers must overcome a number of challenges to replace wood fiber—ranging from the problems of storing annual crops to high lignin contents and fiber consistency in the future. Nevertheless, environmental restrictions may make nonwood fiber sources more attractive. China uses nonwoods for 80% of its pulp needs, indicating that under the right conditions nonwood fiber is potentially viable. However, the Chinese pulp and paper industry is a notorious polluter due, in part, to its use of nonwood fiber.

ENVIRONMENTAL THREATS AND OPPORTUNITIES IN THE PULP AND PAPER INDUSTRY

Environmental pressures and evolving industry dynamics are creating a series of strategic threats and opportunities for paper producers:

- Pressure to recycle has forced rapid change on the paper industry. In many segments, such as newsprint, powerful buyers were instrumental in forcing recycling upon a less than enthusiastic paper industry. If the public turns a critical eye towards the paper industry's forestry practices, the company with poor forest management practices may be punished by the market with loss of market share, particularly if such pressures are brought to bear during a cyclical industry downturn.

- The power of paper buyers is so great that if just one global packaging giant, such as Procter and Gamble, decided to use certified paper products, the action could create a certified market within a year.

- Even though barriers to entry in the paper industry are overwhelming, the enthusiasm in some markets for an environmentally differentiated product may open up an opportunity for a first-mover into this field.

- The nascent certified movement has attracted the attention of large European, particularly German, publishers in the past few years. Their buying clout combined with the interest in certified paper from other large users, such as Swedish furniture giant IKEA's catalogue group, could lead to brisk new markets for certified paper products in the next two to three years.

Paper Opportunities

A handful of paper industry leaders have already organized their strategies around sustainable forest management issues—and some are reaping competitive advantage:

- Three top Swedish paper producers, SCA, Stora, and Assi-Doman, the world's largest paper maker, are members of a Forest Stewardship Council (FSC) working group, which is creating a Swedish certification standard. If, as most observers expect, large tracts of forest are soon certified under an FSC system, significant volumes of certified Swedish paper and wood products will come to market in the next two years.

- The Rock-Tenn Co., one of the largest producers of recycled paperboard and folding cartons in the United States, has successfully implemented a strategy of vertical integration, high technology and extensive use of waste fiber. The company's emphasis on recycled paper has given it flexibility in managing costs, and helped keep earnings variability lower than that of most of its competitors.

- Fort Howard, the world's leading maker of commercial and industrial tissues, pioneered recycling technology that has enabled the company to use low-cost mixed office wastes as a raw material and maintain the highest profit margins in the tissue business. Most of the company's products are made from 100% recycled waste paper, and are processed with a proprietary de-inking technology.

- Brazil-based Aracruz Celulose, the world's largest producer of bleached hardwood market pulp, has established itself as one of the lowest-cost, highest-quality producers of market pulp. In the process of developing its premium product, Aracruz has forged ahead of competitors in eucalyptus forestry practices and made environmental performance a cornerstone of its strategy.

• Some industry observers expect the pressure for more sustainable forest management to provoke radical changes in the paper industry. As long time industry expert Robert W. Hagler of Wood Resources International sees it, "...*we are at the point of fundamental change in the way society values and manages its forest resources...A restructured wood fiber equation will mean new players, new products, and new rôles for traditional participants. Adaptation will be crucial for survival.*"

Indeed, some analysts think that forest supply issues may trigger a wave of mergers and acquisitions to secure wood supplies for the future. Consolidation would boost the industry's overall financial performance by helping to stabilize erratic price swings and avoid periods of overexpansion.

Panels and Engineered Wood Products

Most engineered wood products are panels, basically wood fibers that are reconstituted into flat, thin sheets. In the last ten years, however, producers have leveraged technologies developed for panels to create a variety of new products that substitute for solid lumber.

Panels are used principally in the construction and furniture industries. The amount of industrial roundwood used for the production of wood based panels topped 300 million cubic meters in 1995, about 9% of world roundwood production. Due to a number of considerations discussed below, customers are not yet asking for certified engineered wood products and panels. Opportunities to market certified products do exist, however, particularly in the furniture segment, although no producers have yet capitalized on them.

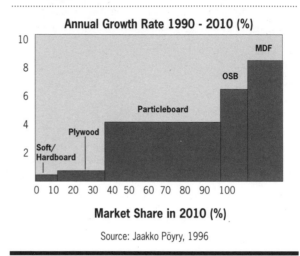

Global Demand for Wood Panels Market Share and Growth Rates - 2010

Annual Growth Rate 1990 - 2010 (%)

Source: Jaakko Pöyry, 1996

Graph 13

NEW PRODUCTS GROWING DRAMATICALLY

Demand for the new class of panels—MDF (Medium Density Fiberboard), OSB (Oriented Strand Board), and particleboard—will soar over the next fifteen years. The products' ability to use many kinds of wood fiber, readily available and constantly advancing technology, and attractive economics make these panels a segment to watch in the future.

MDF will achieve the highest growth rates, becoming over the next fifteen years a high-end panel suitable for a wide variety of applications (see Graph 13). MDF's relatively high cost compared to other panel products will, however, limit its overall market share. Particleboard's attractive economics, by contrast, will earn it a significant market share of the worldwide panels industry. The difficulty of obtaining appropriate logs to use for plywood will cause growth in this product category to stagnate. Gradually, OSB will become the preferred substitute for plywood.

STRUCTURAL PANELS

Thirty years ago "structural panel" meant plywood. Two decades ago a new structural panel, OSB, was developed. OSB is made through a process of aligning "strands" of wood in a multilayered mat with the layers of strands aligned perpendicularly. Glued together under heat and pressure, the perpendicular alignment of these many stranded mats give the board a uniform, consistent strength. OSB has become something of a darling in the industry because it does not require high quality fiber for raw material. The product uses aspen, birch, and other "weed" species common in secondary forests that have little alternative use, and are therefore inexpensive.

OSB Substitutes for Plywood

The plywood industry is robust in much of the world, although it is being rapidly replaced by OSB in North America and Europe. The peeler logs which make up the raw material for plywood are becoming increasingly scarce in Asia and elsewhere as the natural forests containing such high quality timber are cut down. As a result, over the next two decades, OSB and other panels will gradually substitute for plywood in many applications. Meanwhile, the cost of OSB plants is dropping considerably faster than is the cost for plywood mills.

As Graph 14 indicates, a tremendous amount of OSB capacity will come on line in North America over the next fifteen years. Meanwhile, plywood capacity is expected to plummet. The same tendency is evident elsewhere. The United States will have more OSB capacity than it needs over the medium term, although North American trade associations are undertaking ambitious export promotion programs with a goal of exporting nearly 5 billion square feet per year (4.425 million cubic meters).

Overcapacity on the Horizon

Looming overcapacity in both plywood and OSB is prompting companies to look for opportunities in new and specialized product niches. Under pressure

North American Structural Panel Capacity

Source: RISI, 1995

Graph 14

from OSB, plywood producers are seeking out specialty niches and nonconstruction markets. Meanwhile, OSB manufacturers such as Georgia-Pacific, Louisiana-Pacific, and Boise Cascade, are finding new ways to leverage their engineered wood capabilities for the production of I-Joists, laminated veneer lumber (LVL), and parallel strand lumber.

Environmental Forces, OSB, and Plywood

Since OSB's primary source of fiber consists of commercially unattractive species often found in younger, second-growth forests, the industry is protected against the loss of fiber resources by environmental restrictions. Indeed, the difficulty of obtaining the higher quality logs required for plywood from old growth forests is the catalyst behind OSB's displacement of plywood.

Although, the OSB industry and other engineered wood products incur high costs to comply with emissions restrictions that regulate their heavy use

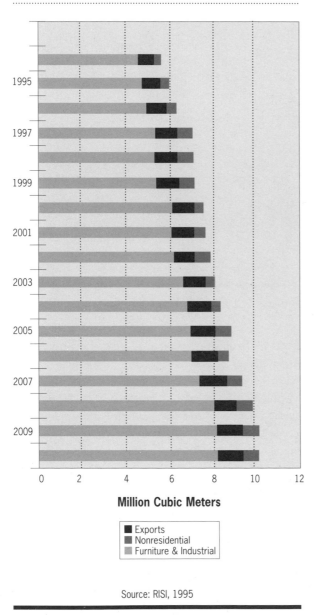

U.S. Particleboard Consumption by Total End-Use Category

Million Cubic Meters

■ Exports
■ Nonresidential
■ Furniture & Industrial

Source: RISI, 1995

Graph 15

of adhesives, compliance with these regulations will not undercut the growth in OSB at the expense of plywood.

The exploitation of forests for plywood production has had an enormous environmental impact on forest resources in Southeast Asia. Plywood has been among the most important exports from this region, particularly to Japan, and the huge export markets have fueled the industry's expansion.

PARTICLEBOARD

More particleboard is produced than any other panel in the world. Under heat and pressure small flakes of wood fiber are bonded together, usually with a urea formaldehyde-based adhesive, to create an economical panel that is heavily used in furniture making, cabinetry, floor underlayment, and door cores. Demand for particleboard has increased dramatically worldwide in recent years as the market for Ready-to-Assemble (RTA) furniture, which uses a great deal of particleboard, has taken off. Particleboard competes with energy applications for its fiber sources in the United States, typically sawdust and plywood shavings, and historically has had no supply difficulties. Relatively local sources of supply are critical to this sector given the low value, high transport costs, and relative bulkiness of both the raw material and final product. Because the product is heavy relative to its value and does not transport well, international trade is limited, so the industry will continue to be chiefly regional.

Worldwide particleboard consumption will grow briskly over the next fifteen years, driven by technical innovations that improve the product's structural characteristics and lower its cost.

Industry Trends

Industrial uses of particleboard, including furniture applications, will account for most of the industry's growth in the United States during the next fifteen years (see Graph 15). Similar growth patterns are expected elsewhere for several key reasons:

- Just as in other forest products sectors, the particleboard segment is moving toward manufacturing cut-to-order pieces and other differentiated products.

- Producing these value-added products enables manufacturers to operate more efficiently and serve customers better.

- No substitutes are in sight with similar performance and cost characteristics. Particleboard's rising popularity in the residential and nonresidential construction sectors, as well as predicted strong growth in furniture and store fixtures, bode well for the future of this category.

Environmental Forces and Particleboard

Although the particleboard industry is concerned about environmental restrictions reducing fiber supplies in the future, air pollution and emissions are the more important, immediate issues. The Clean Air Act is forcing the industry to incur significant costs to cut its emissions of volatile organic compounds (VOCs). Typically, VOCs are eliminated through thermal oxidation, or burning at high temperatures. This process, however, requires expensive capital investments and can add significant fuel costs, particularly when the heat generated by burning is not efficiently recovered by the mill. While mills continue to make incremental improvements in VOC emissions, the regulations will no doubt increase costs for the industry in most developed countries with little or no additional value added.

MEDIUM DENSITY FIBERBOARD

MDF is similar to particleboard, except that in the manufacturing process the raw material, typically wood chips, is reduced to individual fibers before being formed into boards with pressure, heat, and urea formaldehyde. MDF is sought after because it is easy to machine, has smooth surfaces, and is homogeneous. Although it is used primarily for furniture and cabinetry today, the product is making inroads into the thin plywood market, door

---

Gridcore Systems International: An Innovator in Engineered Wood Products

Gridcore is a three-dimensional structural material made from virgin wood and recycled fibers originally developed at the U.S. Department of Agriculture's forest products laboratory in Madison, Wisconsin. The product, created by pressing pulp into a honeycomb core that has a smooth face, is suitable for a wide range of applications—everything from wall panels and ceiling tiles to partitions and furniture cores.

The company has inked marketing and distribution agreements with Weyerhaeuser and Laird Plastics. Recently, it completed a multi-million dollar round of private equity financing, and is expanding its production. The company is riding the wave of several new trends in the building and forest products industries, including:

- Use of a recycled resource as the principal raw material.

- The substitution of alternatives for more wood-intensive products.

- Products engineered specifically for various applications.

---

skins, and millwork applications. In the future, as MDF with special characteristics—water resistance and thin sheets, for instance—develops, the product will become even more popular.

Adaptability to Supplies

The adaptability of fiber supplies for MDF bodes well for its long term viability. According to Sunds Defibrator, a major supplier of MDF production systems, global raw material consumption by the

MDF industry in 1993 was 50% softwood, 20% hardwood, 20% a mixture of the two, and 7% bagasse and cotton stalks. In Asia, MDF uses rubberwood, eucalyptus, and bagasse, which are all expected to be readily available well into the future. MDF's fiber adaptability is undoubtedly a factor in its rapid growth. The product's popularity has exploded during the 1990s, with capacity almost doubling in the United States, output increasing fivefold in Europe, and capacity expected to triple in Asia between 1994 and 1997.

MDF Manufacturing Capacity in Southeast Asia

	Capacity (1000m^3)	
Name of Company	1994	1997
China*	216	216
Indonesia	0	956
Malaysia	400	1,163
Thailand	340	660
Others	105	105
Total	**1,061**	**3,100**

No estimate was made for growth in Chinese capacity, though all observers agree it is quite significant.

Source: Owen F. Haylock, O.F. Haylock Consultants Ltd.

Table 2

Other factors, however, are propelling the growth of MDF as well. The depletion of natural forests which contain the trees required for alternative panel products is also contributing. So is the acceptance of MDF by the furniture industry, which is adopting the product because it is easy to coat with natural-looking surfaces.

IKEA Teams up with a Malaysian Producer

IKEA, the Swedish manufacturer renowned for its trendy, economical RTA furniture, has forged a joint venture with Malaysia's Golden Hope Plantations Bhd. According to *Tropical Timbers*, this Malaysian company is developing manufactured products based on its extensive rubberwood plantations. Golden Hope is building a 100,000 cubic meter plant that will make furniture components and manufacture MDF furniture ti IKEA designs. IKEA, the leader in RTA particleboard furniture, will handle the MDF furniture for all markets outside Malaysia. Although the production is not currently marketed as a "green" product, given the nascent demand for such furniture and IKEA's leadership in the market, an opportunity to do so may exist.

The joint venture highlights several critical trends that are transforming the forest products industry:

- The expansion of fiber producing areas into higher value-added products.

- Efficient use of unattractive species like rubberwood in engineered wood products.

- Globalization of the industry, with designs from Europe manufactured in Southeast Asia for distribution around the world.

- Plantation forestry in the Southern Hemisphere meeting the fiber needs of a global marketplace.

Environmental Forces and MDF

MDF stands to benefit from the same environmental forces that are buffeting other sectors of the industry:

- MDF uses a wide range of raw materials for fibers, including annual crops, which means it is less exposed to supply difficulties than other sectors. That gives MDF an advantage over competing panel products.

- The ability of MDF to use many kinds of fiber sources may permit the industry to position the product as a "green" solution to many furniture and construction needs.

- The explosive growth of MDF capacity in Asia illustrates two powerful trends reshaping the forest products industry: that processing capacity is following wood supplies, and that Asia is fast becoming a hub of furniture production for export. Because much of Asia's furniture is exported to Europe and the United States, where concern over forest management is highest, positioning a company's products as "green," or even obtaining certification, would appear to be prudent for MDF manufacturers that export to Europe, if only as a defensive move.

- Several players dominate the office segment of the furniture market, including Knoll, Herman Miller and Steelcase. At the same time, Fortune 500 companies account for a large percentage of office furniture consumption. Given this dynamic, and the desire of large corporations to be environmentally correct, in the near future these customers may begin to push their large suppliers to offer sustainably produced furniture products.

- MDF, which is based on relatively new technology, is one of the most successful new product categories in wood industry history. The rise of this product illustrates how powerful the impact of new technology can be in creating new opportunities for sustainable forestry in the future.

An "Urban Forest" MDF Mill

Garbage disposal is a high-profile issue in the United States. In Oregon, many forest product processors face extremely tight wood supplies caused by cutbacks in harvests on federal lands. Willamette Industries MDF plant in Eugene, Oregon, has found opportunity in these two environmental realities. The company set up seven sites to collect wood along the 275-mile corridor between Eugene and Seattle. From these locations it gathers over half of the plant's raw material in the form of clean lumber waste, material from construction sites, and other industry sources.

According to *Wood Technology*, the MDF plant was originally built as a particleboard facility. "Our experience using urban wood in particleboard convinced us this was a viable raw material source, but it's better suited for a higher-value product like MDF," reported the mill's general manager David Smith. "The fiber quality available with urban wood is similar to that of pulp chips, and far superior to what we can get out of shavings." With such high quality material on tap, the plant switched over to MDF production after a multi-million dollar renovation, which included designing an original chip washer and rebuilding the mill's fiber preparation area.

"We must make a very clean board from urban wood. That's why the chip washer is essential," Smith said. "Customers are wary of board made from urban wood." In fact, the shift to MDF from particleboard was essential. It costs a great deal more to gather, clean, and process urban wood than to process shavings. The higher value-added MDF can support the higher raw material cost.

VENEER

Veneers are typically used as decorative covers for furniture, wall paneling, and other premium applications to give the appearance of natural wood. According to the United Nations Food and Agricultural Organization (FAO), the worldwide production of veneers reached nearly 14 million cubic meters in 1993. The U.S. veneer industry is relatively fragmented, with capital required to build a plant hovering in the low seven figures. Many veneer companies are family-owned and run, and have been in the business for a long time. However, requirements for higher capital investments are forcing some consolidation. Larger firms with good access to roundwood supplies and the capital available to invest in closer tolerance, highly efficient equipment are increasingly dominating the market.

Global Product

Veneer is among the most "globalized" of forest products. The product's relatively high value and light weight, the industry's ability to customize and differentiate products at a reasonable cost, and wide knowledge about the product among furniture designers make veneers well suited for international trade. Indeed, exports account for nearly 50% of U.S. production. They are expected to remain strong over the next three to five years as European consumers shy away from tropical veneers. The greatest threat to veneers, according to the Miller Freeman annual report "Panels: Products, Applications and Production Trends," is MDF, which is expected to be available soon in high-quality overlays or prints that look just like hardwood plywood.

Worldwide, the industry is moving towards stricter uniformity in manufacturing processes. Customers increasingly demand, for instance, that the car they purchase be exactly the same color as the one whose picture they saw in an advertisement or brochure. In response, manufacturers strive to make all "identical" products as identical as possible. Under such circumstances, temperate, lighter colored veneers that can be stained to a uniform color are preferable to tropical veneers. For similar reasons, MDF may also be preferable.

Environmental Forces and Veneer

Several powerful trends will cause demand for veneers to rise significantly over the medium and long term. The precious few hardwood trees that can provide furniture and panels with a rich, natural look and feel are becoming increasingly difficult to secure—and cost more when available. As larger dimensional wood becomes more expensive, more manufacturers are substituting engineered wood products, even though few meet consumers' aesthetic requirements. Manufacturers are then turning to veneers to cover their engineered wood products, which provides customers with products that meet their aesthetic requirements at a reasonable cost.

Veneer is one product where high quality forest produces a high quality veneer. Although veneer represents less than 1% of total global roundwood production, it is an extremely high value product. A top-quality veneer log can sell at a 600% premium over alternative uses for the log.

Certified Veneer on the Horizon?

The structure of the veneer sector appears to favor opportunities for certified products for several reasons:

- Much of the export market is to Europe, so producers are already tied in to the strongest certified market in the world.

- Veneer producers act more like "job shops" than most other branches of the industry. Therefore, custom orders and chain of custody issues involved in certified products are less burdensome.

- Some producers, such as Curry-Miller in Indiana, already use bar-coding systems to track logs and their products from logyard to customer because veneer buyers often want to purchase entire sliced logs. This type of tracking system is likely to make it easier for veneer producers than other sectors in the industry to adapt to third party certification schemes that require monitoring certified wood from the forest floor to retailers' shelves.

- Long term, the greatest challenge for veneer makers will be the availability of quality hardwood roundwood. Certified forests require more careful, long-term management and the higher-value logs they produce can yield high profits when directed towards the right market. These forests would appear to be ideal candidates for veneer production.

PANELS AND ENGINEERED WOOD PRODUCTS—ENVIRONMENTAL OPPORTUNITIES AND RISKS

Panels and engineered wood products are filling niches created by the increasingly scarce supplies of peeler logs and by the deteriorating quality of available wood, which are both linked to deteriorating forests. OSB, MDF and particleboard all use non-traditional trees or by-products, and these sources have been the platform for growth.

Engineered panels do not depend on high-quality raw material, either, but on the by-products of other elements of the forest products industry, or on low-grade timber, instead. The sector will continue to demonstrate a remarkable resiliency in its ability to adapt to supply variables. "Urban wood," annual crops, and other fiber sources will fill any gaps in traditional fiber supplies. In Asia, producers have turned to annual crops as a source of fiber for particleboard, particularly flax and hemp, bagasse, and cotton stalks.

Several significant obstacles exist to the development of a viable "certified" panel market, particularly a market fueled by consumer demand:

- Since the raw material for many engineered wood products is a by-product of other wood processing industries, and is widely sourced, it presents difficult chain of custody issues.

- The industry is relatively fragmented, produces relatively undifferentiated products, and has no dominant players exercising market leadership.

- Buyer power is widely dispersed among tens of thousands of builders, remodelers, furniture manufacturers, and others who use panels as a component in their final products.

- The ultimate arbiter of demand, individual consumers, are in all cases at least one link away in the value chain from the decision on what particular type or brand of panel will be purchased.

Most engineered panels will probably lag behind other forest products in the development of any large-scale certified products market. However, well positioned engineered wood and panel producers stand to benefit from the environmental forces currently affecting the industry, particularly those firms that develop innovative fiber supplies. Opportunities appear to be strongest for those panel products used in furniture, a sector traditionally more open to "green" and certified products than the construction industry.

Sawnwood

The section that follows analyzes the hardwood and softwood sawnwood segments of the forest products industry, identifying how the forces reshaping the industry, including environmental pressures, will affect the production of temperate hardwoods, tropical hardwoods, and softwood.

Softwood accounts for just over 300 million cubic meters of lumber per year, most of it harvested from native forests in temperate countries (see Graph 16). Softwood lumber, the most important traded solid wood product, made up 25% of world softwood lumber consumption, or nearly 80 million cubic meters, in 1993. Trade flows primarily from Canada to the United States, the western Pacific of

World Softwood Production—1993

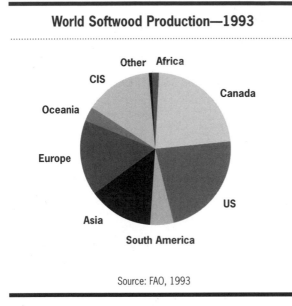

Source: FAO, 1993

Graph 16

U.S. Harvest Trends—Softwood 1962 and 1986

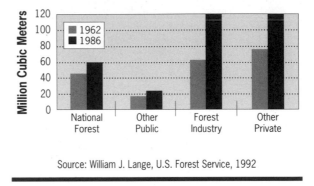

Source: William J. Lange, U.S. Forest Service, 1992

Graph 17

North America to Japan, and Canada to Europe. With the exception of a rise in trade from Oceania and South America to the Pacific Rim, according to Jaako Poyry, these patterns are expected to remain relatively unchanged over the next decade.

Traditionally, the Canadian industry has depended heavily on government-owned lands for its supplies of softwood, with provinces such as British Columbia producing almost 90% of its harvest from public lands. The United States, which has developed strong production of loblolly pine in the South in recent years, is much less dependent on government lands for softwood production than it has been in the past (see Graph 17). Nevertheless, the large public ownership of softwood forests in both countries makes the industry more vulnerable to shifts in public priorities and policies toward forest management than would be the case if these lands were primarily privately held.

CONSUMPTION OF SOFTWOOD LUMBER

The construction industry consumes the largest proportion of softwood lumber, primarily for home building. The United States dwarfs all other countries in its use of softwood. But U.S. consumption of softwood lumber is expected to remain flat over the next fifteen years. During that time slight increases in demand for repairs and remodeling will be offset by a continued dip in demand for the lumber in industrial and residential construction, as indicated by Graph 18. Wood usage rates have slumped in recent years due to dramatic increases in softwood lumber prices, and total U.S. consumption is not expected to reach the peak levels of 1987. In recent years, softwood lumber has also lost market share to steel, engineered wood products, and other substitutes. As government or environmental forces have pulled public forest out of the market, driving prices up, contractors have increasingly turned to wood substitutes to reduce their exposure to sudden increases in prices. At the same time, the deteriorating quality of lumber, closely linked to forest degradation, is hurting the ability of softwood to retain market share.

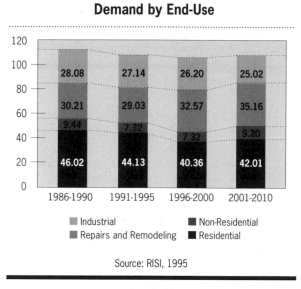

U.S. Domestic Softwood Lumber Demand by End-Use

Source: RISI, 1995

Graph 18

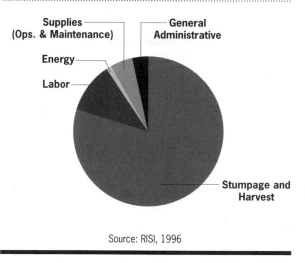

The most important criteria for purchasing paper...

Source: RISI, 1996

Graph 19

In Japan, the world's second largest softwood lumber market, new housing dominates consumption. The Japanese tend to tear down and rebuild older houses rather than repair and remodel them, because they lack available land for new housing starts.

Industry Structure

The softwood lumber industry is a relatively low value-added commodity business. As Graph 19 shows, stumpage, or the value of a tree in the forest, and harvest costs overwhelmingly dominate all other costs, which makes softwood lumber prices extremely sensitive to changes in value of the forest itself. Producers depend on sophisticated milling equipment, including log scanners that automatically size and grade wood to optimize cuts, close tolerance saws, efficient planers and sanders, and skilled workforces to remain competitive.

Industry Trends

A number of forces are affecting the softwood market:

- As second growth and plantation fiber replaces old-growth trees, the quality of softwood is declining. This means that higher quality grades of wood are becoming more expensive, and quality products more difficult to obtain.

- Engineered wood products, such as more competitively-priced composite I-beams, are substituting for large dimension lumber in many applications. In the coming years, expect price changes to also favor LVL for selected applications, predicts James B. Wilson of Oregon State University.

- Builders are becoming increasingly efficient in their use of lumber, and building codes are beginning to reflect these changes, permitting more efficient construction.

- New technologies are emerging that will enable sawmills to "scan" lumber and grade it according to strength. Currently most lumber is graded by dimension, so that all lumber is assumed to have the lowest common denominator in terms of strength. The new technologies will permit much more efficient use of wood in selected applications.

- Users of softwood lumber will increasingly look for substitutes to protect themselves against price swings caused by supply difficulties. As with many industries, substitutes will make inroads only when prices are particularly high. Nevertheless, even though prices may drop off after a price shock, most substituting technologies will maintain the market share they gained during the interim.

- As sawmilling technologies improve, distribution becomes more efficient, national and international trade equalizes price differences across markets, and sawlog supplies become scarcer, a greater proportion of the final value of softwood products will reside in the forest.

Environmental Forces and Softwood Lumber

Because much of the world's softwood forests are government owned, the management of these resources comes under greater public scrutiny than might be the case if the land were privately held. The availability of government forests for logging can change dramatically when political pressures increase, as has happened in the Pacific Northwest in recent years. Although the market has already largely absorbed the curtailing of harvests forced by the spotted owl controversy, dependence on government land will be a strategic issue to the industry well into the future. Because most softwood is produced in developed countries where environmental concerns are high on the public policy agenda and effective regulatory systems are in place, softwood lumber production is especially susceptible to increased government regulation.

Even though environmental forces have had enormous impact on the softwood market, several factors would appear to hinder the development of a market for certified softwood, including:

- A highly fragmented market of end-users.

- Multiple steps between final consumers and forest managers.

- The difficulty of developing differentiated products in a low-value-added commodity business.

- Historically, the construction industry has not proactively sought solutions to environmental issues.

Several circumstances could mitigate these constraints, however:

- If, as expected, large European publishers push for certification of the paper they purchase, large tracts of Scandinavian forests will be certified. These forests not only produce pulp for paper, but also large volumes of softwood lumber and sawnwood products. The Nordic producers will undoubtedly seek to distinguish their products to gain a competitive advantage, or try to create one.

- In addition, given the amount of softwood lumber coming *out* of environmentally important areas like British Columbia and *into* environmentally sensitive markets like California, it is reasonable to expect that the current situation will provoke change at some point.

- The consolidation among the do-it-yourself retailers, an avowed interest by some chains in certified forest products, and the close attention forest products companies are paying to certification.

HARDWOOD SAWNWOOD

Hardwood lumber production worldwide declined to 123 million cubic meters in 1993, down from 134 million cubic meters in 1990. Most hardwood lumber is consumed in the country in which it is produced, so trade accounts for a small percentage of the world hardwood lumber consumption. Major trade flows from the United States, Southeast Asia, and Brazil to Europe, and from Southeast Asia and the United States to Japan. These flows are expected to increase over the next decade, particularly to Japan. West Africa is expected to become a more important player in trade with the Pacific Rim as those markets push for increased production out of Africa.

TEMPERATE HARDWOODS

The United States which possesses vast hardwood resources, is the world's largest producer of temperate hardwoods. Most U.S. hardwoods are produced by nonindustrial private forest owners. Millions of these landowners harvest over 100 million cubic meters of hardwood per year. Hardwood forest ownership, indeed the existence of these forests, is a relic of the decay of the agricultural economy in the eastern United States. As farms were abandoned over the last century, they reverted to forest. Most of the forest is owned today in relatively small parcels that were inherited, sold, or somehow passed along from their agricultural past. Industry sources report that little consolidation is taking place in privately owned hardwood forest land in the United States, and that the predominance of small and medium nonindustrial land holdings is likely to continue. These owners often derive recreational benefits from their land, and keep it as a long term investment.

In the past forty years, U.S. hardwood growth has exceeded harvest. Nevertheless, a combination of passive management practices, withdrawals for environmental reasons, such as wetlands preservation and poor stocking, will probably keep a lid on production over the next twenty-five years.

U.S. Estimated Hardwood Lumber Production and Usage 1995-2005

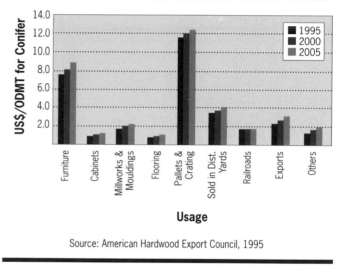

Source: American Hardwood Export Council, 1995

Graph 20

U.S. Hardwood Consumption

The health of the economy determines U.S. hardwood consumption. As Graph 20 indicates, the pallet and crating industries are the leading consumers of hardwood in terms of volume, followed by the furniture industry. The furniture, cabinets, and flooring markets, however, are the leaders in terms of sales. No major shifts in consumption are anticipated. The pallet and crating industry, however, is under great pressure to reuse its product because end users wish to avoid high disposal costs and pallets are threatened by substitution if hardwood prices rise considerably.

Processing of Temperate Hardwoods

The sawmills that process temperate hardwoods are typically small- to medium-sized businesses. They tend to serve a nearby geographic area and depend on that area for their timber supply. The leading hardwood producing state of Pennsylvania, for example, has 578 sawmills. Several factors are driving consolidation in this link of the hardwood value chain:

- Midsized mills with sales between $1 million and $10 million are being squeezed because they do not possess the technology - high-tech CNC milling machines and scanners, specifically - to maximize milling yields.

- More and more mills are trying to produce higher-value-added products. Medium to large mills are more likely to be able to meet the requirements for technology, investments, marketing know-how, and product development demanded by these ventures than smaller mills.

- Successful migration into these value-added products also requires new organizational skills, technologies, and links with markets that many smaller mills do not possess.

Environmental Opportunities in Temperate Hardwoods

U.S. exports of hardwoods grew from $400 million in 1988 to $2 billion in 1996. U.S. producers profited from the growing popularity of light woods, many of which come from the United States, and the reluctance among European buyers to purchase tropical hardwoods for environmental reasons. The growth in exports is expected to slow from its recent explosive levels over the next several years. But continued concerns about tropical hardwoods and diminishing supplies coupled with robust forests in the United States, and an increased familiarity among customers with U.S. hardwoods, should favorably position U.S. exports for additional growth, particularly for wood of matched lengths and widths, and other value-added products.

Despite these favorable conditions, Michael Buckley, director of the U.S. Hardwood Export Council, among others, cites several factors as obstacles to the growth of U.S. exports:

- Most buyers seek wood cut to metric measures. U.S. mills are accustomed to cutting to Imperial measures: The resulting waste can be significant for the purchaser.

- U.S. hardwoods are graded according to U.S. standards, which do not always coincide with European standards.

- U.S. mills dry hardwoods to U.S. specifications, whereas European hardwood purchasers often ask for wood with a higher moisture content.

Certification and Temperate Hardwoods

Conditions exist that would seem to foster greater interest in certification among U.S. hardwood producers. Several U.S. companies such as Kane Hardwood, MTE, and Seven Islands have found that certification of their products has given them access to market opportunities in the more developed European certified product market. As Europe evolves towards an even "greener" market, more opportunities will undoubtedly follow.

Many U.S. nonindustrial forest owners possess some of the characteristics found in the ideal candidate for sustainable forest management:

- They often hold their forest land as a long-term investment, typically one generation, and do not necessarily seek an annual return from their holding.

- Owners are often already interested in managing for multiple objectives, such as hunting and hiking. These individuals are already committed to a fundamental tenet of sustainable forest management—that forests should be managed to maximize several objectives, not just timber production.

- Many owners hold land because of a general interest in nature, and indeed often do not seek to maximize their financial returns.

- Most harvesting practices are already selective, a key component to sustainability.

• Hardwood is a product that has "direct communication" with its consumers—they can feel and touch the wood's grain, and more directly associate the product with the forest.

Although the fragmentation of small hardwood forest owners presents difficult chain of custody issues, they are a group that might be predisposed to buy into sustainable forest management practices more readily than other industry players.

TROPICAL HARDWOODS

Tropical hardwoods constitute the most controversial segment of the forest products industry. Western developed countries increasingly view tropical hardwood forests as the reservoir of the planet's biodiversity and its most important "carbon sink", both critical factors to the future of human well-being. Developing tropical countries, by contrast, see their hardwood forests as a resource to spur economic development. Most resist the notion that more developed countries can tell them what to do with their forests, particularly since the developed countries have, for the most part, already cut down their own virgin forests.

As indicated in Table 3, relatively little of the world's tropical sawlogs and sawnwood makes it into the world market. Plywood is extremely important to the world trade of tropical hardwoods, with trade accounting for over 80% of plywood production in some years. Although total volumes are a relatively low percentage of total production, the extraction of peeler logs is an important forest management issue. The volume of finished product understates the impact of this production on the forest.

Asia dominates tropical hardwood harvests, with production concentrated in Indonesia, Malaysia, and Brazil. According to the International Tropical Timber Organization, Asia and Brazil together account for 70% of logs, 63% of sawnwood, 84% of veneer, and 90% of all tropical plywood.

Tropical Forests Products Exports—1989-1993

	1989	1990	1991	1992	1993
Logs	26,163	25,397	24,056	21,864	16,081
Sawnwood	1,575	7,397	7,391	7,798	7,696
Veneer	578	647	768	1,036	1,121
Plywwod	9,550	9,791	10,626	10,708	11,176
Total	**37,866**	**43,232**	**42,841**	**41,406**	**36,074**

Tropical Forests Products Production—1989-1993

	1989	1990	1991	1992	1993
Logs	150,079	136,625	109,336	135,841	130,279
Sawnwood	42,389	39,357	38,631	39,179	39,203
Veneer	1,212	1,332	1,507	1,786	1,878
Plywwod	12,113	12,375	12,949	13,854	14,756
Total	**205,793**	**189,689**	**162,423**	**190,660**	**186,116**

Source: International Tropical Timber Organization

Table 3

Consumption of Tropical Hardwoods

All of the major importing countries of tropical hardwoods are located in the Pacific Rim region. In all cases, imports are driven by strong economies and a lack of adequate forest production of their own. The biggest importer, Japan, depends on imports from both tropical and temperate sources for some 70% of its wood needs, and as Graph 21 shows, Japan vastly outdistances all other importers.

Trends in Tropical Forest Products

Many European countries have drastically cut their imports of tropical forest products in recent years as retailers, shippers, and governments have responded to pressure from environmental groups, local and national, to reduce imports. These measures have

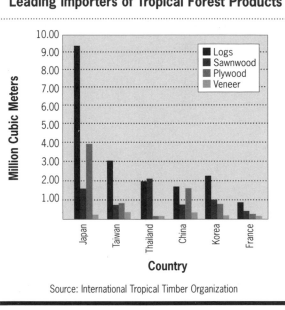

Leading Importers of Tropical Forest Products

Source: International Tropical Timber Organization

Graph 21

largely affected imports from Malaysia and Africa. Some estimates of the impact of European bans, certification efforts, and general skittishness on imports from Malaysia place the drop in those exports to Europe as high as 30%. Other effects of this general unease over the use of tropical hardwoods include:

- A surge in imports of hardwoods from temperate countries, such as the United States, into Europe and other sensitive markets.

- A host of initiatives in tropical producer countries to improve the image of their countries' exports. These efforts cover a range of sincerity; from public relations tactics to serious action to ensure the long-term viability of the industry.

- Some shift of African exports from Europe to the Pacific Rim.

Logging in Asia, by far the biggest producer of tropical hardwood, is dominated by large, vertically inte-

grated timber companies. These companies often produce plywood, paper, sawnwood, and other forest products. Their forests are often granted through government concessions that have served as platforms for diversification into new industries and some reforestation efforts.

As natural forests are exhausted in Asia, companies based in Malaysia and Indonesia have begun to migrate to new regions, such as northern South America and Africa in search of new harvesting opportunities. Although the potential production of these regions will not alter the balance of world trade, the environmental consequences are considerable. Furthermore, the institutional ability of some countries to negotiate equitable deals with foreign companies is questionable.

Environmental Opportunities for Tropical Forest Products

Unfortunately, the overwhelming majority of tropical products are sold in countries where environmental concerns and the issues of tropical deforestation receive short shrift. Imports to these countries are expected to continue, driven by robust economic and population growth. As long as this is the case, tropical forest products will continue to find an enthusiastic market.

However, environmental forces may still open opportunities for more sustainable forest practices in tropical regions for several reasons:

- Few certified tropical forest product companies exist, and all of them are relatively small scale. Nevertheless, many European buyers wish to use tropical woods because of their unmatched characteristics. These buyers complain about a lack of availability of tropical products, and demand for certified products clearly outstrips supply.

- As natural forests diminish around the world, the premiums paid for tropical hardwoods will probably rise. Countries that successfully husband their forests for the long term with the best management, sustained yield management systems, and carefully structured relations with logging companies will be able to generate long-term value for their economies. This is no small task, but certainly a plausible development strategy.

- Building businesses around certain sustainably produced species such as mahogany and teak will become a medium- to long-term opportunity as the market for certified wood matures.

Conclusion

The forest products industry defies facile predictions and explanations. The complex, multifaceted industry has a history of great dynamism. While it would be rash and precipitous to draw fast conclusions about the future of the industry, the intersection of industry change and environmental forces will determine which companies succeed in the next century.

The profitable forest products company of the future will need to have tight integration of a broader range of skills. The winning companies will be those that successfully structure themselves to respond to a dynamic, environmentally sensitive and smaller world.

As Table 4 indicates, the paradigms are shifting. Many companies have already responded, but the migration to the new paradigm still creates opportunity at the leading edge. Environmental forces are determining market

opportunities, supply dynamics, processing technologies, and the regulatory atmosphere. Though the forces are complex and their effects diffuse, they will define, in large measure, the industry's future in the coming century.

Although most sustainable forestry companies have started in the forest, they will not be able to rely upon their environmental prowess alone anymore. Success in the future will require adapting to the many forces transforming the industry. Investors and entrepreneurs will need to tie together the complexities of wood production and the evolving dynamics of the forest products industry to create sustainable new business opportunities.

The Forests Products Company of the Future

Company Name	New Paradigm	Old Paradigm
Fiber Supply	Shifting, source of innovation and competitive advantage	Assured, essentially an issue of cost Local supplies paramount
Processing Technology	Developed in conjunction with suppliers Information intensive	Purchased off shelf with minor variations Labor intensive
Production Focus	Value Driven High value added	Volume Driven Cost reduction focus
Sales and Marketing	Oriented towards final consumer Source of innovation	Sales and volume driven
Environmental Issues	Knowledge of issues may provide early positioning advantage	React to and/or head off regulations
Competition	Global Driven by multiplicity of factors	National or even regional Driven by cost and technology

Table 4

Marketing Products from Sustainably Managed Forests: An Emerging Opportunity

CASE STUDY

PREPARED BY:

DIANA PROPPER DE CALLEJON

TONY LENT

MICHAEL SKELLY

RACHEL CROSSLEY

A Case Study from "The Business of Sustainable Forestry"
A Project of The Sustainable Forestry Working Group

The Sustainable Forestry Working Group

Individuals from the following institutions participated in the preparation of this report.

Environmental Advantage, Inc.

Forest Stewardship Council

The John D. and Catherine T. MacArthur Foundation

Management Institute for Environment and Business

Mater Engineering, Ltd.

Oregon State University
Colleges of Business and Forestry

Pennsylvania State University
School of Forest Resources

University of California at Berkeley
College of Natural Resources

University of Michigan
Corporate Environmental Management Program

Weyerhaeuser Company

The World Bank
Environment Department

World Resources Institute

CCC 1-55963-618-1/98/page 3-1 through page 3-25

Marketing Products from Sustainably Managed Forests: An Emerging Opportunity

CASE STUDY

PREPARED BY:
DIANA PROPPER DE CALLEJON
TONY LENT
MICHAEL SKELLY
RACHEL CROSSLEY

A Case Study from "The Business of Sustainable Forestry"
A Project of The Sustainable Forestry Working Group

Contents

(continued)

Contents

(continued)

Marketing Products from Sustainably Managed Forests: An Emerging Opportunity

Introduction and Executive Summary

Most forest products analysts exploring the market for sustainable forest products have been searching for the green consumer. They have assumed that the well-documented consumer concerns about the impact of the industry on the forest would make consumer demand the dominant force propelling the industry toward sustainability.

While consumers' concerns about the industry's environmental impact remain important, many other, more powerful, forces are at work that will lead to an overall market shift towards sustainable forest management (SFM). These factors are converging to shift environmental attention on the industry from process controls and recycling to the management of forest resources. Today, a greater emphasis on the entire life cycle of forest products is pushing environmental concerns through the value chain from retail stores and pulp mills back down to the forest floor.

This paper assesses the major drivers and pressures on the forest products industry that are combining to bring about more SFM; thereby, significantly increasing the volume of sustainably produced forest products entering the markets.

The paper first looks at "push" drivers—those drivers putting pressure on the industry, pushing it towards greater sustainability. Second, external "pull" drivers are examined. These are incentives that encourage the forest products industry to change its practices and operate more sustainably. The third section describes how these push and pull drivers are converging to gradually create a market for sustainably produced forest products. Finally, geographic and industry structure factors are examined to identify how and where the transition to sustainable forestry is most likely to emerge.

FORCES, IN BRIEF

The forest products industry is one of many industries entering a transition phase in which more sustainable products and practices come to the fore. The move toward sustainability is largely market-driven rather than regulation-driven—as has been the case in the past. However, this transition is not driven, as is often assumed, by large-scale consumer demands. Rather, a variety of forces—so called "push and pull" drivers are combining to create a new market for sustainably produced forest products.

Forces pushing the industry toward sustainability include:

- Environmental groups pressuring the forest products industry to reduce the impact of forestry operations on wetlands, watersheds, and other aquatic ecosystems, and to preserve biodiversity and the scenic beauty of forests.

- The evolution of environmental management systems and forest management certification systems, from within industry and without, that provide mechanisms by which to differentiate the most sustainable companies and products in the marketplace.

- Early adapters, often with a history of superior stewardship in their forest management practices, that see sustainability as a business opportunity. Among the first to change their practices, they become bellwethers of the overall industry direction, and put pressure on others to change.

- Increased government regulation of the industry, which is forcing participants to manage their forests and harvesting operations in a more sustainable fashion.

The forces pulling the industry toward sustainability include:

- Consumer concerns about forest management and demand among consumer intermediaries for products grown and harvested in a more environmentally sound fashion.

- Demand from buyers groups, associations of commercial consumers of forest products that are banding together to create aggregate demand for certified forest products.

- Industrial demand, particularly in the publishing industry, for paper products that carry no environmental risk.

- Government-generated demand from various government entities, primarily local, that are specifying a preference for certified wood products.

- Demand from niche markets including architects, cabinetmakers, and some parts of the furniture, flooring, and hardwood distribution industries.

A DEVELOPING MARKET

The framework presented here for assessing the developing market for sustainable forest products indicates where demand for these products already exists and where it is anticipated to develop. As the analysis shows, the major issue is not whether demand for sustainable forest products will develop, but how fast and in which markets sustainable forest management will become an important market variable.

And while certified products (products that carry an "ecolabel" awarded after an independent, third party has evaluated forest management practices) are the most readily identifiable in the marketplace, certification alone should not be mistaken as the only true indicator of a shift towards sustainability. Certification has catalyzed huge debates around sustainability in industry and environmental circles, and has given a name and process to the development of markets for sustainably produced products. But while certification represents the leading edge of sustainability, many other trends mark a broader, perhaps more gradual, shift to sustainable practices.

The market for sustainable forest products is in its infancy, which cannot be overemphasized. Environmental quality as a differentiating product characteristic is a relatively new phenomenon and certified wood products have existed in name only

in the last five years. Any demand for sustainable and certified products is greatly restricted by the current very low supply of certified wood products on the market—less than 0.60% of world industrial roundwood.

Trends indicate that Europe will lead the development of the market for sustainable forest products for the foreseeable future. Furthermore, demand for environmentally friendly paper in Europe, led by German publishers and members of large buyers groups, may soon shift the current predominance of certified solid wood products to paper. Supply shifts are likely to be first evident in Scandinavia, as other large companies in the region follow the lead of major Swedish forestry companies and become certified.

While the shift towards sustainably produced forest products is likely to be slower and less dramatic in the United States, several major American forest products companies are closely monitoring developments in the sustainability arena. The major U.S. industry association—the American Forest and Paper Association (AFPA)—has established its own "'Sustainable Forestry Initiative" (SFI) to move the industry towards higher operating standards. The formation of the U.S. Certified Forest Products Council and the experimentation with certification by state, and, possibly, federal forestry authorities may soon spur the development of a larger market for certified products. Canada, conversely, has already established an SFM systems program—based on International Organization for Standardization (ISO) environmental management protocols and procedures. One major reason for this head start is that Canada has a greater dependence on exports than the United States does—and demand for certified products from European customers is stronger. Interest among Canadian companies has been strong, indicating the significant volumes of supply may be available in the near future. Customers have also responded favorably too.

Japan and Asia are lagging much further behind, as they currently experience little internal market demand for environmentally sound products. No certified wood is sold as such in those markets. Nonetheless, most Asian countries are experimenting with sustainability initiatives of one sort or another, many with an eye to fulfilling the International Tropical Timber Organization's (ITTO) campaign for sustainable management by the year 2000. While it is not possible to gauge the current and future efficacy of these initiatives, they will presumably focus attention on SFM and produce some improvement in on-the-ground practices.

The rise of sustainability as a more important market variable for the forest products industry, and the increased globalization of the industry and its customers will exert a number of catalytic effects. Many international companies are evaluating the sustainability of their products, tracing them all the way back to the raw material. Should they make changes, this will have an impact on all products and markets. And, ironically, the same factors that have kept such global players out of the sustainable products market in the past are likely to fuel its development once such companies enlist.

Thus, a combination of forces is slowly converging on both the supply and demand sides of the forest products value chain, fueling the development of a market for sustainable forest products. As more industry customers are becoming concerned about the origin of the forest products they consume, and as pressure mounts from environmental groups, governments, consumer and industrial demand, and leaders in the industry, suppliers are searching for ways to improve their forest practices. Forest products companies worldwide are now keenly aware of the market's burgeoning interest in forest sustainability. Some are making significant changes to their practices already; others are biding their time and watching developments closely. Over time, though, the cumulative effects of these changes will make sustainable forest management an integral part of the industry's overall value equation.

Push Drivers

This section reviews four major drivers at work within the industry that are pushing it to develop better forest management practices. These include pressure from environmental groups, voluntary certification initiatives springing up within and outside the industry, early adapters leading the charge in defining and adopting new practices, and governmental regulation.

PRESSURE FROM ENVIRONMENTAL GROUPS

Environmental groups with their conservation agendas, governments geared to political sensitivities, trade associations in defense of their membership, and certified companies with products they wish to position as green are all pushing the forest products industry toward greater sustainability. Although tactics vary widely, from slightly defensive in the case of trade associations to highly aggressive among some environmental groups, the likely net result is greater forest sustainability.

Environmental groups are leaders in propelling the forest products industry toward SFM. The head of the German magazine publishers' association, Wolfgang Furstner, explained the importance of these groups in a speech before the International Periodicals Symposium in mid-1996:

> Today, environmental organizations worldwide are an expression of society's fear and concern, having recognized the consequences of exploitation of nature and its resources when the principal of sustainability is not adhered to. Such organizations are so strong and widely accepted because they represent the guilty conscience of an industrial society which knows that it has to pay the price for its prosperity with the often ruthless exploitation of nature and its resources.

Throughout the century, environmental groups have repeatedly tangled with forest products companies. The U.S. environmental movement can trace its origins, in part, to efforts to protect forests from the cut-and-run practices of loggers early this century. During the 1960s and 1970s, the groups pressured paper mills to reduce effluent discharge. During the 1980s, they concentrated on increasing the content of recycled fiber in paper products. In the 1990s, environmentalists have increasingly pressed the forest products industry to improve forest management practices.

Environmental groups have had an enormous impact on the forest products industry for a number of reasons. They enjoy more respect and trust among the general public and educated elites than does industry or government. For this reason, they have been able to set the agenda of forest sustainability debate, and are often able to influence outcomes even before governments get involved. In addition, because of the respect afforded to them by the public, environmental groups often inform and shape consumer demand. Examples of this trend are multiplying.

The tuna industry learned this lesson in the early 1990s when tuna consumption plunged after environmentalists protested that the industry's fishing practices were killing large numbers of dolphins. The industry was forced to introduce nets that did not catch and drown dolphins.

In Europe, to a greater extent than in the United States, there is widespread acceptance of the idea that the root cause of many environmental problems is over-consumption of the world's finite resources, and that much of the solution therefore lies in reducing consumption. Even though properly managed forests are a renewable resource, the forest products industry has proven vulnerable to this notion, particularly the paper products segment. In the United Kingdom, the World Wide Fund for Nature (WWF), a widely-respected environmental group, pressured major retailers and large forest products buyers in the early 1990s to join a group that pledged to purchase paper and wood

from third-party certified forests. Following the U.K.'s lead, similar groups have since been formed, lead by national environmental groups, in Holland, Germany, Austria, Switzerland, the United States, and Australia. As a result, many forest products suppliers that service these groups' members are in the process of becoming certified in order to retain their business.

According to a study of German consumers by Jaakko Pöyry, a leading forest products consulting company, environmental groups, including Greenpeace, enjoy very high credibility among the general public. That credibility helped Greenpeace successfully lead the effort in the late 1980s to force many European paper companies to produce totally chlorine-free (TCF) paper and to make elemental chlorine-free (ECF) the minimum threshold of environmental acceptability. Most recently, Greenpeace has rebuked the magazine and newspaper industries for the forest management practices of their suppliers and their "excessive" paper consumption.

In the U.S. Pacific Northwest, environmental groups have won several victories over the federal government that sharply limit industry access to forests owned by the federal government. And in Canada, they have forced the national and provincial governments to review forest management regulations and implement stricter laws.

Outside North America and Europe, environmental groups are active but so far have not had sufficient political clout to force significant changes in forest management practices. However, the intensified globalization of the industry is likely to expose forest products companies to more pressure from environmental groups, and bolster their ability to influence individual and corporate consumers, thus bringing about further changes in industry practice to meet market demands for sustainably-produced wood products.

VOLUNTARY SUSTAINABILITY INITIATIVES AND THIRD-PARTY CERTIFICATION

For purposes of this discussion, independent third-party forest certification is considered to represent the leading edge of the effort to bring SFM into the marketplace. No attempt is made here to evaluate the respective merits of environmental management systems-based (proposed by ISO) versus field-based forest management-based certification systems (proposed by FSC). Instead, both are considered steps toward more sustainable forest management.

Over the last few years many initiatives aimed at improving forest management have sprung up, some led by industry, some by environmental groups, and some a combination of both. Only the major initiatives will be discussed here.

Forest Stewardship Council

The Forest Stewardship Council (FSC), a nonprofit organization of international environmental groups, academics, and industry members, has set forth principles by which all forest types should be managed. The group hopes to establish a single globally accepted set of standards for certifiers of forests and wood products. The FSC evaluates, accredits, and monitors independent certification companies. These companies not only evaluate on-the-ground inspections of forest management practices, but also track a series of chain-of-custody protocols to ensure that finished products that carry the FSC stamp of approval are made of wood that comes from a certified forest. FSC was the first certification system in place and has won the greatest credibility among environmental groups.

The FSC national standard setting process involves convening a wide variety of stakeholders in a given country or region who articulate a series of best principles and standards for that area. As of 1997, many states in the United States and countries around the world were developing these standards.

International Organization for Standardization

The International Organization for Standardization (ISO) is a global standards body which in 1993 created its "14000" Environmental Management Standards (EMS) dedicated to developing general environmental management tools and systems. The Canadian Standards Association (CSA) hopes that the standards it is developing will become acceptable to ISO. By year-end 1997, ISO's technical committee is expected to have published and accepted the report prepared by the CSA. In preparation for ISO acceptance of the Canadian-prepared standards, a number of Canadian forest products companies have begun the process of certification. ISO has also attracted the interest of large forest products companies elsewhere, including the United States, New Zealand, and Europe.

Other Initiatives

In addition to the initiatives developed under the auspices of the FSC and ISO, the forest products industry and several important producing countries have set up their own forest management initiatives. In the United States, the AFPA has the SFI, which requires all of the group's members to adopt a set of principles whose goal is ongoing progress toward SFM. SFI's requirements, however, had proven sufficiently onerous to result in the AFPA's losing more than ten percent of its membership by the mid-1990s. Despite the attrition, the organization has pledged to push the initiative forward.

Other national industry associations and governments have created voluntary initiatives to improve forest management and the public perception of the industry's management practices. The ITTO, an association of tropical timber producing and consumer nations, has set a goal that by the year 2000 its member nations will manage their forests in a sustainable fashion. Environmental groups fault most initiatives like those of AFPA and ITTO on the grounds that such programs are self policed, i.e., they do not involve independent third-party verification of forest management practices. Nevertheless, the existence of such initiatives indicates that both industry and government recognize the need to move toward greater sustainability in forest management.

Even though industry-sponsored efforts have their merits, they are unlikely to satisfy a number of the actors pressing the forest products industry for change. The principal problem with industry-verified initiatives is that forest products companies do not enjoy enough public credibility, even among generally proindustry groups. Interviews with numerous industry insiders indicate that the companies that subscribe to the AFPA's SFI are aware of its limitations. While all agree that SFI is a positive step, many company officials question whether the SFI serves their long-term interests. These executives are actively tracking the FSC and ISO certification efforts, and keeping their options as open as possible.

It is far too early to tell which, if any, certification initiative will win the most adherents or predominate in which markets. But undoubtedly the proliferation of these sustainability schemes, and the involvement of industry associations, environmental organizations, and timber companies will push the market in several respects:

- Certification is the bellwether of market change, but does not constitute the entire market. Certification initiatives represent the leading edge of sustainability, but are also an indicator of a trend of rising pressure on the forest products industry to improve the way in which it manages forests—with or without certification.

- As more companies participate in sustainability initiatives, they increase pressure on the nonparticipants.

- The greater the number of companies that adopt sustainability efforts, the more supplies of sustainable product will increase.

EARLY ADAPTERS—BELLWETHERS OF INDUSTRY DIRECTION

As companies with well-known legacies of exemplary forest management, such as Collins Pine Co. in California, formalize their commitment to sustainable forest management through certification, they propel other companies to follow suit and raise their standards. The process has created a significant group of suppliers that is now marketing environmentally differentiated products. Their business rationales for becoming certified provide useful insights into the motives behind the industry's attempts to improve forest management.

Company Positioning

Most certified companies opted to certify their operations as a natural outgrowth of a history of forest stewardship that generally predated the movement toward third-party certification. Seven Islands, Collins Pine, and Menominee Tribal Enterprises, to name several U.S. examples, have long traditions of superior forest stewardship, which readily lent themselves to third-party certification. Executives of these companies view certification as a way to validate their long standing commitment to sustainable forestry and communicate its value to regulators, customers, and neighbors.

Improve or Secure Access to Forest and Other Resources

A number of forest products companies have certified their operations to ensure continued access to forest resources or to improve possibilities of obtaining other resources. In the case of Seven Islands of Maine, the *Journal of Forestry* reported that Seven Islands pursued certification in part to enhance the company's public image and insulate it from forest products critics. At the time, certification seemed politically prudent since major environmental groups were shifting their attention from public lands in the West to private lands in the northeastern United States. Although Seven Islands owns its own forest, the company preferred certification as a kind of self-regulation to the prospect of externally imposed regulations controlling its access to raw material.

In the cases of Portico, FUNDECOR, and Ston Forestal, all in Costa Rica, their decisions to certify were based largely on resource availability. Costa Rica has relatively strict forest management laws and one of the highest percentages of total forest area under certified management. Portico certified its forest operations to help secure a large area of forest. FUNDECOR applied for certification to help it gain access to international sources of debt financing destined for its reforestation and natural forest management programs. In an effort to diffuse criticism from environmental groups, Ston Forestal certified its *gmelina* plantations after a battle with local environmentalists, who claimed that the company's proposed port facilities would damage marine ecosystems. Ston now displays its certification logo prominently in its public relations advertising.

Market Access and Product Differentiation

Certification began at the forest floor—the beginning of the forest product value chain—as an attempt to capture a premium price for superior forest stewardship. The experience of organic food marketers and the makers of green consumer products in the early 1990s, such as natural cosmetics, green specialty housewares, and natural clothing, led some forest products producers to expect that consumers would demand and pay a premium for environmentally-certified wood products—an assumption that has generally not yet materialized. Certification, however, can be an attractive product differentiation strategy. Many forest products are commodities, so certification helps producers distinguish their products from competitors.

Although producers report they rarely receive a premium price for their certified products, certification does help to open up new market opportunities. Colonial Craft, a midsized U.S. manufacturer of window and door grills, picture frames, and molding, features a line of certified products that

executives say has helped the company make inroads into otherwise difficult to enter markets. Aracruz Celulose of Brazil, the world's largest pulp producer, is actively studying the certification of its eucalyptus plantations as a means to help it ensure access to the European market. And nearly one million hectares of forestland in Poland have become certified, in part, to meet demand for certified products from members of the United Kingdom Buyers Group.

Another incentive to become certified is to protect against loss of existing market share for environmental reasons. Consumer concern over tropical forest loss has caused a significant decline in tropical hardwood imports in many European countries in the last few years. Companies that can offer certified tropical products stand a good chance of regaining that market share.

INCREASED GOVERNMENTAL REGULATION OF THE INDUSTRY

Governments, reacting to public pressure for improved forest management, are also pressing the industry hard on forest management issues. In some cases government regulations are taking forests out of production entirely, and sharply curtailing harvesting.

In the United States, stricter wetland regulations are making harvesting impossible in areas that were previously open to logging. In another case, Brazilian regulators in 1996 imposed a two-year moratorium on harvesting mahogany due to over-cutting. In the past few years, many nations have adopted various types of restrictive log export policies, in part to capture a higher share of value-added processing. Other examples that indicate regulators are moving toward greater oversight and restrictions on forests include:

- California and Oregon lead the United States in restrictive forest management regulations based on ecosystem management. Under this system, timber production is only one of several variables that the management system seeks to optimize, which is a dramatic departure from past approaches under which timber production was an almost exclusive goal.

- In 1997 Pennsylvania and Minnesota began large-scale pilot audits involving hundreds of thousands of hectares of land to help gauge the feasibility of third-party certification for all of the wood produced on their state-owned lands.

- Several developing countries, including Brazil, Malaysia, and Bolivia, which have traditionally followed the pattern of developed countries and engaged in "mining" of old growth forests, have begun to mandate greater SFM in their countries' forestry operations. In mid-1997 Brazil imposed a moratorium on mahogany logging to protect that species.

Pull Drivers

The fragmented structure of the forest products industry complicates the articulation of demand for sustainably produced forest products. Millions of forest owners produce wood and fiber for thousands of different end-users. In between is a relatively long value chain of primary processors, secondary processors, distributors, wholesales, and retailers. Demand plays out at various points of consumption, from end-consumers, to industrial customers, to groups organized to stimulate supplies of certified forest products. Initial efforts at generating demand were closely tied to the early adapters of sustainability, generally nonintegrated producers that sold solid wood products. As more industry players on both the supply and demand side of the market equation move toward sustainability, the forces stimulating demand will affect a broader range of forest products, including the paper products industry.

This section reviews those drivers pulling on industry, pressuring it to change. These include consumers' role in stimulating demand, the markets

that have grown around several certified producers, industrial demand, and demand orchestrated by buyers groups and government bodies.

LATENT DEMAND AMONG CONSUMERS

Numerous studies show that European and North American consumers prefer forest products produced in an environmentally friendly way. The European Commission's Eurobarometer study of consumer attitudes on a wide range of issues in 1995 found that 58% of those polled in the United Kingdom and 50% of Germans were willing to pay more for green products. Similar studies in the United States have shown that a significant percentage of consumers express a similar willingness.

As Graph 1 shows, one study conducted in the United States by Vlosky found that only 28% of consumers said they would not pay a premium for certified wood products. The remainder said they were willing to pay premiums between 10% and more than 50%. The problem lies in how to account for the difference between what consumers say they will do and what they actually do.

The authors believe that although consumers are worried about the effects of their behavior, the

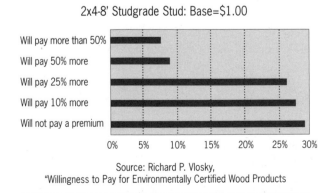

Respondent Willingness to Pay for Certified Wood Products

2x4-8' Studgrade Stud: Base=$1.00

Source: Richard P. Vlosky,
"Willingness to Pay for Environmentally Certified Wood Products

Graph 1

research on "willingness to pay" is overly optimistic. Many in the forest products industry share the view that the consumer, when faced with two similar products whose only differentiating characteristic is a green label, will opt for the less expensive item. Many companies contend that their own studies show that most consumers are not interested in certified or sustainably produced wood. Indeed, several focus groups have indicated that most consumers have no idea what certified forest products are. Home Depot's less than stellar success in the mid-1990s with a handful of certified sawnwood products would seem to support the industry thesis.

But consumers express profound concern about the impact of their purchasing decisions on the environment, and this concern does find its way back to the marketplace through the efforts of the large industrial customers of the forest products industry that possess the wherewithal to make suppliers pay attention.

Consumers Worried about the Forest

The more meaningful finding of the various "willingness to pay" studies is that they have consistently found a level of discomfort among consumers about their consumption of forest products. The movement toward higher recycling content in paper grew out of concerns about the disposal of municipal solid waste, but was greatly assisted by the public's worries over the consumption of forest resources. Despite the industry's attempts to portray itself as an excellent steward of forest resources, consumers are much less likely to believe industry claims about environmental issues than those made by environmental groups. According to a study by the Angus Reid Group, 79% of adults believe all or some of what environmental groups tell them about the environment, and 78% trust the credibility of scientific evidence. Only 37% believe business or industry claims.

Consumer Concerns about Paper Use

Recent surveys indicate that consumers consider paper consumption a significant issue.

A private survey of the Swedish forest industry completed in 1995 showed that 64% of Britons and 60% of Germans had a very or fairly positive perception of Swedish forestry. Nevertheless, 75% of Britons agreed they worried about how much paper they used and tried to use as little as possible, and 62% disagreed with the statement that the world's forests are easily able to meet the world's needs. This is a concern for an industry whose future depends, in part, on the continued consumption of paper products.

Publishers are constantly threatened with substitution by other media, historically radio and television, but more recently by the Internet and other electronic media. They are justifiably concerned that consumer aversion to paper consumption could cause significant impact on the traditional publishing industry above and beyond any market share loss that will take place under the best of circumstances.

Consumer Concerns about Solid Wood Products

The Western Wood Products Association tried to understand concerns about wood consumption by surveying the individuals who often make the purchasing decisions on behalf of consumers. As Graph 2 indicates, the study found that nearly 35% of architects thought that their clients believe they

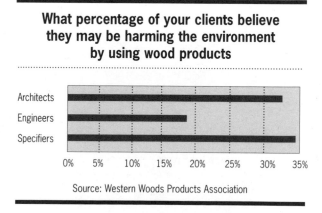

What percentage of your clients believe they may be harming the environment by using wood products

Source: Western Woods Products Association

Graph 2

might be harming the environment by using wood products. An even greater percentage of specifiers (people who specify wood or other product for a given application) and a significant number of engineers said the same about their clients.

As Graph 3 shows, among those who expressed concern about environmental issues, once again forest management topped the list, with over 70% of those respondents who believe wood consumption harms the environment saying that industry is "not taking care of national forests." Furthermore, both engineers and architects reported that their customers were more concerned about forest management practices than they had been in the past.

Consumer concerns are less important in sawn-wood and panel products than for paper products because most consumers actually see and feel relatively little of the wood they consume, and are

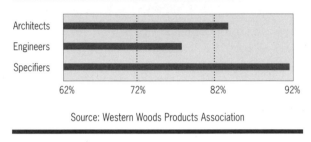

Respondents who say customers would be interested in products endorsed by "third party" certification company

Source: Western Woods Products Association

Graph 4

more distanced from those purchasing decisions. Nevertheless, the fact that the public perceives wood's environmental credentials as weak indicates that there is room to improve the product's image, particularly as substitutes threaten wood's historical dominance in many end-uses.

The same Western Wood Products Association study shows that third-party certification of SFM would help allay some of these concerns. As Graph 4 shows the study found that 84% of architects, 75% of engineers, and over 90% of specifiers say their clients would be interested in wood products endorsed by a third-party certification organization.

Another study by Vlosky examined what kind of entity would be most trusted by architects, building contractors, and home center retailers to certify forest management and harvesting practices. Third-party certification entities emerged as the most trusted, well ahead of the Federal government and environmental groups. From the perspective of what might be considered a relatively probusiness group, third-party certifiers were perceived to be far more trustworthy than the forest products industry (see Table 1).

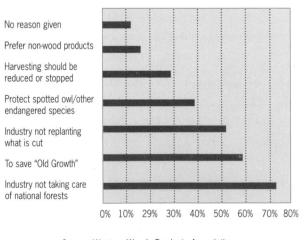

Reasons given for respondents belief that using wood products harms the environment

Source: Western Woods Products Association

Graph 3

Business Customer Perspectives on Forest Management Certification

*Level of Trust to Certify Forest Management and Harvesting
(1=Trust Most to 4=Trust Least)*

	Architects	Building Contractors	Home Center Retailers	Weighted Average
Third-Party Certification Entity	2.0	1.6	1.7	1.7
Forest Products Industry	2.6	2.4	1.9	2.2
Federal Government	2.8	3.1	3.0	3.0
Non-Governmental Environmental Groups	2.5	3.0	3.6	3.1

Source: Richard Vlosky

Table 1

BUYERS GROUPS

Several buyers groups or associations of companies interested in purchasing certified forest products, have organized in Europe and North America. These groups were created primarily because the members had such great difficulty obtaining certified forest products. The United Kingdom and the North American Buyers Groups, the two most active initiatives to date, and their roles in stimulating demand for sustainable forest products, are discussed below.

U.K. and other European Buyers Groups

The U.K. Buyers Group organized in 1991 under the name 1995-Plus Group, after its initial target for members to purchase all wood products from certified sources by 1995. Originally, it consisted of about a dozen members, notably leading do-it-yourself (DIY) chains that were under attack from environmental groups. In 1995 the group moved into high gear after the large DIY chain J Sainsburys joined and the organization began to concentrate on paper products. The current group comprises seventy-two companies, including DIY chains with a combined total market share of 80%,

the number one bed maker, several dozen supply and distribution companies, the two largest supermarkets in the country, the leading newsagent, and the top pharmacy chain.

Because the members include the top-tier players in their respective sectors, the group represents substantial buying power. Table 2 details the estimated total wood product purchasing volume of the U.K. Buyers Group. In the retail sector, however, the estimates are somewhat uncertain since only one company has assessed its consumption rigorously.

Overall, the U.K. Buyers Group represents approximately 14.9% of total U.K. industrial roundwood consumption. While their total forest products consumption is impressive, several factors weaken the group's buying clout. A relatively small proportion of their purchases are products where the retailers have overwhelming market power, notably paper products, such as tissues and diapers, which they

Demand for Forest Products from members of the UK Buyers Group

(Cubic Meter Equivalents of Industrial Roundwood)

SECTOR	Volume
DIY Stores	1,400,000
Retailers - general	6,700,000
Suppliers	1,200,000
Less intergroup Sales	(5,00,000)
Less Recycled Fiber Estimate	(1,300,000)
Total Volume	7,500,000

Source: Delphi International, 1996

Table 2

buy in bulk. Furthermore, while retailers may sell large volumes of newspapers and try to persuade their suppliers to use sustainable timber, they cannot switch suppliers.

Currently available supplies of certified forest products are insufficient even for the relatively small U.K. market. For that reason The U.K. Buyers Group members have made stimulating supply a primary task. In a survey of its members, 38% said they were making considerable efforts in this area. As a group, the organization has been cajoling Scandinavian paper producers to certify their forestry operations. According to Brian McCloy, a marketing specialist with the Canadian Pulp and Paper Association (CPPA), much of the impetus for Canadian ISO certification efforts has come from interest in certified products by British retail and industrial consumers. One major publishing company, BBC Magazines, has joined the group. BBC accounts for 10% of the U.K. magazine market. Although it is committed to certification, BBC is uncertain as to the future source of certified paper.

Several other Buyers Groups have grown up in Europe modeled after the 1995-Plus Group, and spurred by its success. Groups in Austria, Belgium, the Netherlands, Germany, and Switzerland all have slightly different structures and missions, but the same basic goal—facilitating the trade in certified forest products.

U.S. Certified Forest Products Council

Environmental Advantage (EA), an environmental consulting firm, working with several major corporations, universities, and conservation groups, designed and established the Certified Forest Products Council (CFPC). The CFPC held its inaugural meeting in April 1997. CFPC's primary goal and function is to help major buyers of wood products in the United States—the world's largest wood products market—to promote SFM by shifting to buying and using independently- certified forest products. The group intends to set up mechanisms to help facilitate and increase the purchase, sale, and use of these products.

Companies and organizations that buy, resell, or are secondary manufacturers of wood products will be members of the not-for-profit business initiative. Several large companies in the home improvement, self-assembly furniture, wood products manufacturing, construction, and retail segments are expected to become founding members of the CFPC. Smaller furniture manufacturers, architects, specialty product companies, and large paper products users are expected to join as well. CFPC hopes to provide a clearly marked target to which certified producers can market their products. Like its counterpart in the United Kingdom, the CFPC wants to stimulate supplies of certified wood products.

Japan

In early 1996, leading companies and environmental groups in Japan, including Kirin Brewery, Canon, Inc., Sanyo Corp., Sony Corp., Matsusita Electric Industrial Co., the Green Consumer Society, and WWF-Japan, among others, formed the Green Purchasing Network. The coalition is designed to kindle the market for green products. Though the network's goals are relatively modest and by no means concentrated on the development of markets for sustainably produced forest products, the creation of such a coalition, given the importance of its members in the world economy, is yet another indication that forest sustainability issues are being taken seriously by industrial buyers of forest products.

INDUSTRIAL DEMAND

Of the several cords of industrial roundwood that the typical U.S. citizen consumes directly or indirectly each year, precious little of that wood is purchased directly by the consumer. Most of the wood is consumed as paper, as part of a house or a remodeling job, hidden in a piece of furniture, or in some other form. The final consumer is unaware that he or she is purchasing a forest product. For this reason, industrial demand for sustainable forest products will be the determining factor in moving the sustainable forest products market forward. Most forest products flow through industrial channels, industrial players often hold significant leverage in the marketplace, and they generally have no vested interest in the forest management side of the industry.

Several factors are helping to shape the evolution of industrial demand.

Managing Perceived Environmental Risk in Supply Chain

Some large buyers of forest products (such as home improvement stores, large furniture manufacturers, retailers, and publishers), many of whom have joined buyers groups are beginning to pressure their suppliers to guarantee that the forest products they purchase have been produced in an environmentally sound fashion. It is not at all uncommon for buyers to personally inspect forestry operations. In the words of one executive at Procter & Gamble, one of the largest customers of fiber for packaging and diapers, "sustainable forestry should be a given." Procter & Gamble and other large companies, particularly those that sell name-brand products, increasingly want to prevent any environmentally tainted materials from entering their products, even in packaging. They are increasingly asking their suppliers to give them written assurances that, among other things, fiber and paper products were not produced using pulp obtained from tropical forests and that the products come from well-managed forests. In most cases, these assurances consist of negatives—that no tropical forests were cut to make a product, or that no species were endangered.

At the moment, few large purchasers are expressing their risk aversion through positive assurances by stating that a given product was produced in an environmentally sustainable fashion. One of the main goals of the buyers groups springing up around the world is to reverse this situation, so that companies can make positive, reputable claims about the products they sell. As the Magazine Publishers of America's Task Force on The Magazine Industry and the Environment declared in 1996, all paper users, including magazine publishers, may eventually want to certify the sources of the fiber in their paper. This task force included representation from publications as varied as the *New York Times, Readers' Digest, Time,* and *Country Living, Playboy,* and the Conde Nast and Hearst Enterprises magazine groups. Although most of the task force recommendations were relatively conservative, the fact that so many large publishing interests are looking at the issue of forest sustainability indicates the high level of concern among the pulp and paper industry's major customers.

Protecting Brand Equity and Customer Goodwill

Currently, the largest potential market for sustainable products is Europe. In the United Kingdom during the early 1990s, several environmental groups demonstrated with large inflatable chain saws in the parking lots of the leading DIY home improvement product retailers. The action made these companies realize the risks of not knowing the environmental implications of the products they sold. They quickly committed to buying certified forest products and are now doing so in small volumes. Certification gives retailers the tool they need to answer queries about the nature of their products, and avert the risk of damaging their reputations by selling environmentally unfriendly products. Because Procter &Gamble, for instance, sells branded products, it cannot afford to risk the prestige of its brands on an unfriendly forest product used in its packaging. Similarly, British retailers like J Sainsbury plc cannot afford to risk their reputations as trusted retailers over a low-value forest product.

One survey of the members of the 1995-Plus Group found that most retailers cited specific consumer pressure as their initial motive for joining the group. Once members, the retailers came to consider providing environmentally sound products as another opportunity to serve their customers. According to an unpublished study by Delphi International in the United Kingdom, "Two major respondents noted that it [certification] is a matter of building customers' trust in the quality of the company's products and policies. Timber is just another area where they can demonstrate their ability and willingness to respond positively."

Addressing or Anticipating Customer Concerns

Similarly, providing assurances of sustainability can also help suppliers assure their industrial customers of the environmental quality of products. A 1994 internal study by Weyerhaeuser Co. revealed that forestry practices topped the list of issues identified as areas of concern by their industrial and retail customers. Forestry practices were almost twice as important to customers than the formerly predominant issues of chemical pollution and recycling. Forest products companies typically produce relatively undifferentiated products and sell them into price-sensitive markets. In these essentially commodity markets in which quality, price, service, and delivery are often virtually identical, market opportunities would appear to be open to the supplier that positions itself ahead of the competition on these issues.

GOVERNMENT-GENERATED DEMAND

A number of governmental entities, primarily local and municipal, in the United States and in Europe, have passed ordinances designed to make sustainability an important criteria in both their own purchasing decisions and in the markets over which they hold jurisdiction. The General Agreement on Tariffs and Trade (GATT) and other international trade agreements render illegal most regulations that restrict trade based on environmental considerations. This constraint has impeded the development of many trade related initiatives, but by no means all:

- As early as 1988, the European Parliament passed a nonbinding measure which proposed that the European Community (EC) import only tropical hardwood products certified to be produced under forest management and protection programs.

- In 1997 the U.S. Department of Defense submitted a bid package for a large-scale renovation project of one of the world's largest office buildings, the Pentagon. The contracting office is actively seeking sources of FSC-certified forest products and lumber.

- In 1992 Austria tried to increase tariffs on unsustainably-produced timber and require ecolabeling of tropical timber products. Both measures were eventually drastically watered down to make them GATT-compatible.

- A number of local and city governments in Britain, Germany, and the United States have banned the use of tropical timber in public buildings and projects. They are now in the process of changing this policy in favor of certified products.

NICHE MARKETS

The existence of small producers of third-party certified forest product companies has attracted a number of buyers. Architects, cabinetmakers, furniture manufacturers, and flooring companies have been primarily responsible for pulling small volumes of product into the marketplace. While these niche markets are small, they are the most important outlets for the certified wood products to date. Organizations such as the Good Wood Alliance in the United States, which grew out of the Woodworkers Alliance for Rainforest Protection (WARP) were instrumental in vocalizing demand of their membership for sustainably-produced forest products, thus providing a clear market target for early certified producers.

Push and Pull Drivers Slowly Converging to Create a Market

Despite strong pressures on both the supply and demand side of the forest products industry for greater SFM, the quantities of sustainably produced and certified product in the marketplace are still extremely low due to their nascence. This section describes the current level and source of supplies of FSC-certified forest products, examines the reasons why the market is still in the early stages of development, and explains why the introduction of environmentally differentiated products is such a long-term process.

FSC-CERTIFIED PRODUCT

One way to measure of the state of the market for sustainable forest products is by simply quantifying the supply of industrial roundwood currently marketed by FSC-certified companies. Despite many signs of a nascent market for sustainable forest products, supplies of this certified product still represent a tiny percentage of world industrial roundwood—less than 0.60%. Even more surprising, only a small volume of those current supplies are positioned or marketed as certified product. Most certified wood goes directly into the mainstream supply as undifferentiated product.

The task of quantifying the FSC-certified supply of wood is relatively simple, given the small number of producers that have certified their forestry operations to date. Table 3 lists those producers and their production volumes. Most companies direct their production toward solid wood products, producing hardwood and softwood lumber. Analyzing the current certified producers by volume and by land under management, however, reveals another picture. Polish production of certified forest products accounts for nearly half of total production

FSC-Certified Companies and Production and Land Area with Calculation of World Industrial Roundwood Percentage

Name of Company	Cubic Meters	Land Area (ha^2)
Seven Islands	742,922	960,000
Menominee	27,242	89,000
Kane Hardwood	29,412	50,000
Collins Pine	209,862	43,000
Keweenaw Land Association	13,500	62,726
Ulatawa Estates (PNG)	59	60,000
Amacol	30,000	59,000
PORTICO	23,580	10,000
Broadleaf Forest Development	236	25,000
Ston Forestal	3,000,000	14,000
Big Creek Lumber	51,876	2,833
Flor y Fauna SA		3,500
CICOL	1,009	40,000
Perum Perhutani	730,000	2,063,100
Duratex	–	–
SPFEQR	354	–
Bainings Community Project	1,200	12,500
Polish State Forest Service-Gdansk	940,000	294,000
Polish State Forest Service-Szczecinek	2,000,000	622,563
Polish State Forest Service-Katowice	1,648,800	635,000
Durawood Products	1,886	24,850
African Charcoal	–	6,000
UZACHI	19,800	26,000
ISOROY	180,000	210,000
Totals	**9,651,738**	**5,313,072**
Total World Production (1995)	**1,680,000,000**	
Certified as % of World Production	**0.5898%**	

Source: Environmental Advantage

Table 3

worldwide (48%), but less than 10% of Polish certified wood is positioned and marketed as such. The largest producer of chips for paper production, Ston Forestal in Costa Rica, markets none of its product as certified.

CONTRADICTION IN THE MARKET?

None of these certified companies could be considered a global player or one that exercises authority in the markets it serves. The companies produce a broad spectrum of products, from charcoal and plywood to *gmelina* chips. Yet, few have been successful in marketing their products as certified. How does one then make sense of an apparent contradictory set of circumstances?

- Major intermediate and final customers are concerned about the industry's forest management practices and want to be assured that companies are adequately managing forest resources.

- Buyers who are concerned about environmental issues and specifically interested in purchasing certified forest products have organized.

Yet,

- Suppliers are producing certified forest products that, for the most part, they are unable to sell as such.

- This contradiction accounts for much of the skepticism about future market for certified wood products.

COMPLEX INDUSTRY MAKES MARKET CREATION DIFFICULT

It is understandable why confusion exists over the future of certified forest products, or that some skeptics dismiss the market for sustainable wood as illusory. The market is evolving slowly because there are multiple steps between a tree in a forest and a finished product. Forest products can flow from landowner to logging company to sawmill to broker to secondary processor to wholesale distributor to retailer before reaching the end consumer. The hundreds of thousands of landowners who produce wood for the industry, as well as hundreds of thousands of intermediary and final customers, complicate the process. The fact that the industry is also built around its ability to pull specific species of trees and specific products through the value chain efficiently, and that orders for forest products are made for specific times and quantities only adds to the complexity.

PUTTING SUPPLY AND DEMAND TOGETHER

Certified producers are faced with the difficult task of finding interested customers for their relatively undifferentiated products in the sea of possible purchasers. Customers interested in certified products, be they retailers, individual consumers, or intermediate consumers, have the unenviable task of finding, from a limited pool of certified producers, those that can fill orders for the species, grade, dates, and volumes they want. When primary and secondary processors and distributors and current chain-of-custody requirements are thrown into the mix, the task becomes even more daunting. With current supplies low and demand concentrated in a few markets, it is usually happenstance that brings supply and demand together.

For the market to develop, sustainable forest owners and environmental consumers at opposite ends of the value chain must come together across a considerable distance (see Diagram 1). Owners of nonintegrated forests who do not own secondary processing facilities—most currently certified producers—are at best familiar with companies in the

Two Disparate Ends of Value Chain for Certified Forest Products

Pioneers and Early Adapters ▼ Certified Forest Products Companies

Forest Products Value Chain— Brings Together Producers and Consumers of Forest Products. Facilitator or Barrier?

Industrial/ Consumer Demand ▼ Retailers and Intermediate Consumers of Forest Products

Diagram 1

value chain immediately ahead of and just below them. The challenge, therefore, is how the two ends of the value chain can or will meet to form a vibrant market for certified and sustainably produced forest products.

The following section outlines how the combination of push and pull drivers and industry structure will influence the development of the market for sustainable forest products.

Geographic and Industry Structure Factors

Traditional consumer demand for certified forest products has not played an important role in developing this market. Consumers are not walking into their neighborhood DIY store or the local lumber-yard and asking for certified wood products, nor is that a likely scenario. But then most markets for new goods do not develop this way. Most products are developed by companies that successfully anticipate coming consumer preferences and trends in specific markets, and then create new products and adapt production systems in response to them. With this in mind, it makes sense to analyze trends and industry structure in specific geographic markets for insight into how markets for sustainable products are likely to develop.

EUROPEAN MARKET MOST HIGHLY DEVELOPED

As described earlier, it is clear that Europe will lead the developments in the market for sustainable forest products for the foreseeable future. Historically, Northern Europeans have been more concerned about environmental issues than their counterparts in other developed countries.

Market for Paper in Europe May Shift Emphasis from Solid Wood Products

Although almost all FSC-certified forest products on the market today are solid wood products, an analysis of the European market suggests that the next significant arena for certification will be in the forests of Scandinavia, driven both by demand from European Buyers Group members, and by the European publishing industry.

As indicated in Diagram 2, the pulp and paper industry's value chain is relatively short. In Europe publishers commonly purchase directly from their suppli-

---

**Sustainable Forest Management Along the Value Chain
Pulp and Paper Industry**

Consumers
- Generally concerned about environment
- Expect at minimum that products be environmentally OK
- Very small niche markets and disperse demand means consumer will not drive sustainability

▼

Publishers and Printers
- Susceptible to environmental group pressure
- Publishers very sophisticated and politically astute
- Face severe substitution threats from other media
- Exercise enormous buyer power

▼

Traders and Distribution
- Little exposure to SFM and certification issues
- No vested interests in either direction
- Much more important in US than in Europe

▼

Primary Producers
- SFM implementation depends on them
- Generally produce commodity-like products
- Heavy exposure to buyer power

Diagram 2

ers, rather than use the broker-mediated purchases that are commonplace in the North American market. Consumers are concerned about paper consumption and expect environmentally sound forestry practices from the publishers they trust. This makes publishers, generally a fairly politically astute group, susceptible to pressure from environmental groups. Because publishers have no real stake in the status quo, and are eager to appear to be on the side of consumers and environmental groups, they are predisposed to exert pressure on paper producers with respect to sustainability issues.

When confronted with customer pressures, the paper companies have good reason to respond. Paper prices are determined by the normal "paper cycle" that dominates the industry, which periodically drives prices up before dramatically lowering them when new paper mills come on line. Large orders can mean the difference between running a paper mill at capacity and shutting it down—and capacity utilization is the key to profitability.

German publishers and other customers have already forced Scandinavian paper producers to abandon some raw wood suppliers in Russia, and are spearheading the drive to certify vast areas of forest in the region. One study of German purchasing managers and other individuals involved in paper supplies for the publishing industry found that "environmentally-sound production" was one of the most important criteria for selecting suppliers (after price and quality, often identical in the commodity market for newsprint).

Northern European publishers are increasingly making it clear that they will not purchase paper that they think may be environmentally-tainted, though they have not offered to pay a premium for it. But even without the inducement of a price premium, the following conditions are likely to encourage paper producers to offer sustainably produced supplies.

- During the periodic downturns in the industry, discounting often becomes the norm as suppliers try to move product in saturated markets. During such a down cycle, certified producers would maintain access to premium markets while undifferentiated producers might have to accept discounts.

- Publishers not only possess the buying power in a commodity market, they dominate the most desirable top end of this market.

- Publishers are rarely backwards integrated into the paper business, and have no stake in current forest management practices. Furthermore, several major consumer countries in Europe do not have strong local industries to oppose SFM or to defend current practices.

- Purchasing managers in the publishing industry are generally conservative and reluctant to expose their companies to risk over forest management issues.

Other European newspaper and magazine companies are following the German publishers' lead. U.K. publishers are beginning to prod their suppliers to certify their forestry operations. The European publishing industry is also lobbying for a speedy resolution of the FSC-ISO debate to protect itself against claims from environmentalists and consumers that the industry is harming the environment.

At the same time, the 1995-Plus Group is pressing forest products companies to supplement currently minuscule supplies of certified product available to them. The group purchases only a tiny fraction of its forest product needs from certified producers. To reach its goal of 100% certification by the year 2000, supplies need to increase dramatically.

Scandinavia Leads Supply Shifts

In the face of strong interest from their major customers in the United Kingdom and Germany, Scandinavian forest products industries have begun to reexamine their previous opposition to third-party certification. Sweden, the most advanced country, was working on the details of a country-wide certification process in early 1997. According to the United Nations Food and Agriculture Organization (FAO), Sweden produces approximately 58 million cubic meters of industrial roundwood, and if the country certified its production en masse, the Swedish volume alone would boost worldwide supplies of certified product by a huge multiple.

Although Finland's effort lags behind Sweden's, it has already constituted a working group on certification made up of industry, environmentalists, academics, and government representatives. The group intends to implement a Finnish national system that would also be compatible with the FSC and ISO systems. Large-scale certification of Finnish forests could begin as early as 1998. The addition of a significant percentage of Finland's 35 million cubic meters of industrial roundwood would represent another significant boost in certified supplies.

Several characteristics of the Scandinavian forest products industry increase the likelihood for large-scale certification.

- The Scandinavian forest industry is historically export-oriented and has learned to be highly sensitive to customer and market demands.

- Implementation of FSC certification would be easier and less costly in Scandinavia than in many other regions. Scandinavian companies do not log old-growth forests because they were harvested long ago. Forestry practices are among the most advanced in the world. Much harvesting is done during cold winter months when erosion problems are not as severe and harvesting equipment can be mobilized with relatively low impact on the frozen ground.

- Detractors of certification in Europe claim that proponent companies are simply looking for a nontariff barrier to protect their share of the European market. One company's barrier, however, is another's competitive advantage. Clearly,

Scandinavian producers would welcome the opportunity to compete on environmental grounds.

- Sweden is home to IKEA, the world's largest furniture maker, which has a reputation for environmentally friendly policies. "Clean, green, and Swedish" summarizes its product positioning strategy.

- Scandinavian forestry prides itself on world leadership in the industry, and some industry players see certification as another opportunity to exercise this leadership.

Although Scandinavian interest in certification is driven primarily by German publishers and the 1995-Plus Group, wholesale certification would yield volumes of certified product far larger than these two markets could absorb. Presumably Scandinavian producers would try to differentiate their green products from those of competitors' to gain a more favorable market position, which would further stimulate demand for certified products.

U.S. MARKET SHIFT SLOWER AND LESS DRAMATIC THAN EUROPE'S

The U.S. market has not yet embraced third-party forest certification as strongly as has its European counterpart. A number of factors have hindered its development.

- By introducing its own certification system, the AFPA has undercut the FSC. This has muddied the issue with industrial customers, who need to be informed to be able to identify environmentally differentiated products.

- The U.S. industry depends less on exports than do European and Canadian producers. Weaker links to the European markets where environmental concerns are strongest has enabled U.S. producers to avoid the close scrutiny of their forest management practices that suppliers to the European market face.

- U.S. imports of tropical forest products have been traditionally low, and, as a result, have drawn less attention from environmental groups than have tropical imports to Europe.

- The United States depends less on public lands than does Canada for wood, and has a tradition of vigorous defense of private property prerogatives.

These conditions make U.S. industry less receptive to ideas and criticisms about its forest management practices. Nevertheless, developments underway indicate that U.S. industry's reluctance to embrace outside monitoring of its forest practices is changing.

U.S. Companies Closely Monitoring Situation

Since many U.S. companies have extensive operations in Canada, they are keenly aware of the Canadian industry's new forest certification initiatives. Indeed, several leading players in the Canadian initiative are owned by or closely affiliated with U.S. companies. According to a senior executive at Weldwood, a subsidiary of Champion International, the company "is positioning itself for registration to the CSA standard at its Hinton Forest Management Agreement area, which comprises approximately one million hectares with an annual allowable cut of two million cubic meters. The CSA process clearly benefits all of our stakeholders." Weyerhaeuser Canada is also participating in the CSA process, as are several other U.S.-based companies.

Most U.S. companies, however, have adopted a wait and see approach while participating in the AFPA's SFI program. They prefer to let the AFPA lead the counteroffensive against FSC-style certification.

U.S. States Undertake Pilot Certification Projects

Two states are experimenting with large-scale pilot certification projects in state-owned forests. Pennsylvania has contracted with Scientific Certification Systems (SCS), an FSC-recognized certifier, to review a 1.2 million-acre sample of the state's 2.1 million acres of forestland in north cen-

tral Pennsylvania. In discussing the initiative, the Secretary of Conservation and Natural Resources commented that "ecosystem management and sustainable forestry practices are basic to our mission, and will be the focus of our next fifteen-year management plan for our forests beginning in 2000."

Under a pilot program in Minnesota, 614,000 acres of county and state lands are being assessed by Smartwood, another FSC certifier. The project team intends to assess each agency's forest lands based on timber resource sustainability, forest ecosystem management, and socioeconomic issues. Minnesota implemented the test primarily because "the public wants to be reassured" that public forest lands are being properly managed, said John Krantz of the Minnesota Department of Natural Resources. If the initiative were fully implemented across state-owned lands, certified wood supplies would increase considerably.

In 1997, preliminary discussions took place in Washington D.C. among certification proponents, policy makers, and officials of the U.S. Forest Service with respect to the possibility of certifying U.S. government-owned lands. Some observers think that if third-party certification were implemented on U.S. public lands, and if the industry and environmentalists reached a credible compromise over forest use, the industry could regain access to public forests that have been off-limits to logging in recent years. If successful, these talks may eventually lead to increased supplies of certified forest products.

U.S. Pulp and Paper Market

In the United States, industry structure, a less confrontational environmental movement, and lower awareness of environmental issues have reduced pressures on suppliers to improve forest management. Nevertheless, some companies are taking steps to minimize their risks over forest management issues.

- Procter & Gamble has slowly backed out of the pulp and paper business and forest land ownership. Some observers see this as a move to reduce the company and its flagship brands' exposure to risks related to forest management and pulp production.

- Currently, most major U.S. companies with either high brand-equity or a base of environmentally-concerned consumers protect themselves against some forest management risk issues by requiring that their suppliers provide evidence of SFM or that they avoid certain practices, such as tropical forest harvesting.

- A number of U.S. companies have stopped buying paper from MacMillen-Bloedel in response to the company's poor sustainability record in British Columbia. To date, most sustainability-related issues in the United States have been negatively defined this way; that is, pushing suppliers not to engage in certain forest management practices.

Although the situations in Europe and the United States differ, the markets operate under similar dynamics of buyer power, commodity products, concerned consumers, and activist environmental groups. The shift toward higher recycled content may be a harbinger of the evolution toward more SFM.

History of Recycling May Foreshadow a Shift in Forest Management

The United States and European paper supplies shifted from just 16.5% of recycled fiber content in 1985 to a projected 37.5% in 1998. This sea change in the industry required billions of dollars in capital investments, the assembly of complex collection and distribution systems, the repositioning of paper products companies to seize new opportunities, and changes in consumer behavior.

Recycling gained momentum from both ends of the product life cycle. In the mid-1980s, many people believed that the United States was rapidly running out of landfill space to house paper garbage.

Underlying the concern about waste disposal lay a broad preoccupation about the ability of forests to continue to produce all of the fiber consumed by the nation. Statistics about how many acres of forest were required for a single edition of the Sunday *New York Times* were repeatedly bandied about. Recycling was perceived as an antidote to the scourges of deforestation, clogged municipal landfills, and a lack of stewardship of the world's resources. Industry fought recycling content initiatives for a period of time, but gradually succumbed to a series of regulations, pressure from buyers pushing for more recycled content, and consumers' concerns. Today hundreds of millions of consumers carefully package up their recyclables each fortnight and deposit them for pick up to be recycled—at no small cost in time and energy expended and living space devoted to recycling storage. Fifteen years ago, most observers doubted the feasibility of such a rapid change.

Some would suggest that the wholesale adaptation by the U.S. marketplace to sustainable forestry presents an even greater challenge since consumers have only recently become accustomed to the requirements of a recycled paper supply. Not so. If anything, the rapid innovations in recycling and the industry dynamics which lent velocity to the process probably make changes in the marketplace toward sustainable forestry even more likely. Although green consumerism has not become as powerful a force as some of its earlier proponents once predicted, polls consistently show that environmental issues are at the top of any list of social concerns.

CANADA EMBRACING ISO-14000 STANDARDS

The dynamics that are beginning to fuel demand for sustainable forest products are behind Canada's early certification initiative. Canada ships over 50% of world exports of newsprint, 34% of wood pulp exports, and contributes nearly 15% of world trade in printing and writing papers, according to the CPPA's annual statistical review. For over a decade, the industry has been under fire by environmental groups for its continued logging of old growth forests. Because trees grow more slowly in Canada's harsh climate, more land is required per unit of roundwood production, which makes high-quality land management all the more important. The combination of the Canadian industry's international exposure, environmental pressures, and the need for sustainable production led the Canadian forest products companies to develop what they hope will become an internationally-acceptable set of standards for Canadian forestry based on the ISO system.

Canadian interest in the market for sustainable products was initially piqued by the 1995-Plus Group and focused on solid wood products. When paper emerged as a key variable in the debate, Canadian companies took action. Canada's export tradition has made the industry accustomed to adapting practices to customer requirements. As one executive at a large Canadian paper company put it, "When our customers came to us with concerns about chlorine bleaching, they made us become much more strict with ourselves than the regulators—to stay in touch with the market, we have to be at the leading edge."

Significant Supplies of Certified Product Available in Short Term

The CPPA reports very strong interest among its members for the Canadian certification initiative. Because the standards were only promulgated in late 1996, most interested companies are in the process of doing "gap" analyses to determine where their management practices need to improve. Nevertheless, supplies should be on the market by 1999. According to the CPPA, thirteen organizations expressed "very strong" interest in certification and were set to begin the process in 1997. These companies and groups of private landowners alone account for some eight million hectares of forest and ten to twelve million cubic meters of industrial roundwood.

Canadian View of Market Premium

The Canadian industry has not adopted chain-of-custody protocols or created an ecolabel for its certified products. In designing their certification system, the Canadians have worked under the assumption that sustainability will become a general market condition, something that customers simply expect from their suppliers. The CSA standards were created to provide such assurances. The standards were not formulated with the idea of creating market opportunities for niche players servicing ecosensitive markets. FSC certification, with its logo system and procedures for chain of custody, on the other hand, is better suited for the creation of differentiated products.

Indeed, Scandinavia's embrace of certification lends urgency to the Canadian effort, since Canadian suppliers do not want to be left behind in their ability to demonstrate the sustainability of their forest management practices.

Initial Canadian Customer Response Favorable

Customers have responded positively to the Canadian certification initiative. The director of the Newspaper Society, a group of 1,300 regional and local newspapers in Britain, said of the initiative, "The Society supports the CSA standard. Newspapers need assurance that forests are managed to serve both environmental and industry needs." In Germany, Wolfgang Oberress of Haind Papier GmbH endorsed the CSA system saying, "Our company welcomes the Canadian initiative to produce uniform international standards for the certification of ecologically-based forest management. Canadian forest interests have played a leading role worldwide in this field and have worked consistently over three years to develop the certification standards. We now have a considerable achievement before us."

JAPAN AND ASIA

No certified wood products are presently sold as such in Japan. Although some wood products make environmental claims, few are substantiated. The mere existence of such claims may indicate some interest among Japanese consumers in the environmental attributes of the products they consume. By far the most important importer of tropical forest products, Japan has the greatest potential to exert pressure on its suppliers to improve their forest management. Historically, however, Japan has not taken such actions, and the prospects for the development of a market for sustainable forest products there are rather remote.

The large Asian producers have taken considerable umbrage at efforts by European countries to boycott Asian tropical hardwoods over concerns about forest management practices. In 1996 Malaysian producers even talked about organizing a counter-boycott of German equipment. European concerns have undoubtedly cost Asian producers market share in Europe, though it is difficult to quantify the loss. According to Aw Beng Peck of Asian Timber, however, among pulp and paper producers, the demand within Asia for paper products is so great that the industries are, so far, unaffected by market access issues in Europe.

Nonetheless, in what some see as a defensive maneuver, most Asian countries have undertaken sustainability initiatives of one sort or another, many with an eye to fulfilling the ITTO's campaign for sustainable management by the year 2000. It is beyond the scope of this discussion to gauge the current and future efficacy of these initiatives, but presumably by focusing attention on SFM, these countries are likely to produce some improvement in on-the-ground practices.

GLOBALIZED MARKET FOR SUSTAINABLE FOREST PRODUCTS

The rise of sustainability as a more important market variable for the forest products industry, and the increased globalization of the industry and its customers will exert a number of catalytic effects. McDonald's, for instance, recently convened its major paper suppliers and asked that each one do an in-depth presentation on the sustainability of its forest management practices. McDonald's then proceeded to award more business to the company that it thought was the best steward of forest resources. In many cases, it will be easier for global customers to implement such policies for all supplies than just those for specific markets. Few companies would want to explain that sustainability is only a condition for certain markets; most would like to attribute their interest in forest sustainability to forward-thinking and progressive environmental policies.

Until now, such a commitment as McDonald's would have been risky because the supplies of sustainable products were simply unavailable and the potential to overstate commitments would have been high.

Ironically, the same factors that have kept some global players out of the sustainable products market in the past are likely to fuel its development once such companies enlist. This is the case with Manadnok Paper Mills of New Hampshire. The company reports that its line of wallpaper products may be excluded from the U.K. market if it is unable to deliver certified paper to its customers. If suppliers do not clear the certification hurdle, the company believes that Scandinavian producers will quickly fill the gap. For that reason, Manadnok is actively trying to locate certified suppliers.

On the supply side of the equation, the catalytic effects are potentially equally dynamic. Most major forest products companies are multinational players, with operations in many countries, even different hemispheres. Champion, a U.S. company with significant eucalyptus plantations in Brazil, is watching to see if its competitor Aracruz, the Brazilian pulp producer, moves ahead with certification. Blandon Paper Company, a Minnesota-based lightweight coated paper manufacturer is owned by Fletcher Challenge, a New Zealand company with significant operations in Canada. According to a company official, Blandon is monitoring the evolution of the Canadian ISO-compatible system because its New Zealand-based owners think that ISO certification may offer the company some long-term advantages.

Conclusions

A combination of forces is converging on both the supply and demand sides of the forest products value chain to fuel a market for sustainable forest products. More industry customers are becoming concerned about the origin of the forest products they consume. Suppliers, in response to pressure from environmental groups, governments, consumer and industrial demand, and leaders in the industry, are searching for ways to improve their forest practices. Forest products companies worldwide are now keenly aware of the market's burgeoning interest in forest sustainability. They are seeking the appropriate vehicle to communicate their efforts and concerns over forest management.

Over time, the cumulative effects of these changes will make SFM an integral part of the industry's overall value equation. The variables of quality, price, reliability, and service will continue to dominate most purchasing decisions, but sustainability will also become an essential variable. And as sustainability is integrated into production and purchasing decisions, the standards that suppliers must meet will rise.

The pace of change in the industry will be largely determined by geography and industry structure. Europe is clearly the first region where sustainability is taking a firm hold in the marketplace. In the next five years, the authors predict that third-party certification will emerge as a major market force in Europe. So much so, in fact, that any producer interested in selling into premium markets will need to be certified, making certification a general market condition for suppliers aiming at the upper ends of the market.

If certification takes a firm hold in Europe, any countries interested in the European market will have to respond to remain competitive. Canada is already doing so. If just one or two major U.S. companies become certified, as appears likely, large sectors of the U.S. forest products industry will open up to third-party scrutiny of its forest management practices. Once large supplies of certified product are available, sustainability may well become a characteristic of products that premium customers simply expect.

References

Canadian Forestry Certification Coalition. 1997. Bulletin, Vol. 3, No. 1, January.

Canadian Pulp and Paper Association. 1996. *Reference Tables 1996,* July.

Crossley, Rachel. 1995. A Review of Global Forest Management Certification Initiatives: Political and Institution Aspects. October, p. 16.

Environmental Advantage database, 1997.

FINNPAP. 1994. Internal study of 300 German consumers.

FINNPAP. Interviews with 30 German individuals involved in purchasing paper for German printing and publishing industry.

Food and Agriculture Organization of the United Nations, Rome, 1995. *FAO Forest Products Yearbook, 1993.*

Furstner, Wolfgang. 1996. *The German Magazine Publishers Ecological Vision and Strategy.* Paper presented at the International Periodicals Symposium, June.

Johnson, Brad, Procter & Gamble. 1996. Personal communication, August.

Juslin, Hikki, coordinator of Finnish Certification Initiative. 1997. Personal communication, January.

Koski, Teppo, senior consultant for environmental strategy, Jaakko Pöyry, 1997. Personal communication, January.

Krantz, John, Minnesota Department of Natural Resources. 1997. Personal communication, February.

Laishley, Don, director of forest strategy, Weldwood of Canada, Ltd. 1997. Quoted in *Sustainable Forestry Bulletin,* January.

Lapointe, Jerry, Canadian Pulp and Paper Association. 1997. Personal communication, February.

Maezawa, Eishi, World Wildlife Fund Japan. 1996. Personal communication, December.

Magazine Publishers of America and American Society of Magazine Editors. 1996. The Magazine Industry and the Environment— Findings and Recommendations of the MPA/ASME Task Force, September.

Mansley, Mark. 1996. The Demand for Certified Wood Products. Unpublished draft.

McCloy, Brian, Canadian Pulp and Paper Association. 1997. Personal comunication, January.

McCoy, John, Blandon Paper Co. 1997. Personal communication, February.

McNulty, John and Cashwell, John. 1995. The Landowners Perspective on Certification. *Journal of Forestry,* April.

Peck, Aw Beng, Asian Timber. 1997. Personal communication, February.

Pennsylvania Department of Conservation and Natural Resources. 1997. Press release, January.

Pöyry, Jaakko. 1996.

Taylor, Donald, Champion International. 1997. Personal communication, January.

Vlosky, Richard P. and Ozanne, Lucie K. 1996. Willingness to Pay for Environmentally Certified Wood Products: The Consumer Perspective, June.

Vlosky, Richard P. 1996. Forest Products Certification: The Business Customer Perspective, August.

Western Wood Products Association. 1993. Research Shows Lumber Retail and Wholesale Customers More Concerned about Environmental Impacts of Purchasing Decisions, April.

Weyerhaeuser internal study. 1994. Used with permission.

Emerging Technologies for Sustainable Forestry

CASE STUDY

PREPARED BY:
CATHERINE M. MATER

A Case Study from "The Business of Sustainable Forestry"
A Project of The Sustainable Forestry Working Group

The Sustainable Forestry Working Group

Individuals from the following institutions participated in the preparation of this report.

Environmental Advantage, Inc.

Forest Stewardship Council

The John D. and Catherine T. MacArthur Foundation

Management Institute for Environment and Business

Mater Engineering, Ltd.

Oregon State University
Colleges of Business and Forestry

Pennsylvania State University
School of Forest Resources

University of California at Berkeley
College of Natural Resources

University of Michigan
Corporate Environmental Management Program

Weyerhaeuser Company

The World Bank
Environment Department

World Resources Institute

CCC 1-55963-619-X/98/page 4-1 through page 4-27

Emerging
Technologies for
Sustainable Forestry

CASE STUDY

PREPARED BY:
CATHERINE M. MATER

A Case Study from "The Business of Sustainable Forestry"
A Project of The Sustainable Forestry Working Group

Contents

Emerging Technologies for Sustainable Forestry

Introduction

As experience grows with sustainable forest management (SFM) practices throughout the world, one single factor continues to emerge as noncontrovertible: *SFM practices do appear to cost more to implement in the forest.* It is this factor that continues to drive the debate over whether SFM practices are economically feasible for the forest products industry.

If SFM proponents fail to recognize the importance of helping industry to increase the higher value of wood produced with equal or less resource use, then incentive-based efforts to infuse SFM practices and certified wood product development into accepted industry standards will not succeed.

Finding ways to foster the adoption of emerging technologies that enable the forest industry to accomplish better bottom-line results could prove to be of significant benefit to fast-tracking the implementation of SFM practices worldwide. Identifying these emerging technologies, however, and providing a pathway for easier entry into the market is no simple task.

This Emerging Technologies note highlights some of the most promising technologies, techniques, and strategies that may foster the implementation of SFM practices by offering improved environmental and bottom-line results to the forest products industry.

SFM Technology Objectives

Typically, there is an ever-increasing offering of new traditional technologies that help forestry and wood processing operations increase the rate and volume of their output using traditional resource standards, and optimize machinery and manpower efficiencies to achieve those goals. SFM technologies and techniques, however, might be defined as those that help to create a better balance between sustaining natural resources and economic development. Although the objectives and intended results between traditional and SFM technologies have similarities, they also have clear differences as indicated in Figure 1.

While increasing the rate of raw resource output or production is a traditional industry objective, the technologies highlighted here address opportunities to increase the readiness of operations to process a variety of raw resource grades or quality. Because the consistency of log supply is a major constraint for many wood processing operations, those facilities that can purchase and profitably process a range of log quality often have a more direct and consistent access to supplies of raw resource. Similarly, while increasing the volume of raw resource production is a traditional industry objective, the technologies identified here focus more on increasing the value of wood processed through effective waste recovery, value-added production, and the development of custom grades of wood.

The value of SFM technologies is created through their ability to enhance value-added processing, which means maximizing the resources available through SFM practices. The benefits of enhancing processing that adds value to wood resources are clear when you consider the following data produced by the Oregon Wood Products Competitiveness Corp.

For every 1 million board feet of wood processed:

- Approximately three full-time, family-wage jobs are created during primary processing, which converts logs to commodity lumber.

- Another twenty full-time, family-wage jobs can be created from further processing that same 1 million board feet of lumber into component parts such as furniture blanks and turnings.

- Up to eighty additional full-time, family-wage jobs can be created by converting that same 1 million board feet represented in component parts to high-end furniture ready for consumer use.

Difference between Traditional and SFM Objectives and Results

Traditional Objective	Intended Results	SFM Objective	Intended Results
Increase rate of raw resource output	Faster processing of raw resource done over time	Increase readiness of operation to process traditional and custom grade wood output	Increased access to and reliability of wood supply due to ability to process both high and low grade material
Increase volume of raw resource output	Higher volume of raw resource processed over time	Increase value of output using existing wood volume input	Increased dollar value per unit of resource processed due to waste recovery, value-added, and custom grade development
Optimize machinery efficiencies for producing traditional wood grades	• Decrease downtime • Decrease wood waste • Increase volume of resource output • Decrease number of employees and labor cost	Optimize machinery efficiencies to produce both traditional and custom grade material	• Decrease downtime • Decrease wood waste • Increase value of resource output • May actually increase number of employees
Optimize manpower efficiencies for producing traditional wood grades	• Increase employee safety • Increase worker training skills • Decrease material production time	Optimize manpower efficiencies for producing traditional and custom wood grades	• Increase employee safety • Increase worker training skills • Provide more stabilized job security

Source: Mater Engineering

Figure 1

Equally important, many of these value-added products can effectively use underutilized species, smaller pieces of wood alone or joined together, and wood with "character" that is often considered waste or defect material (with lower value) in traditional wood processing, material that would otherwise be of lower value or treated as waste.

INDUSTRY NEEDS AND CONSTRAINTS

Identifying the most promising emerging SFM technologies requires understanding the critical needs and constraints that players throughout the chain of forest products development face. Issues that are of key concern to the forestland owner are likely to be significantly different than those that directly affect the decision making of loggers, or wood products manufacturers. Figure 2 identifies some typical needs and constraints experienced by the landowner, the logger, and the wood products producer.

By identifying specific bottlenecks within the listed constraints of each category of major industry player, the specific types of solutions that emerging SFM technologies need to offer emerge, as Figure 3 illustrates.

Targeted Needs and Constraints

	Key Needs	Key Constraints
Landowners	• Access to information systems and services providing sustainable forest management practices	• Land management systems and services difficult to access and often too expensive for the smaller landowner
	• Identifying and accessing knowledgeable logging contractors	• Systems not adaptable for use in isolated locations
		• Must rely on word-of-mouth recommendations for logging contractors
Loggers	• Decreasing the cost of doing business (increase profits)	• Increase in worker's compensation costs
	• Complying with environmental regulations for logging in the forests	• Lack of information on new logging technologies to match environmental concerns
Wood Producers (Primary and Secondary)	• Decreasing wood waste during production	• Lack of information on affordable value-added production options
	• Increasing the value of product per unit produced	• Lack of information on affordable waste reduction options
	• Stabilizing raw resource supply	• Lack of access to capital for systems improvements
	• Accessing affordable chain-of-custody solutions for moving certified wood	• Lack of information on chain-of-custody tracking options save for barcoding and separate production runs

Source: Mater Engineering

Figure 2

Bottlenecks and Solutions

	Key Bottlenecks	Targeted SFM Solutions
Landowners	• GIS capabilities not adapted to personal computer technology used by consumers	• New GIS software especially designed for personal computer use
	• Neighbor referrals for logger contacts unreliable	• Logger's certification program
Loggers	• Log harvest machines of the size, weight and function which create heavy impact in forest ecosystem	• Low-impact harvesting technology
	• Increased worker's comp. costs	• Logger's certification program
Primary Producers (Lumber)	• Slabs, trim ends, and lumber shorts recovery and use in product development	• Scrap Recovery System • Trim Block Drying Rack System
	• Conversion of lower volume sawdust and chips to value-added product (especially for smaller producers)	• Sorbilite Systems
	• Waste factor due to defect cutout in production	• New scanning technologies

Secondary Wood Product Manufacturers	• Utilization of short pieces in production development	• Fingerjointing Technology:
		  • Greenweld
		  • Soybean-based adhesives

	• Lack of wood resources with preferred characteristics for product development	• Wood-hardening technology
	• Waste factor due to defect cutout in production	• New scanning technology

Source: Mater Engineering

Figure 3

Based on an analysis of the industry's needs and constraints, ten emerging SFM technologies were selected for discussion. The criteria for selection was, in part, determined by the following guidelines:

- The technology or innovative practice/technique must have already gone through product/service testing producing documented positive results.

- It must have just entered the marketplace or be about to do so.

- The technology needs higher visibility to gain market acceptance.

- It has been in the marketplace in other regions throughout the world, but lacks significant market presence in North America.

The technologies selected were those that targeted key SFM opportunities such as converting wood waste into wood profits, making more wood product with equal or less resource use, and being adaptable for use by both large and small wood product manufacturers worldwide. Some of the technologies discussed are one-of-a-kind offerings. Others represent a type of technology that may be produced by more than one manufacturer, but still lack significant market presence to further SFM practices.

Promising SFM Technologies

The emerging SFM technologies selected for coverage include:

User-Friendly GIS Software:	*ForestVIEW 96*
Innovative Logger Programs:	*Certified Logging Professionals (CLP)*
Low-Impact Harvesting:	*Ponsse Systems*
Fiber Wood Waste Conversion:	*Composite Product Technologies (Sorbilite)*

Solid Wood Waste Conversion:	*Scrap Recovery Systems (Auburn)*
New Drying Technology:	*Trim Block Drying Rack System*
Short Piece/Scrap Utilization:	*"Wet" Wood Fingerjointing Technology:* *a) Greenweld Process* *b) Soy-based Process*
New Scanning Technology:	*Robo-Eye*
Wood Hardening Technology:	*Indurite* (TM)

NEW USER-FRIENDLY GIS SOFTWARE: FORESTVIEW 96

Geographic Information Systems (GIS) and Global Positioning Systems (GPS) technology have existed for several years and already have many useful applications. Natural resource management, precision agriculture, forestry, environmental applications, utility inventory, and city planning are just some of the areas where this technology can be applied.

GIS and GPS systems have been invaluable technologies in helping the forester gain access to information that is vital for the practice of SFM. A variety of GPS collection units gather and enter field information about the land under assessment. GPS units collect control points that become the data that can be used to create images that can produce inventories of timber and identify wetlands, recreation trails, habitats for wildlife, vegetation, and forest landscape schemes. This data can then be gathered and plotted.

The maps created using these GIS systems, however, can be complex and expensive. Usually, the landowner brings the information he has collected in the field to a company that performs the GIS services. This information can be in the form of topographic maps, satellite images, quadrant maps, or even city maps. The GIS service company then generates specific maps of the area being assessed or inventoried. The service is expensive, and the technology has not been readily available or affordable to the small landowner or individual consulting resource manager.

Although technological advances in GIS and GPS systems are evolving quickly, hardware and software systems that are affordable, user-friendly, and can be adapted to standard computer technology that consulting foresters or individual forest owners are likely to use have been largely unavailable.

Program Capabilities

ForestVIEW 96 breaks both those availability and affordability barriers. Introduced in June 1996, the program is a powerful PC-based, desktop package that gives the small landowner and consulting resource manager user-friendly, affordable tools to create maps and perform land assessment and inventory independent of expensive outside services.

The software package, designed for use on standard consumer computer systems, costs at least 50% less than the cost of the outside services that provide the same land-mapping results. The hardware used in the collection of the field data is functional, user-friendly, and portable, which makes it a technology suited for use in isolated forest regions.

ForestVIEW 96 promises to increase dramatically the access to information and knowledge that forestland owners and resource managers need to assist them in sustainably managing their forestlands. The system is a standardized software package that enables users to create accurate maps on a PC using the Windows 95 operating system. It can integrate a variety of data from GIS and Computer Aided Design (CAD) sources. Scanned images, such as aerial photos and quadrant maps, as well as GPS data can be entered into the software. This system can store multiple layers of data, and recalculate figures from those layers. The entire software package sells for $3,195. ForestVIEW 96 contains many important features that if purchased individually, would exceed $7,000 in cost.

Three separate programs can be linked to the ForestVIEW 96 system:

- The *SuperEasy* program allows the user to enter data directly in the field, as well as edit it there, instead of waiting to get back to the PC. It corrects errors in the field and can kick out all reports.

- *SuperAce,* a PC processing package, takes the field data, and computes volumes of sustainable harvest based on the multiple layers of data.

- The data from *SuperAce* is then hooked to *FLIPS 96* (Forest Level Inventory Planning Systems). This program performs growth calculations, generates projections (future inventory), and performs general resource appraisal. Instead of the user building inventory separately, and then entering it into another program, *FLIPS* interacts with the *ForestVIEW 96* program.

- Finally, *FieldNotes* is a pen-based software package that allows for field data collection that can be plugged into a PC running *ForestVIEW 96*. *FieldNotes* contains a Microsoft Windows Interface, and supports a variety of industry standard formats.

All someone needs to operate *ForestVIEW 96* is an adequate computer that uses Microsoft Windows 95 operating system, a scanner to scan in photos, maps, and the like, and a color printer for maps. A GPS unit might also be needed to provide control points for the image data.

ForestVIEW 96 also includes two valuable special features, rubber sheeting and heads-up digitizing, as discussed below.

Rubber Sheeting

Rubber Sheeting is a term used to describe the process by which imported elements, such as a photo, a satellite, and a map, are arranged to fit together, or align. For instance, if the user scans in a map, he or she can identify points of reference and then scan in a photo of the same location. But the exact location of the points on the map will not automatically be matched by the photo, due to distortion, film speed, or source. Rubber sheeting stretches the image to fit the map; it arranges the pixels so the different images are aligned. It eliminates such things as elevation differences and radial distortions. This replaces the process where the user would have to "eyeball" the different images to get a match.

Heads-Up Digitizing

Heads-up digitizing is a term used to describe the process of digitizing directly on the computer monitor as opposed to sending the image to the digitizing department and having technicians digitize the image by "eyeballing." Digitizing involves entering points and lines or adding other information such as roads, streams, and timber stands. Using heads-up digitizing, the user can bring the image (map) up on the computer monitor and draw lines directly on the screen using the mouse. The user's head is looking up at the screen, not down at the keyboard or a hard copy of the image, hence the term "heads-up." People in the digitizing department would add the information, enter the points and lines, while eyeballing the image and using a grease pencil on the hard copy. Heads-up digitizing is a streamlined, more accurate way of adding additional information to the image on the screen.

Technology Contact:

> Mr. Toby Atterbury
> Atterbury Consultants, Inc.
> 3800 SW Cedar Hills Boulevard
> Suite 280
> Beaverton, Oregon 97005
> Phone: 503/646-5393 (Main office)
> Phone: 503/520-3150 (Forest shop)
> Fax: 503/644-683

INNOVATIVE LOGGER PROGRAMS: CERTIFIED LOGGING PROFESSIONALS PROGRAM

While not actually a technology, the Certified Logging Professionals Program (CLP) initiated in the state of Maine is an unusually successful emerging technique that helps to sustain resources and development. In Maine, wood fiber production constitutes a large part of the overall economy as shown by the following statistics.

- Maine has 19.7 million acres of forestland.

- Maine's forest products industry employs over 30,000 people and produces 26.7% of the states' total manufacturing output, or almost $9 billion dollars of the gross state product.

- Maine is the nation's largest supplier of paper products, housing sixteen paper mills and fifteen pulp mills that produce approximately twelve tons of paper daily.

An Answer to Key Concerns

The logging industry initiated CLP in 1991 to address four key concerns of the logging profession:

- High injury rate.

- Aging workforce; lack of young workforce selecting logging as a profession.

- Loss of high school and vocational training programs that were natural avenues for training loggers.

- Threat of state regulation of the environmental practices of loggers in the forest.

While many states have programs that assist the professional forester in engaging in more SFM practices, few house professional programs for loggers to accomplish that purpose. Maine's CLP program is unique in its ability to achieve economic benefits to the logger in tandem with ecosystem benefits to the forest.

The CLP was designed to achieve these key objectives:

1) To decrease injury rates among loggers.

2) To give logging a professional status that would encourage the development of skill, knowledge, and pride among loggers.

3) To promote sustainable forestry.

4) To begin with a program that had a few good people who could set the stage for creating a successful broad-based training approach.

Ultimately, however, the program was designed to create a voluntary process by which a nongovernmental agency or association would recognize and "license" individuals who met certain predetermined qualifications specified by that agency or association.

Program Requirements

Currently, the program requires individuals to attend an initial intensive hands-on training program that includes the following curriculum areas:

- Sustainable Forestry Principles
- Forest Management
- Safe and Efficient Harvesting
- Business Logging
- Preserving Fish and Wildlife
- Production Felling Training
- Mechanical Harvesting
- First Aid

Specific curriculum components incorporated into the CLP training program cover a wide range of elements necessary for increasing worker safety and sustainable management practices in the forest. Through the use of instructional materials developed by Cornell Cooperative Extension and the University of Maine Cooperative Extension, effective forest management skills are taught that incorporate many SFM practices, such as preserving wildlife diversity.

Other program requirements include:

- Documenting six months of paid experience using the techniques and practices taught in the certification program.

- Passing an in-field inspection/interview to evaluate how well the individual applies the learned techniques in the field.

- Attending recertification training courses the first year after the initial certification and every two years thereafter to ensure continuing education and direct application in the field.

- Passing recertification field inspections the first year after the initial certification and every two years thereafter to ensure that certified loggers continue to apply the required techniques in the field.

Program Results

The CLP program has generated impressive results for both the environment and the logging business as noted below.

1) The program has been so successful that a special insurance code classification for certified loggers has been created to separate certified loggers from noncertified loggers. The insurance code is 2702 for noncertified loggers, and 2721 for CLP loggers.

2) Certified loggers have helped to reduce the injury rates of Maine's loggers. In 1990, Maine loggers experienced 20 injuries per 100 full-time employees (FTEs) compared to the state's average of 14.5 injuries per 100 FTEs. By 1994, in part due to the CLP program, loggers' injuries dropped to 8 per 100 FTEs compared to the state's average of 10.5 per 100 FTEs.

3) Responding to decreases in injury rates, workers' compensation rates for CLP-certified loggers also dropped. Between the inception of the CLP program in 1991 and February 1996, the workers' compensation rate for CLP-certified loggers was 10% lower than for noncertified loggers. By March 1996, the workers' compensation rate for CLP-certified loggers was 27% lower than for noncertified loggers, and by May 1997, it was 53% lower.

4) The number of days of work lost for certified loggers has dropped from 13 days per 100 FTEs in 1990 to 4.8 days per 100 FTEs in 1994.

5) Because of the SFM training offered by the CLP program, certified loggers are now preferred as loggers by large forest products producers in the state. This year, for example, International Paper announced a goal of obtaining CLP training for all its logging contractors.

6) Since 1991, the number of trained loggers who remain active in the state has continued to grow. Almost 2,000 loggers have become CLP-certified since the program's inception.

Trained loggers are issued certification cards documenting their performance in the program and are given an authorized certification number. When they complete the initial certification process, trained loggers sign a "Code of Ethics for Certified Logging Professionals" that requires them to follow seven key principals of professional certified logging including "...protecting the natural environment and enhancing the natural resources of the state by operating in accordance with sound forest management and environmental principles."

The program has created an unusual yet successful partnership between forest industry representatives, the Maine Tree Foundation, OSHA, and the Maine insurance industry.

Program Contact:

> Mr. Mike St. Peter
> Program Manager
> PO Box 557
> Jackman, Maine 04945
> Phone: 207/668-2851

LOW-IMPACT HARVESTING: PONSSE SYSTEMS

Alternative logging practices are a major component of SFM. Practices that maintain the visual quality of forests, log harvesting systems that do less damage to the forest floor than conventional cutting techniques, and systems that have the capability to perform selective cutting or thinning are all needed to foster SFM.

Low-impact log harvesting is not necessarily a new technology concept. Designs for equipment that pro-duce less damage to surrounding forest ecosystems during harvesting continue to improve. However, within North America, especially for smaller-scale log harvesting operations, awareness and use of this low-impact harvest technology remains low. For that reason, systems like the Finnish-designed Ponsse log harvesting units for low-impact forest thinning operations deserve consideration here.

An Additional Source for Quality Wood

Generally speaking, trees that are often "thinned" from the forest under SFM are small in diameter and considered of lower grade because they have a variable diameter and defects in the log. In standard practice this material is typically viewed as pulpwood. Because the entire pulpwood log will be chipped for pulp, the logger has no incentive to harvest any quality wood that might be obtained from that log. A classic example may be a small diameter log with a heavy taper at the top. The lower portion of the log, although smaller diameter, may include quality wood that could be used in shorter-length logs called boltwood that can made into many value-added products, such as furniture and flooring.

The objectives in the design of the Ponsse log harvesting system included:

1) Designing a log harvesting and forwarding system that can do cut-to-length processing for first and second forest thinning, a process required to achieve SFM status, while doing minimum damage to the surrounding plant life and forest ecosystems as the machinery travels over it.

2) Achieving ease of maneuvering so that the equipment can effectively work above, around, and in-between logs intended to be left after harvest.

3) Designing a portable system that would decrease lost production time due to site transfers and set up.

4) Achieving the first three objectives without loosing speed and efficiency of production.

Ponsse has achieved a worldwide reputation for accomplishing its technical objectives. Beginning in 1983, the company faced strong environmental concerns from the Finnish public and government over the intensive logging practices used in Finland. The company was compelled to plow more research and development dollars into designing harvesting systems that better protected the environment. That effort produced the first in a long line of log harvesting equipment that, according to the company brochure, "...takes environmental protection seriously."

Well-Tested Design

During the 1980s, the Finnish government tested various log harvesting machinery to determine which caused less damage in environmentally-sensitive forest areas. The test results proved so lopsided in favor of the Ponsse design, that the tests were repeated. In both tests, the Ponsse machines out-performed all competitors by 300%.

In particular, the Ponsse HS 10 log harvesting unit, provides excellent performance in first and second forest thinning conditions while doing minimal damage to the forest. The system harvests the whole log and initiates a highly efficient cut-to-length process while still on the forest site. Loggers scan an individual log at harvest time to determine the amount of sawlog or boltwood to be extracted from the pulpwood volume of the tree. Simply put, the system cuts the log on stump, de-limbs the log while scanning, then cuts the log according to the sawlog grade or boltwood grade anticipated from the log scan. The entire process is accomplished within a few minutes from the time the tree is first cut from the stump. For first and second forest thinning of trees between a 7'-8' diameter size at breast height, about 50% of the wood is converted from pulpwood to more valuable log grade for wood product manufacturing by using the system.

Small U.S. harvesting businesses, such as Brown Trucking and Logging of Rhinelander, Wisconsin, that have purchased a Ponsse system report a 20% increase in the value of logs they harvest. They give two reasons for this—the increased volume of boltwood that is converted from the traditional pulpwood harvest, and the increased access to forest areas due to the maneuverability of the unit over more difficult terrain, as well as its low-impact on the forest floor. Although the Ponsse HS 10 system has a $435,000 price tag, Brown Trucking reported that the machine earned a payback on its initial investment within the first two years of operation.

Technology Contact:

> Mr. Hannu Fautari
> President
> Ponsse, Ltd.
> 2310 Peachford Road
> Atlanta, Georgia 30338
> Phone: 770/454-7799
> Fax: 707/454-9415
> In Finland:
> 74200 Vierema, Finland
> Phone: 358/77/768-461
> Fax: 358/77/768- 4690

WOOD FIBER WASTE CONVERSION: COMPOSITE PRODUCT TECHNOLOGIES (SORBILITE)

Converting wood waste in the form of sawdust, chips, and the like into composite products has become a valuable and growing technology in the wood products industry in the last decade. Composite board production, in particular, has registered substantial growth, especially with the improvements and rising demand for medium density fiberboard (MDF) and oriented strand board (OSB). Efficient technologies that can convert fiber waste to composite products, however, have generally been developed for large-scale wood products producers.

Appropriately sized and priced composite product technologies that adapt to smaller-scale production and still allow for increased value-added processing has been consistently lacking, especially in North America. Sorbilite technology, introduced in Europe and marketed in other regions—almost 70% of the existing customer base is in South

America and Asia—provides some innovative technology that could be profitably adapted for North American applications.

Small-Scale Use Can Bring Big Benefits

The Sorbilite composite molding system was designed to provide cost effective, efficient compression molding technology that can be adapted to smaller-scale use, and produce a high-quality composite product using a lower volume wood fiber waste than other comparable systems. As such, the Sorbilite system offers substantial benefits to the environment—and communities. It uses wood fibers that might normally be chipped or landfilled, and converts that fiber into a high value-added product. In the process it creates additional full-time, family-wage jobs for rural communities. It helps to conserve forests by converting wood waste from existing processed material into valuable product, thus lowering the need to harvest more wood from the forest. The technology allows as little as one pound of fiber to be processed into value-added product at any given time, making it tailor-made for small-scale producers.

The system has other advantages. It uses substantially less energy in production than other composite board production technology and can manufacture compressed molded products that apply to a variety of value-added product sectors including:

- casket production
- upholstered furniture parts
- full-sized raised panel doors
- decorative molding
- building parts
- floor tiles
- cabinet doors
- chair parts
- fireplace mantles
- toys
- frames and wall plaques

The Sorbilite system also generates a number of specific performance benefits.

- The Sorbilite compression molding process allows for the use of a combination of fibrous material in product development. Along with wood fiber, other materials such as carpet fibers, paper, and peanut shells can all be added to a special bonding agent to create products that are as strong and dense as wood, but cost less to produce.

- The Sorbilite Membrane Press generates intense, even pressure up to 725 psi. The vertical pressing power is five times greater than any competitive machine currently on the market with a low operating cost and a 3.5 hp motor. The press weighs only a fraction of most competing machines, so no special building preparation or site is required.

- Estimated calculations for composite product manufactured by the Sorbilite process indicate substantial profit advantages. Using cabinet door production as an illustration, Figure 4 shows a gross profit per door produced of $2.85, with a daily gross profit of $7,900.

- Additional cost comparisons further indicate the economic benefit of using the Sorbilite process. To manufacture a wood back for a secretary chair, plywood costs about 90 cents per chair. The Sorbilite process can mold that chair back for about 30 cents.

- With an estimated upfront purchase price of $350,000, payback on investment appears to be between twelve and thirty months, depending on the volume and products being produced, and the location of the production facility in relation to markets. This is a substantial difference in cost compared to traditional composite panel production equipment that can cost well over a million dollars.

- The mechanical characteristics of the Sorbilite product, compared to other traditional panel

Sorbilite Profit Calculation (1995 Values)

Production Capabilities

For 12"X 15" (1.25 sq. ft.) cabinet doors (two sides); and assuming a total raw materials cost of $1.87 per sq. ft.

Press time:	2 minutes
Doors per press cycle:	15
Doors per hour:	400
Doors per 7 hour day:	2,800
Labor:	4 men at $10.00 per hour: 10¢ per door
Overhead, energy, etc.:	25¢ per door
Total cost with the Sorbilite Membrane Press:	$2.22 per sq. ft.
Estimated sale price per sq. ft.:	$4.50
Gross profit per door:	**$2.85**
Gross profit per day:	**$7,900**

Source: Sorbalite

Figure 4

products such as particleboard, receive high marks. The Sorbilite molded product is almost three times as hard and outperforms particleboard in holding screws by 45%.

- The Sorbilite process uses chemical bonding of the fiber combined with compression pressure (up to 2,000 tons), instead of cooking the entire product to create the compressed molding products. Conventional press technology usually heats or cooks the entire panel during processing to activate the bonding agent mixed with the fibers. As a result, the Sorbilite process saves energy during product processing.

- The Sorbilite process emits very little steam or noxious gas during processing. The small compact systems can also be located where fiber waste is created, which is a real advantage in helping to create jobs in many rural and isolated parts of the world.

- The process can use wet wood fiber with up to a 50% moisture content, unlike traditional compression technology that requires 8% or less moisture content.

Technology Contact:
> Ms. Deanne Beckwith
> Sales Manager
> Sorbilite
> 5721 Bayside Road
> Virginia Beach, Virginia 23455
> Phone: 757/464-3564
> Fax: 757/464-6959

SOLID WOOD WASTE CONVERSION: SCRAP RECOVERY SYSTEMS (AUBURN)

The recovery and use of wood residue and scraps from traditional wood processing is not new to the industry, but technologies that effectively and affordably combine scrap recovery process options with the ability to process smaller pieces are new and merit attention. Auburn Machinery of Auburn, Maine, manufactures one of these technologies. The line of Yield Pro machines offers several processing options:

- *Yield Pro-2:* Many machines on the market have only one head. This is a 2-head machine that can machine materials with a flat bottom face and one good straight edge. It is used for process ripsaw edgings, short cut-off blocks, mismachined components, and other small pieces that are usually relegated to the waste bin.

- *Yield Pro-3:* The 3-head machine processes materials with a flat bottom face. It is used for mill slabs along with edgings, cut-off blocks, mismachined components, etc.

- *Yield Pro-4:* This 4-head machine can process materials with a flat bottom face and one good straight edge. It produces ripsaw edgings, short cut-off blocks, mis-machined components, etc.

Even with the strong emphasis on maximizing raw resources and value-added production in today's forest products industry, many operations still treat wood trim ends, lumber shorts, lumber jackets, edgings, mis-machined material, and low-grade or defect cuts as scrap material to be chipped, burned, or landfilled. As a rule of thumb, a good way to evaluate good waste recovery in an operation is to spot-check the conveyor system feeding the wood waste pile of the chipper or hog to see how much "valuable" scrap is ending up as chips or waste material for the landfill.

Technology exists to improve scrap recovery within operations, but several factors limit that technology. It tends not to allow single-pass processing of lumber slabs and short wood pieces, or have the ability to process shorter-length wood waste material. To perform such operations with existing technology would typically require more than one pass on the machine, which is inefficient.

An Efficient Solution

The Auburn system was designed to address some of these technical constraints by:

1) Converting both softwood and hardwood material to resource that can be used in value-added wood product production.

2) Targeting a more cost-effective single-pass process adapted to difficult-to-process material such as lumber slabs, odd-shaped pieces, and trim ends and shorts.

3) Accomplishing the above with adaptable technology that can accommodate hand-infeed as well as automatic infeed, both wet and dry wood processing, and multiple feed rate options for wood infeed handling.

4) Combining scrap recovery processing with innovative small wood piece drying to address a major technology need in the industry.

5) Creating an environmentally responsible technology that creates a higher return on value from every tree cut.

The Yield Pro line accomplishes its objectives by providing a technical option for both large- and small-scale producers to convert:

- random short cut-off blocks to uniform fingerjoint blanks;

- ripsaw edgings to moulding blanks and glue blocks;

- edger strips to kiln sticks and grade stakes;

- mis-machined parts to usable materials;

- live-edge squares into smaller-edged units; and

- high defect material to pallet stock and crating parts.

The process technology also allows more value-added product to be made with the existing resource and more value-added jobs to be created in the region where wood resource extraction occurs. Specific examples of its processing performance include:

- Machining a 1-1/2" reject square piece of wood, which has a rounded edge down to a useable 1" dimension piece in only one pass.

- Converting a debarked pine sawmill slab destined for the chipper, which is worth $35/mbf chip value, to a machined 1" by 4" joint blank valued at approximately $750/mbf.

- Converting single and double live-edged boards into valuable component parts.

- Converting an assortment of waste trim ends and wood shorts into valuable blocks used in finger-jointed lumber.

The Yield-Pro system handles multiple wood scrap sizes with thicknesses of 2" to 2", widths of 3/4" to 6", and can process pieces as short as 6". The process handles both wet wood and dry wood scraps, and has variable operating speeds of 20 lineal feet per minute for manual feed and 100 lineal feet per minute for speed feed.

Widespread adoption of the technology is contingent upon development of new technology to effectively and affordably dry wet lumber trim ends and shorts. Auburn has now teamed up with the developers of that technology, Carter and Sprague, to offer a full product line (see below for discussion of Trim Block Drying Rack system).

Primary wood product producers that were considering the purchase of this technology were contacted in late 1996 for their projections on how must waste they might recover in their operations if they used the Auburn scrap recovery system. The manufacturers estimated a total percentage added recovery of 8% of their volume, with 2% of that recovered volume coming from slab recovery alone. A maximum two-year return on investment was projected, with initial capital investment costs ranging from $60,000 to $100,000 for the purchase of complete systems.

The Auburn technology is well adapted for use in primary production mills as well as secondary wood processing operations.

Technology Contact:
> Mr. Thom Labrie
> President
> Auburn Machinery, Inc.
> PO Box 3065
> Auburn, Maine 04212
> Phone: 207/784-4244

WOOD DRYING TECHNOLOGY: TRIM BLOCK DRYING RACK SYSTEM

The Trim Block Drying Rack system is a one-of-a-kind emerging technology. Designed and manufactured by Oregon-based Carter and Sprague, specialists in the design and development of thermal products for the forest products industry, the Trim Block Drying Rack system provides a solution to the long-standing problem of how to dry wood trim ends and shorts for conversion into value-added products.

Until recently, the wood products industry lacked the economic incentives and pressures of diminishing wood supplies to pay close attention to this type of waste recovery. The lack of an affordable drying process further thwarted any efforts to recover slab, trim ends and lumber shorts. The new Trim Block Drying Rack system facilitates bottom-line driven yet environmentally appropriate change in manufacturers' production behavior and practice. Combined with effective scrap recovery processing equipment such as the Auburn system, and existing and new fingerjointing technology, industry no longer needs to treat green trim ends that are wet, and shorts as wood waste, or convert 100% of its green trim ends and shorts to chips or boiler fuel.

Technical Objectives

Specific technical objectives for this emerging technology included the ability to:

- Convert wood waste into resource for value-added wood product.

- Create more full-time family-wage jobs in areas close to where timber is extracted.

- Make more product using equal or less volume of resource.

- Have a drying technology that could be adapted for both hardwood and softwood mills.

- Reduce energy consumption during the wood drying time, but also provide uniform drying throughout the wood load.

- Reduce defects during the dry time.

- Make a system that was easy to handle physically during storing, loading, and unloading.

Test results conducted on the Trim Block Drying Rack system support manufacturer claims:

1) The system was tested on sugar pine, ponderosa pine, Douglas fir, and white pine commodity lumber trim ends. Results showed that it reduced moisture contents down to 19% to 7%.

2) In early 1997 testing was completed for selected hardwoods.

3) The system reduced labor costs over a nonrack method for drying trim ends and shorts.

4) The system eliminated kiln sticks and sticker stain on the wood, which resulted in fewer documented defects in the dried material.

5) The quality of drying was improved due to increased "air-to-board" contact and reduced moisture removal from ends.

6) The system used less energy to dry boards than conventional drying technology due to more air-to-board contact and the hollow aluminum support structure, which increases the uniform transfer of heat through full load.

7) The system was easy to store, load, and unload.

Looking at the estimated lost opportunity of not using the trim block dry rack for both hardwood and softwood manufacturing helps put the importance of the technology into perspective. Lost opportunity analyses for softwoods, as shown in Figure 5, indicate that by using the system a manufacturer could recover a net value per year of over $1 million.

As part of the investigation conducted for this technical note, softwood producers that recently purchased the Tim Block Drying Rack system were interviewed about their estimated recovery rates, return on investments, increased job generation, and added wood product volume based on recovered waste. One softwood manufacturer processing approximately 60 million board feet of lumber annually provided the following data:

Softwood Net Recovered Value for Wood Waste Conversion to F/J Stock

1) Assume 80 mmbf/annual production.

2) Rough green trim ends = 8% of total production or 6.4 mmbf/yr.

3) 60% of that rough green trim end volume can be converted into fingerjoint (F/J) stock or 3.84 mmbf/yr.

4) Calculation assumptions:

- Sales value of F/J stock   = $480/mbf
- Labor to load   = $55/mbf
- Drying cost   = $30/mbf
- Softwood chip value   = $125/mbf

5) Calculated recovered net value using the Trim Block:

- Sales value of F/J   = $1,843,200
- Less chip value   =   ($480,000)
- Less cost to recover and dry =   ($326,400)

Total Recovered Net Value = $1,036,800

Source: Mater Engineering

Figure 5

- The manufacturer projected an estimated annual increase in wood volume production of 3.6 million board feet as a result of the Trim Block Drying Rack installation.

- The operation would go from a one-shift production to a two-shift operation with a 20% increase in total manpower.

- With the ability to effectively dry trim block, the operation would not only convert its own waste into product, but also be able to purchase green and partially-dried trim ends and shorts from other wood processing operations in the area for processing.

- The capital investment made in the purchase of the Trim Block Dry Racks would be recouped in the first year.

Hardwood Net Recovered Value for Wood Waste Conversion to F/J Stock

1) Assume 10 mmbf/annual production.

2) Rough green trim ends = 5% of total production or 500 mmbf/yr.

3) 100% of that rough green trim end volume can be converted into finger joint stock.

4) Calculation assumptions:

• Sales value of F/J stock	=	$1,250/mbf
• Labor to load	=	$65/mbf
• Drying cost	=	$40/mbf
• Softwood chip value	=	$50/mbf

5) Calculated recovered net value using the Trim Block:

• Sales value of F/J	=	$625,200
• Less chip value	=	($25,000)
• Less cost to recover and dry	=	($52,500)
Total Recovered Net Value	**=**	**$547,500**

Source: Mater Engineering

Figure 6

For standard hardwood mills, the calculated total recovered net value per year is over $500,000, as shown in Figure 6.

Technology Contact:

Mr. Lyle Carter
President
Carter and Sprague
PO Box 6206
Beaverton, Oregon 97007
Phone: 503/848-8478
Fax: 503/848-9508
Phone: 207/784-4244

SHORT PIECE AND SCRAP UTILIZATION: "WET" WOOD FINGERJOINTING TECHNOLOGY

The wood products industry has been developing the ability to join pieces of wood together in a fingerjoint fashion for the past decade. Fingerjoint technology offers clear environmental benefits by recovering wood that would otherwise end up as waste or in low-value uses. It offers compelling financial benefits for both hardwood and softwood operations, as indicated in Figure 7. Fingerjointed softwood moulding made from ponderosa pine, for instance, is worth some $1,250/mbf, while the same material used for pallet stock is worth just $160/mbf. With hardwoods the difference in value is similar: Fingerjointed hardwood moulding blanks can sell for as much as $1,350/mbf, while the same material used for pallet stock sells for $200/mbf.

Market demand for fingerjointed products is evidenced in many products—most particularly paint grade clears for mouldings, facings, door and window parts, including paint grade clears and core stock materials, vertical use only studs, and full structural rated products, such as truss cords, floor joists, and other horizontal or tension loading uses.

Fingerjointed material is capturing a commanding share of some targeted markets—especially moulding and millwork products. Some producers report up to 30% premiums paid for fingerjointed product. The reported dollar values for premiums paid range from a few dollars/mbf up to $50/mbf.

From the perspective of product buyers, purchasing fingerjointed material over solid wood product provides several benefits. Most important, fingerjointed material can maintain its dimensional integrity when stored. Because of its long continuous grain, solid lumber tends to cup and warp when "yarded" over time. Export customers in Japan have experienced as much as a 32% loss in lumber stores in yards due to cupping and warping. Fingerjointed material performs much better in this arena: It has an average yard loss of only 3-4%.

Although fingerjointing is not an emerging technology, adhesive applications that allow for fingerjointing of wet wood and even frozen wood is. In the industry, it is standard practice when dealing with fingerjointing wood first to dry all of the wood, which is a costly process, then cut out any defects, then fingerjoint it.

Fingerjoint Technology Conversion Data

For Softwoods:

Item	$/unit (mbf)	% Value Increase/unit from next lowest value
Boiler Fuel	$14/ton or $14/mbf	baseline
Paper Chips	$125/ton or $125/mbf	+793%
Pallet Stock	$160/mbf	+28%
Block for fingerjoint	$480/mbf (ponderosa pine)	+200%
Fingerjointed softwood moulding blanks (preripped)	$1,250/mbf (ponderosa pine)	+160%

For Hardwoods:

Item	$/unit (mbf)	% Value Increase/unit from next lowest value
Boiler Fuel	$12/ton or $24/mbf	baseline
Paper Chips	$25/ton or $50/mbf	+108%
Mulch	$54/ton or $108/mbf (limited species)	+116%
Pallet Stock	$200/mbf	+85%
Block for fingerjoint	$500/mbf	+150%
Fingerjointed hardwood moulding blanks	$1,350/mbf (poplar)	+170%

Source: Mater Engineering based on available pricing data

Figure 7

Technical Objectives

The technical objectives for producing a system that can process wet fingerjoint material are as follows:

- To allow for fingerjointing of wet softwood trim ends and shorts that have a moisture content of 30-100% to produce defect-free lumber that can then be dried for value-added product use. Drying after fingerjointing requires a drop in moisture content to 9-12%.

- To allow for fingerjointing of partially-dried softwood lumber trim ends and shorts with an 13-20% moisture content to produce defect-free lumber for further drying as required for value-added product use.

- To reduce processing costs by drying less waste and concentrating costly drying time only on defect-free fingerjointed wood.

Benefits

The development of new adhesive technology and processing of wet wood fingerjointing has several notable environmental and economic benefits:

1) Defects are removed before drying, which can save up to 50% of the drying costs normally incurred by traditional milling operations. Between 30-50% of kiln dried lumber currently used to produce fingerjointed product has defects that must be cut out and discarded.

2) The capacity of the kiln is maximized since all dried timber is salable.

3) Short lengths of wet lumber can be upgraded into marketable product instead of being dumped or converted into chips. One 1994 example documents that wet wood trim end waste was converted to fingerjointed lumber with a market value of $825/mbf;

4) The process has a faster curing time than standard dry wood fingerjoint adhesive technology—five minutes compared to twenty-four hours for conventional adhesives.

When evaluating wet wood fingerjoint processing, two emerging adhesive technologies, the Greenweld Process and the Kreibich System, appear to be leaders.

Greenweld Process

Greenweld Technologies, Ltd., a wholly-owned subsidiary of the New Zealand Research Institute Ltd. located in Rotorua, New Zealand, has pioneered the efforts in fingerjointing non-dried, even frozen wood. With plants now located in New Zealand, Australia, and the United Kingdom, they have just opened their first U.S. operations in Oregon through a partnership with WTD, one of the ten largest volume lumber producers in the U.S. The intent is to further introduce the technology to North American markets.

Testing has been completed for Greenweld application on multiple medium-density softwoods and hardwoods throughout the world with favorable results. Testing has proven to evaluate the possibility of mixing species in fingerjointed product development such as fingerjointed lumber made from combinations of black spruce, jack pine, and balsam fir. With the exception of more dense hardwoods with high oil content found in the tropics, the Greenweld technology has demonstrated performance to required building code and lumber standards in multiple countries including the U.S.

Continuing technological advances have improved the Greenweld process in several critical areas:

- While meeting U.S. performance standards for green fingerjointed lumber use on a side surface (vertical position) as of July of 1997, the company received certification of its green fingerjointed lumber for use on its face (horizontal position) in October, 1997. This greatly expands the use of the product to new applications and allows further waste reduction during the fingerjoint processing.

Instead of the typical 1 1/8" fingerjoint typically required for horizontal use, Greenweld requires only a 7/8" joint resulting in an additional savings of approximately 1,500 lineal feet of wood waste per eight-hour shift.

- Instead of the typical 7 to 8 days required before quality control product testing can be achieved for traditional dry wood fingerjoint technology, the Greenweld process reduces the wait time down to just 24 hours, with in-process problem checking conducted every hour, 10 minutes after the board is produced. Problem areas are quickly detected during, not after, production with final product ready for shipment after 24 hours.

- The process produces approximately 80 blocks per minute and can process piece variations between 5" to 30" in length, 3" to 6" in width, and up to 4" lumber thicknesses.

Financial results for companies selecting to invest in the Greenweld technology are already bearing notable bottomline results:

- Sales prices for green (wet) fingerjointed studs match or exceeds green (wet) solid wood studs. At the writing of this technical note, the sales price was at $350/mbf. Premium green fingerjointed studs (producing a clear, non-seam wood look) was selling at $375/mbf.

- The net profit per mbf of green fingerjointed product ranged, at the time of this analysis, from $125/mbf to $130/mbf. (This assumes a $150/mbf cost to purchase and transport wood waste to a Greenweld site, and an additional $120/mbf cost to produce green fingerjointed product.)

- The cost differential between purchasing green fingerjointed lumber and dry fingerjointed lumber is $35/mbf, providing a sizable bottomline benefit to building contractors who are likely to be attracted to not only the lower cost, but the increased nail-entry ease without splitting concerns often experienced with dried wood.

- According to companies selecting to invest in the Greenweld technology, with an up-front technology purchase cost of approximately $3.5 million (assuming no stud mill or fingerjointing technology exists on site), realized payback on investment has been within industry acceptance levels of 18-24 months.

Use of the Greenweld technology by wood product producers can result in significant wood waste reductions during processing of commodity lumber while adding valuable, community-sustaining jobs to a state's economy. According to data provided by the American Forest and Paper Association, the total hardwood and softwood lumber production in the U.S. for 1995 was 44.6 billion board feet. Using a 5% waste recovery factor and a factor of 3 jobs created for every 1 million board feet of lumber produced, the Greenweld technology could help realize approximately 2.3 billion board feet of wood waste recovery throughout the U.S. annually and could help generate almost 7,000 new fulltime, family-wage jobs. All of this without increasing timber harvest levels and by converting wood waste into wood profits.

Other documented benefits of using the Greenweld adhesive technology include:

- Successful testing on species up to a 200% moisture content;

- Successful testing on frozen wood;

- Uses standard fingerjointing machinery typically used in dry wood fingerjointing. No new machinery required.

Noted constraints associated with the use of this technology include:

- As with many engineered wood product technologies, the process does use phenol-resorcinol formaldehyde (PRF) resin (toxic);

- The resin accelerator used in the adhesive product produces some fumes in the work environment that must be removed with installed ventilation systems or mechanical cooling of the accelerator;

- Waste water generated from the fingerjoint process must be treated to produce "land fillable sludge".

- A royalty payment of $7/mbf of production is required to be paid back to the New Zealand Research Institute in order to use the Greenweld technology.

Technology Contacts:

Mr. Sean Clearkin
Chief Executive Office
Greenweld Technologies, Ltd.
PO Box 6230
Rotorua, New Zealand
Phone: 647/347-5850
Fax: 647/347-5331
E-Mail: greenweld@rfri.cri.nz

Mr. Jeffrey W. Garver
Vice President
Greenweld North America Co.
5700 SE Reservoir Avenue
Corvallis, Oregon 97333-2936
Phone: 541/758-1023
Fax: 541/754-6134
E-Mail: J.Garver@greenweld.com

Soybean-Based Adhesive (Kreibich System)

Developed through a joint effort sponsored by the United Soybean Board, a new soy-based adhesive targeted for the wood fingerjoint industry was announced in 1996.

Documented benefits for employing this adhesive product include:

- The process was introduced to the market in September 1996 through a joint effort between an ex-Weyerhaeuser employee and the United Soybean Board.

- Successful testing has been completed on multiple North American softwood species.

- Successful testing has been completed on species exceeding 150% moisture content.

- Requires *no* high-technology equipment or electricity. Glue can be applied by hand with a brush; end pressure can be applied by simple lever action or even with a sledge hammer, which is a benefit to underdeveloped countries.

- Purported cost savings over other wet fingerjointing technologies is fairly significant.

This adhesive application process cannot be used with standard fingerjointing technology but must have new technology adapted for a "honeymoon" application. The process also uses phenol-resorcinol formaldehyde (PRF) resin (toxic).

Technology Contact:

Dr. Charles Grabiel
Omni Tech International
2715 Ashman Street
Midland, Michigan 48640
Phone: 517/631-3377 (x 304)
Fax: 517/631-7360

Dr. Roland Kreibich
Kreibich Associates
4201 South 344th Street
Auburn, Washington 98001-9545
Phone: 206/838-0964 or 206/924-6010
Fax: 206/924-4239

WASTE REDUCTION: NEW SCANNING TECHNOLOGY (ROBO-EYE)

Computerized scanning systems have been integrated into the forest products industry for many years. The systems have historically been used to help mill managers figure out how to cut the highest value of wood from a log based on characteristics such as the size of the log and the grade of log. These characteristics are often determined by the amount of defect present in the log such as knots, splits, and fluctuating grain configurations. Wood processing operations have documented increases in log values of between 20-50% by effectively using appropriate scanning systems. Realizing these increased values is especially important in several areas:

1) For hardwood operations, purchasing logs represent up to 75% or more of lumber production costs. Gaining the maximum value from a log when it is processed is crucial.

2) For both hardwood and softwood operations, tighter restrictions on gaining access to raw material has heightened the need to maximize the use of those resources that are available for processing.

3) Increasingly, value-added production operations throughout the world are emphasizing the use of lessor-known and underutilized species in their product development. Achieving greater production and processing efficiencies through color and defect scanning systems can help raise the confidence of buyers in product quality and consistency, which are two high risk areas for product buyers in dealing with nontraditional wood species.

4) Because wood supply in many regions throughout the world is coming from smaller diameter logs (12"-18" in diameter), gaining the maximum value from each log processed is important.

Ironically, heated conflicts between industry and environmental groups have also helped to improve log scanning technology. The rising incidence of log sabotage in forests where tree sales are controversial, in part, has led to the development of better x-ray scanning systems. These more sophisticated systems can detect foreign objects such as ceramic spikes, nails, barbed wire, and shot glass that have been inserted into the trees, and which pose serious processing hazards to loggers and mill workers.

While much of the scanning technology has been dedicated to scanning logs, there continues to be increased demand for better scanning systems that can be used in lumber production and value-added wood product manufacturing.

Individual scanning systems that are currently viewed as leading-edge technology for lumber and value-added wood product production typically use either color imaging sensors, laser-based ranging sensors, or, to a more limited extent, x-ray scanners. The importance of these technologies for increasing the value from each piece of lumber processed is based on their respective beneficial functions.

Color Scanning

These systems use color line scan technology which is particularly useful for detecting features such as knots, holes, splits, checks, stains, and decay in the wood being processed. Color scanning is also effective in evaluating color in the wood, a function most important for lumber sorting and creating custom grade material based on aesthetic wood qualities.

Laser-Based Ranging Systems

This technology is based on high-speed pixel array cameras that scan at a rate of 384 frames per second. The profiling capability created from this technology is especially useful for detecting lumber shape characteristics such as wane, thickness, splits, checks, holes, and various types of lumber warping.

X-ray Scanning Systems

This x-ray technology is similar to that used in airport systems that scan luggage. It is well suited to detecting accurately the full extent of knots and decay in lumber. X-ray systems are also good at finding internal features such as knots, honeycomb, mineral streak, and gum pockets which cannot be detected by color or laser scanning systems.

Robo-Eye Combines All Three

The technology challenge for scanning systems has been to develop a reasonably priced system that incorporates all three functions into one system. Although still not commercially produced, a team of Virginia Polytechnic & State University researchers in electrical engineering, wood science, and forest products, and the U.S. Forest Service Southern Research Stations are advancing such technology. A new multiple sensor machine (MSM) under development at Virginia Tech—coined

"Robo-Eye" by CNN—is setting the stage for the next generation of integrated scanning systems in lumber production. In it's final stage, the MSM incorporates color, laser, and x-ray scanning functions into one machine. The general purpose machine should be able to automatically check lumber for surface features (knot holes, splits, decay, color, grain orientation), for board geometry (warp, crook, wane, thickness variations, voids), and for internal features (honeycomb, voids, decay).

Although the technology, which has a patent pending, is not yet ready for market, a completed MSM system has been built by an equipment manufacturer and was being tested in a plant in 1997. Testing was scheduled to be completed by year-end 1997. The total cost for the completed MSM unit is expected to be $250,000 to $500,000.

Technology Contact:

Dr. Richard Connors
Virginia Polytechnic & State University
Dept. of Electrical Engineering
340 Whittemore Hall
Blacksburg, Virginia 24061
Phone: 540/231-6896

Philip A. Araman
USDA Forest Service/Southern Research Station
Brooks Forest Products Center
Virginia Tech
Blacksburg, Va. 24061-0503
Phone: (540) 231-5341
Fax: (540) 231-8868

WOOD HARDENING TECHNOLOGY: INDURITE (TM)

This new wood hardening process, originally developed by researchers at the New Zealand Forest Research Institute in Rotorua, New Zealand, was initiated in response to concerns regarding the overharvesting of the tree species Rimu, found in the native forests of that country. Rimu, a hardwood with a beautiful reddish-brown color and grain, is used in many value-added products such as furniture, paneling, and flooring. In 1994, however, due to concerns of overharvesting, Rimu production plunged from 120,000 cubic meters of wood harvested to approximately 20,000 cubic meters. The decision to invest in developing technology that could harden radiata pine, an abundant softwood plantation species in New Zealand, so that it had comparable performance and visual qualities to Rimu, appeared to be a rational option for the industry and an appropriate response to environmental concerns.

The process has created both a new technology and a new product that merits discussion. The process allows faster-growing softwoods that are typically not used in many value-added products to replace more valuable, and increasingly threatened hardwoods in those applications, which has both environmental and economic benefits.

When confronted with limits on the harvesting of a particular species, the forest products industry typically adopts one of three options to relieve the immediate impact of the species restriction:

• Identify another species of similar type and characteristics that can be substituted in product development. In many cases, this may mean searching for similar wood resources in other regions of the world that pose less stringent harvest practice requirements for access to the resource.

• Convert the material used in product development from an all-wood based to a partial-wood based or nonwood based material;

• Convert from the use of a hardwood for product development to a softwood. When North American hardwood prices escalated during the 1980s, for instance, the upholstered furniture frame industry switched from using hardwood in furniture frames to good quality softwoods.

In this case, a segment of the New Zealand wood products industry opted to research ways to improve the mechanical characteristics of an abundant and replenishable softwood that does not come from native forests, so that it resembled the desirable mechanical characteristics of the restricted hardwoods. The technical objectives were specific:

• Create an affordable, marketable replacement for Rimu using treated radiata pine.

• Change the characteristics in radiata pine without sacrificing its desirable wood processing features, such as ease of machining and dimensional stability.

• Develop the first wood hardening production facility in New Zealand selling this new hardened pine, with intentions to license the technology for offshore production in Malaysia, China, and Australia.

Today, Evergreen Indus-Trees Limited of Auckland, New Zealand offers the trademarked Indurite process to wood product manufacturers worldwide. An associated company is currently constructing a processing plant in Tauranga, New Zealand, to process radiata pine for remanufacturing into flooring for the Japanese market. A similar plant is being established to convert Fijian-grown Caribbean pine into glue-edged laminated panels in New Zealand, as well.

Treatment Process

The Indurite process essentially increases the density of the treated lumber by impregnating a largely cellulose-based formulation into the wood. In effect, the process pours wood into wood. The Indurite process, which can be applied to all species that accept water-borne pressure impregnation, has been successfully applied to all other pines, aspen, birch, poplar, coconut, and eucalyptus species.

After combining six components in a batching tank, lumber that is pre-dried to 30% moisture content is then impregnated in a vacuum pressure cell for approximately one hour. During this time, the weight of the lumber increases up to 50%. The lumber is then dried in a standard medium-temperature kiln in line with the normal drying schedules for appearance grade lumber.

Results

Test results show that impregnated radiata pine becomes harder than oak, teak, and African mahogany. And, as noted in Figure 8, other characteristics of the wood such as stability, gluability, and machinability—all critically important in the manufacture of value-added products—are also greatly enhanced over nontreated wood. Before processing, the softwood radiata pine is excluded from consideration for product development. Afterwards, the Indurite-treated lumber becomes a high-quality, lower-cost resource option than hardwoods for use in the manufacture of flooring, furniture, doors, joinery, and a variety of other interior finishing and decorative applications.

The Indurite process offers other attractive benefits to wood product manufacturers. During impregnation, a stain can be introduced throughout the lumber permitting machining to occur after the treatment, which is an important benefit considering that the wood has enhanced machining characteristics. The production of Indurite-treated veneer is also possible since the slicing process ruptures cells in the wood, which enhances the wood's ability to absorb the formulation.

Bottom Line

On a per unit basis, the cost of introducing the Indurite treatment to softwoods and lower-quality hardwoods runs approximately $250 per thousand board feet of lumber processed. This means that, coupled with the normal purchase price of dried lumber, the economic benefits to the value-added wood product manufacturer in using Indurite-treated lumber can be substantial. As Figure 9 indicates, when comparing the cost of Indurite-treated pines from multiple locations across the United States to the cost of traditional higher-quality hardwoods that are preferred in product development, an average 40% cost savings could be realized by using the Indurite treated wood. Coupled with its enhanced abilities to accept coating and staining, the Indurite-treated lumber can be an effective,

---

Performance of Pine Hardened by the Indurite Process

Desired Characteristic	Performance
Hardness	• Average hardness increased by 52%. • Treated pine hardness exceeds oak, teak, and African mahogany hardness.
Gluability	• Achieves higher shear strength than untreated pine.
Machining Properties	• *Superior* results noted in cross-cutting, turning, moulding, shaping, morticing, and drilling. • *Inferior* results in fingerjointing and nail splitting.
Coatability	• Stain coatings go on evenly and lightly. • Top coats produce superior finish.
Appearance	• Natural wood color retained. • Dyes and stains are easily introduced. • Reduced "yellowing" of wood is noted when exposed to sunlight.
Environmental Impact	• 95% biologically-based. • Requires little energy consumption. • Is non-toxic.
Fire Resistance	• Has higher fire-resistant properties over non-treated pine.

Source: Independent testing commisioned by Evergreen Indus-Trees Ltd.

Figure 8

---

economical substitute for both lighter-colored and darker-colored native hardwoods.

Technology Contact:

Mr. Peter McArthur
Chief Executive
Evergreen Indus-Trees, Ltd.
PO Box 2142
Tauranga, New Zealand
Phone: 647/578-5990
Fax: 647/578-9550
E-mail: evergreen.ind.@xtra.co.z.

Price Comparisons for Indurite— Treated Softwoods Compared to Traditional Hardwoods (1 x 4) (1997 $ Values)

Species	Standard $/mbf	Premium $/mbf
Softwoods*:		
Idaho White Pine	719	1,066
Ponderosa Pine	686	1,223
Southern Pine	612	922
Sugar Pine	828	1,427
Eastern White Pine	689	990
Hardwoods:		
Morthern Ash	785	1,110
Northern Red Oak	1,235	1,625
Northern Hard maple	1,480	2,275
Appalachian Cherrt	1,195	2,595
Appalachian Walnut	1,060	2,030

* Includes $250/mbf for Indurite treatment.

Source: Mater Engineering

Figure 9

Barriers to the Adoption of SFM Technologies

Given their great promises, why do so many emerging technologies that can enhance SFM lack visibility and marketshare? The answer lies in the multitude of constraints that hamper the adoption of these technologies and techniques by the forest industry.

SIZE AND GEOGRAPHY

First, for those technologies already out in the market, making those technologies visible to small and medium-sized wood product producers located in rural or isolated geographic regions is an ongoing challenge. Yet, these producers have significant impact in the forest products industry. In the United States, for example, small producers of twenty employees or less constitute almost 80% of all wood products operations, including pulp and paper operations. Often located in rural areas with limited access to traditional industry trade association activities or publications, this audience often misses information on new technology offerings.

VOLUME VS. VALUE

Second, innovations in new techniques and technologies in the forest products industry are most often developed from two primary sources: academic institutions, and small private organizations and individuals. For academic institutions, the research tends to be concentrated on technologies that are more likely to benefit the major wood product producers in the solid wood and pulp and paper industries. This is, in large part, due to the funding these producers provide to conduct the research. While the focus of research differs from institution to institution, it is not unusual to find research targeted on technology that increases the volume instead of the value of wood produced, or to find more research weight given to primary wood product production over value-added wood production. Academic institutions also have long lead times to complete research, which makes it difficult to produce market-timely results.

Technical innovations that come from small organizations or individuals also face a variety of constraints that limit their access to the market:

- Limited financial capability to adequately market the technology.

- Lack of marketing acumen to sell the technology to industry.

• Desire to retain ownership of the technology and to sell the product produced from the technology. This often results in overselling the technology/product without appropriate production capability. (Note: Industry rule of thumb #1: it's usually the second owner of the technology that makes the technology work financially, even though the original designer got it to the table.)

• Inability to offer customer servicing, including necessary parts and repairs.

WHO GOES FIRST

The mindset of the traditional wood products industry is another major market access constraint for emerging technologies. As an industry rule of thumb no one wants to adopt a new technology first, unless they have designed it themselves. This industry mindset is often expressed in several ways in relationship to the purchase of new technology.

1) Any performance improvement that new technology delivers must be marketed to a daily production mindset. For example, a new dry kiln technology may provide a 35% reduction in material lost due to traditional overdrying. But for the production facility to invest in the new technology, the performance must reach 50% waste reduction to allow a second shift to be added to the mill. A 35% reduction does not increase the ability of the mill to process more volume on a daily basis; a 50% reduction could allow the facility to double its daily production.

2) In general, return on investments for new technology need to be within 18-24 months with rates of return in the 22-25% range.

3) Many wood product producers conclude that if a new technology is not offered by major traditional equipment manufacturers, it can't be any good. However, before traditional equipment manufacturers will invest in new technology, they need to see documented interest in the technology from wood product producers. The technology designer is forced to serve as a middleman to between the supplier and producer—a process that requires expertise, time, and money, which are usually in short supply.

4) The lack of computer-based skills in the industry is a significant detriment to investments in SFM technologies. With much of new processing technology computer-based, the average employee for a typical wood products operation often lacks the computer training and skills to adapt willingly to the new technology. Mill managers, who are also often deficient in computer skills, complain of too much downtime when working with new technology. The impact of this one constraint alone to the acceptance of SFM technology should not be viewed lightly. The problem highlights a substantial need not only for increased employee training, but also for employee ownership of new technology ideas. Ease of technology application and user-friendly functions also hold high market value.

R&D FINANCING

The need for emerging technologies to initially focus on large-producer sales to quickly capitalize research and development (R&D) investments from both public and private sector funders can produce a no-win option for the SFM technology designer. A technology better adapted to value-added process improvements may find a more ready audience with small- to medium-sized product producers, but it would require a few major producers to invest in the technology in order to meet the payback periods required to secure the initial R&D financing. Traditional equipment manufacturers targeted to represent the technology may, in addition, be unwilling participants unless a major producer is involved.

Finally, the lack of availability and access to research regarding market opportunities for underused and lessor-known species in product development, lower grade lumber use in higher-end product manufacturing, and "character wood" developed from custom grades for higher-end product development is an additional constraint to emerging technologies because it fosters a business-as-usual approach to production.

Solutions That Invite Change

To reduce the constraints to market entry for SFM emerging technologies, the issues of credibility and visibility must be directly addressed.

GAINING CREDIBILITY

For an SFM emerging technology to establish credibility and leverage, support from the industry and the environmental community can be crucial. Both are currently difficult to access.

For industry, gaining credibility for emerging technologies often rests in having access to independent testing of the technology by an accredited organization or institution. This independent testing can prove enormously helpful in moving the technology to market. But it can also be costly, and few programs exist to help designers of technologies that can be adapted to smaller-scale uses, which are most needed in the world's rural regions. The total estimated annual public and private investment in forestry and forest products R&D in the United States is only about $1.3 billion. For forest products research, public funding contributes approximately $40 million, while private funds equal about $825 million.

Industry and universities have a solid track record in engaging in cooperative research projects, but most university links are with large corporations whose research interests are either focused on biological problems, genetics and vegetative management, or physical problems, such as soils and water quality. In 1987, fifty-one U.S. universities had research cooperatives of this nature with wood-based industry. The cooperatives were investing approximately $5.2 million in R&D, over half provided by industry.

Production as Part of SFM Strategy

In the United States, federal programs do exist to help introduce new technologies to market, but access to these funds often has additional requirements. For example, the Small Business Innovative Research (SBIR) program has a three-step process that usually spans several years before funding is provided to actually introduce a new technology to the market. Expanding the SBIR program to include a Small Business Emerging Technology (SBET) arm may be an attractive addition to the existing SBIR program which was slated for Congressional review in 1997.

Similarly, universities currently engaged in offering new sustainable forestry programs and curriculums may find unusual opportunity to collaborate in obtaining both public and private R&D dollars to provide independent testing for SFM emerging technologies. As part of university technology testing, SFM technology rental programs could be established that would allow industry to test a technology on-site to verify its performance before investing in it. The rental program would also give the university the opportunity to test the technology in a laboratory setting, yet monitor and record its performance in an industrial environment.

Obtaining credibility within the environmental community may require a significant shift in thinking within environmental organizations and the funders that support them. If one key objective of SFM is to help sustain communities in balance with forest ecosystems, environmental organizations are missing an important link. For example, can the passage of a new forestry law in Bolivia that dramatically limits the annual volume of mahogany to be harvested from Bolivian forests prove to be a sound policy when a substantial amount of the new allowable cut is wasted each year during the production process? Inappropriate wood drying techniques and lack of access to technologies that allow the use of short pieces of wood are just two of many production issues in Bolivia which leave many forest products experts speculating that 30-50% of hardwoods initially processed in the country end up as wood waste. These production issues are equally relevant in other regions, including many rural areas in the United States.

Environmental organizations no longer have the luxury of looking only in the forests when considering how to successfully implement SFM policies. Yet, few of these organizations or those that fund them, are making production a part of their sustainable forest ecosystem strategy. Consider: In 1988, research grants from private foundations for natural resource conservation activities throughout the world outpaced those grants made for forestry and forest products R&D by almost a 3:1 ratio. Aside from small-scale "boutique" product manufacturing, evidence of environmental support for sustaining communities through more efficient, effective forest products manufacturing is notably missing. The same condition holds true for other issues, such as defining market opportunities for products made from lessor-known species and for certified wood products.

As a matter of policy and funding, encouraging the global environmental community to become actively engaged in evaluating "up-stream" production solutions that can help stabilize community impacts when implementing forest conservation efforts should rank as a top priority with those who provide financial support to environmental groups.

Identifying the top ten emerging technologies each year that can help foster SFM practices may be a valuable activity for environmental organizations. It could encourage unusual politically and financially powerful partnerships between environmental groups and important segments of a forest products industry, who are both better known for sitting at opposite ends of the table on these issues.

GAINING VISIBILITY

Gaining visibility for emerging technologies usually requires two marketing pathways that need to be followed simultaneously. The first path is to get the technology introduced to the market; the second to get the new product manufactured by the emerging technology accepted in the marketplace. It does little good to introduce a new wood hardening technology if the newly created hardened wood is not recognized as acceptable under traditional building codes. Gaining visibility and access to the market in these areas can be daunting, yet very little assistance is available to the technology producer.

Forging Financial Partnerships

From a standpoint of marketing technology, there are some bright spots, even if they are generally isolated events. During the early 1990s, for example, Key Bank in Seattle, Washington, earmarked a $50 million dollar commitment-to-loan for value-added wood products technology that is to be marketed offshore. A first of its kind for the banking institution, the program, which emphasized conventional value-added technology, underscored the reluctance of financial institutions to work with new, *emerging* or *nonstandard* technologies backed by the assets of smaller-scale producers or individuals.

A comparative review of the U.S. environmental products and services industry illustrates the point. In 1991, the industry consisted of 207 public companies each averaging $198 million in annual revenue, compared to 58,700 privately held companies each averaging $1.3 million in annual revenue. Smaller companies have a harder time getting financed—especially for export markets. Further, the lack of up-front capital may be exacerbated by the lack of focus in the United States in general, and in the environmental services segment in particular, on export market opportunities. The 1992 data for the environmental products and services industry show that only 5% of the total U.S. environmental services industry production is exported, compared to 24% for Japan and 31% for Germany.

Under typical lending guidelines, nonstandard technologies are often classified as too high-risk for funding consideration, especially when backed by small-company assets. Moreover, the technology generally lacks testing in the market, which further deters banks from supporting technology developed by small companies.

Even so, a partnership between a traditional banking institution such as a Key Bank, which has taken steps to focus on the export of value-added wood product technology, and the World Bank could prove to be effective in making emerging SFM technologies visible and accessible in world markets.

Private nonprofit foundations could also play a critical role as a catalyst for financial partnerships that could help foster the adoption and development of emerging technologies. They could help bring the various parties together, foster an exchange of information among different players in the product distribution system, and highlight the importance of considering the entire SFM chain of production—from the forest to the end-user—when targeting investments in emerging technologies.

LINKS ARE KEY

This emerging technologies document attempted to illustrate the critical nature of this linked logic. It is, for example, a much more difficult task to market a Trim Block Drying Rack system that efficiently dries short pieces of lumber *unless* it is linked with a short-piece lumber recovery and sizing technology such as the Auburn system.

Marketed separately, these emerging technologies offer the product manufacturer no more than a technology *option*. Combined they offer the manufacturer a technology *solution*. The difference is substantial—it can mean success or failure in marketing these technologies.

Aracruz Celulose S.A.
and Riocell S.A.:
Efficiency and
Sustainability on
Brazilian Pulp
Plantations

PREPARED BY:
WORLD RESOURCES INSTITUTE

A Case Study from "The Business of Sustainable Forestry"
A Project of The Sustainable Forestry Working Group

The Sustainable Forestry Working Group

Individuals from the following institutions participated in the preparation of this report.

Environmental Advantage, Inc.

Forest Stewardship Council

The John D. and Catherine T. MacArthur Foundation

Management Institute for Environment and Business

Mater Engineering, Ltd.

Oregon State University
Colleges of Business and Forestry

Pennsylvania State University
School of Forest Resources

University of California at Berkeley
College of Natural Resources

University of Michigan
Corporate Environmental Management Program

Weyerhaeuser Company

The World Bank
Environment Department

World Resources Institute

CCC 1-55963-620-3/98/page 5-1 through page 5-31

Aracruz Celulose S.A. and Riocell S.A.: Efficiency and Sustainability on Brazilian Pulp Plantations

CASE STUDY

PREPARED BY:
WORLD RESOURCES INSTITUTE

A Case Study from "The Business of Sustainable Forestry"
A Project of The Sustainable Forestry Working Group

Contents

(continued)

Contents

Aracruz Celulose S.A. and Riocell S.A.: Efficiency and Sustainability on Brazilian Pulp Plantations

Introduction

Tightening global timber supplies and rising concerns over the loss of natural forests are fueling investment worldwide in plantation forestry. Plantations, which produce relatively standardized wood fiber, offer a variety of advantages. Typically, they are more predictable, reliable, and malleable than natural forests. The Southern Hemisphere, with its relatively favorable growing conditions and labor and lands costs, has emerged as a center of plantation fiber production. Some of the most impressive plantation forestry is taking place in Brazil, under the management of that country's growing pulp and paper industry. As this comparative examination of two Brazilian pulp producers, Aracruz Celulose S.A. and Riocell S.A., demonstrates, companies face a wide range of environmental, social, and economic forces as they try to balance efficiency and sustainability in plantation forestry.

Aracruz Celulose S.A.– Business Description

Robert Day, World Resources Institute

Headquartered in Rio de Janeiro, Brazil, Aracruz Celulose S.A. was founded in 1967 as a bleached market pulp manufacturer. The original capital included equal parts private international funding and funding from public and private Brazilian interests. Norwegian-born Erling Lorentzen, a successful businessman and former World War II resistance fighter, spearheaded the project. Lorentzen and his partners saw the advantages of Brazilian tree plantations for fiber export, and decided that domestic conversion to pulp made good business sense. With help from favorable national government policies, the company purchased 30,000 hectares in southeastern Brazil from the state government at a very low price. On this land they planted eucalyptus seeds imported from Australia and began construction of a pulp mill and port.

Thirty years later, Aracruz Celulose is the world's largest producer of bleached eucalyptus pulp. The company's Espirito Santo manufacturing plant produces over one million tons of pulp annually, almost all of which is exported. Total revenues in 1995 reached US$796 million. The company's forests now total 203,000 hectares, of which 132,000 hectares are sustained-yield eucalyptus plantations and 57,000 hectares are set aside as natural forest reserves.

In a commodity industry, Aracruz is the world's low-cost producer, which helps the company avoid the low margins characteristic of the industry. In 1995, the company achieved a pre-tax margin of 52.3%. Under the guidance of Lorentzen, Aracruz is also a leader in the World Business Council for Sustainable Development (WBCSD) and other business-driven environmental initiatives. Aracruz Celulose, through sustainable yield forestry and top quality manufacturing, has become a global leader and a model for industrialization within Brazil.

BUSINESS SUCCESS—PUBLIC CONTROVERSY

Aracruz aspires to be a leader in its market, and in its environmental and social performance. Brazil's favorable climate and Aracruz's technological superiority in cloning and growing trees, have given the company a competitive advantage in its industry that financial analysts estimate would take a well-financed competitor over two decades to match. Aracruz consistently generates positive cashflow, no easy feat in a commodity market with substantial and rapid price swings.

Aracruz's environmental and social performance, however, have come under fire. Critics argue that the company is insensitive to the rights and needs of local communities, that its vast eucalyptus plantations damage the soil and deplete vital water supplies, and that pollution at the pulp mill is at unhealthy levels.

Aracruz and its supporters respond that their plantations should be considered crops like corn or wheat, and that the biodiversity of its land should not be compared with that of native forests, but instead to its prior use. Most of Aracruz's eucalyptus is planted in previously cleared areas. Moreover, the company preserves 57,000 hectares of native forest interspersed with its eucalyptus plantation. On the social side, Aracruz contributes significantly to the economic development of that region and further contributes healthcare, housing, education, and other benefits to the local communities.

The divergent views that Aracruz and its critics hold of the company's business illustrate the conflicts that surround plantation forestry:

- Critics point out that plantation forests are less biologically diverse than natural forests, and eliminate biodiversity on vast tracts of land. Aracruz responds that its plantations, planted on degraded agricultural land, provide valuable forest cover for far more species than they used to, and, more important, take economic pressure off natural forests.

- Critics charge that eucalyptus consumes substantial amounts of water and deprives communities and farmers of a much needed resource. Aracruz counters that eucalyptus does not consume more water than other species of trees if its growth rate is considered: the fast growing trees actually increase land productivity and reduce the cost of fiber.

- Critics maintain that cloning trees creates monocultures that are more vulnerable to disease and pests. However, genetically similar trees provide more uniform wood fiber, which produces a premium quality product. Aracruz points out that they use over 100 genetic varieties of eucalyptus clones, and that interspersing plantation and native reserves counteracts the homogeneity, helping to ensure an ecological balance, which helps control pests and diseases.

- Critics accuse Aracruz of occupying 13,000 hectares of land belonging to indigenous groups. Aracruz, however, legally acquired the land.

In 1992, with the world's attention drawn to Brazil during the Rio environmental summit, the international activist group Greenpeace cooperated with several local nongovernmental organizations (NGOs) to publicly lambast Aracruz Celulose. With its "eucalyptus monoculture [Aracruz] destroyed 10,000 hectares of Atlantic rainforest," according to Greenpeace documents. By blockading the company's port with its ship Rainbow Warrior II, Greenpeace held the spotlight on many of the problems that have been blamed on the giant pulp producer.

The contrast between the company's business success and its controversial public image highlights the confusion and tension surrounding Aracruz Celulose. On the one hand, the company has undeniably achieved a cost advantage within its industry through what its supporters call sustainable forestry. On the other, Aracruz is the lightning rod for many of the environmental and social controversies surrounding the Brazilian forest products industry.

The irony, however, is that Aracruz exemplifies plantation forestry, which many environmentalists argue is an answer for sustainably meeting the world's fiber needs. An examination of Aracruz's financial performance, and the conflicting interpretations of its eco-social performance, illustrates both the promises and pitfalls of this type of forest management.

THE VOLATILE PULP MARKET

Aracruz Celulose's emphasis on eco-efficiency—the efficient use of raw materials through waste minimization—as well as its sustained yield forestry are vital for its success in the volatile and difficult global pulp industry. Access to stable fiber supplies will be increasingly important.

In 1991, the global production of pulp reached 429 million cubic meters, according to the United Nations Food and Agriculture Organization (FAO). This amount does not include waste from solidwood production, which can be a source of as much as 40% of the fiber in the United States. Pulpwood production represents about one quarter of industrial roundwood production. The pulp industry accounts for about $390 billion in annual sales.

The pulp and paper market is expected to expand in the near future, as demand for paper products continues to rise. North America and western Europe together currently account for around 60% of the world's fiber consumption. The FAO projects that from 1991-2010 paper and paperboard consumption will increase an average of 3.2% annually. Developing countries, however, will average a 5.3% uptick in demand each year. Even though non-wood and recycled fiber will meet some of this growing demand, the FAO predicts that global pulp consumption will rise 2.9% annually over this period. Meanwhile, wood pulp production is projected to increase only 1.1% a year, according to the *Sustainable Paper Cycle* published in 1995 by the WBCSD. These demand pressures should bring significantly higher prices for pulp producers fortunate to have a stable fiber supply.

Producers are not entirely in control of when they harvest fiber and produce pulp because the growing cycles for pulpwood are so long. These circumstances make the pulp industry highly cyclical, driven by high capital intensity and uneven supply. Improvements in technology, transportation, and communication have allowed for increased globalization, so that producers can sell their pulp anywhere in the world. Consequently, the pulp cycle is now characterized by steeper, more frequent price swings, and margins are low in the industry. This unstable environment threatens to drive many producers out of the market. To mitigate some of the cyclical effects on earnings, most pulp producers are vertically integrated, combining their pulp production with paper product manufacturing.

Several different categories of pulp are used to produce a wide variety of products ranging from fine writing papers and tissue to corrugated cardboard. Differences in the pulping processes, and the contrasts between softwood and hardwood pulp create the various categories of pulp. Softwood fibers are longer, which makes them stronger. But they are not as smooth, opaque or malleable as the shorter hardwood fibers. Each grade of pulp is treated as a commodity. Despite these differences, the various pulp categories are interchangeable to some extent and are often combined for particular paper products.

Aracruz produces only bleached hardkraft pulp (BHK), no secondary products, and supplies 21% of the world's bleached eucalyptus pulp demand, which makes it the industry leader. Aracruz has captured a significant portion of the U.S. market for bleached eucalyptus kraft market pulp (BEKP)—a 61% share. Overall, Aracruz supplies 3% of the world's total demand for pulp. Worldwide, 2.86 million tons of additional BHK capacity are projected to come on line by 1999, mostly in Asia, where Aracruz sells 17% of its product. This significant increase in supply should further squeeze BHK prices. Aracruz has concentrated on expanding in the European and Asian markets, and in 1994 held market shares of 12% and 19%, respectively, in the BEKP markets in those regions .

In 1995, 48% of Aracruz's pulp was used for tissue, with the remainder used primarily for writing and specialty papers. According to industry analysts, the eucalyptus pulp made by Aracruz and a handful of other southern producers has earned a reputation as a "premium" grade, well suited for the production of tissue and uncoated paper. Prices for Aracruz's pulp have gyrated from $641 per ton in 1990 to a low of $366 per ton in 1993, back to a high of $810 per ton in 1995, then plunging to $469 per ton in 1996 as part of the cyclical nature of the market.

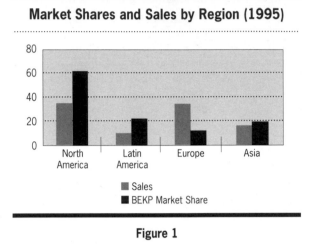

Market Shares and Sales by Region (1995)

Figure 1

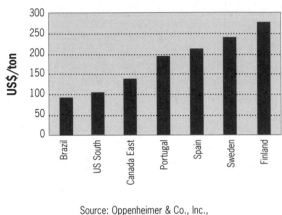

Regional Fiber Costs per Ton

Source: Oppenheimer & Co., Inc.,
International Research Latin America, Klabin, 4/9/96

Figure 2

According to Aracruz's forecast, totally chlorine-free (TCF) technology has established a small, stable market centered in western Europe. The 1993 estimated global consumption of 3 million tons of TCF is projected to rise to between 5 million to 6 million tons in 2000, according to the 1995 report on the paper industry by WBCSD. Demand for "environmentally-friendly" pulp and paper products such as elemental chlorine-free (ECF) and effluent-free pulp is expected to continue growing, driven by consumer demand and government actions. As an early adopter of these technologies, Aracruz is well positioned to take advantage of any expansion in these markets. Aracruz's European customers currently include Jamont, Arjo Wiggins Appleton, Sappi, Inveresk, Tullis Russell, and UK Paper (see Figure 1 for a geographical breakdown of Aracruz's sales). The high quality of Aracruz's pulp has also attracted major U.S. customers such as Kimberly-Clark and Procter & Gamble. Although prices in this market are highly cyclical, demand should grow steadily.

Low Costs/High Margins

Aracruz Celulose is already a financial success. It consistently returns a positive net cashflow in a highly cyclical industry. Appendix A gives the

company's Consolidated Statement of Operations for 1993-1995. In 1995 sales totaled $796 million, but dropped to $516 million in 1996, which reflects the downturn in the pulp market.

By producing high quality pulp in a commodity market, Aracruz benefits by attracting important customers who, by signing long-term agreements, help the company maintain sales volume and market share. But Aracruz is still a price-taker. The market price is mainly determined by Northern Hemisphere pulp producers that account for most production. The company's financial success, therefore, is tied primarily to its ability to control costs. In this, Aracruz has a natural advantage based upon geography and the species of genetically improved trees that it grows. Aracruz is the world's top producer of bleached eucalyptus pulp, with 1996 capacity at 1,025,000 tons and plans for another 20% expansion, while its closest competitor, Georgia-Pacific, has a 765,000-ton capacity.

Eucalyptus is a highly efficient fiber producer. Native to Australia and some parts of Indonesia, its

Rotation Age and Average Productivity of Major Eucalyptus Growing Regions

Region	Species	Rotation (yrs)	m³/ha/yr
Brazil (Aracruz)	hybrid	7	44.4
South Africa	E. grandis	8-10	20.0
Chile	E. globulus	10-12	20.0
Portugal	E. globulus	12-15	12.0
Spain	E. globulus	12-15	10.0

Source: Aracruz, Facts & Figures, January 1996

Table 1

Brazilian Pulp & Paper Industry: Metric Tons Produced Per Employee, 1995

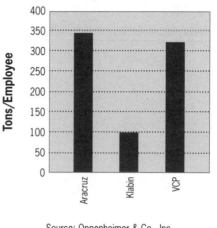

Source: Oppenheimer & Co., Inc.,
International Research Latin America, Klabin, 4/9/96

Figure 3

high productivity has made the tree a top choice worldwide for plantation forestry. Eucalyptus, however, grows best in the triangle of southeastern Brazil, Uruguay, and Paraguay, which gives Brazil a cost advantage in fiber production over other major pulp producing regions of the world (see Figure 2). And as Table 1 illustrates, Brazil leads all other major eucalyptus-growing regions, producing over twice as much per hectare as the next most productive region.

By focusing on costs, Aracruz is now the world's low-cost producer of bleached eucalyptus pulp. Eco-efficiency improvements in the forest and mill contributed significantly to this success, but nonenvironmentally related efficiency gains, such as downsizing the workforce and efforts to outsource operations, were also key. With 2,600 employees in 1996, Aracruz has achieved labor productivity levels of 407 tons of pulp per employee, which helped to keep administrative costs at 9.5% of sales that year. Even within the low-cost region, Aracruz is more efficient per ton of pulp produced than two local competitors, as Figure 3 shows.

Costs of sales in 1995 totaled $295.5 million, which yielded a production cost per ton of pulp sold of $301. Aracruz estimates that annual production costs range between $260 per ton to $310 per ton, which

matches independent estimates. Since the price of pulp realized by Aracruz has not recently dipped below $364 per ton, Aracruz has maintained a consistent positive net cashflow from a low of $74 million in 1993 to a high of $524 million in 1995. (Source: AR 1995.) Even in the cyclical downturn year of 1996, Aracruz achieved a net cash flow of $193 million. In 1995, for instance, Aracruz achieved a pre-tax margin of 52.3%. This performance compares favorably to top North American pulp and paper companies, whose margins in that year ranged from 8%-24%, and to other Brazilian eucalyptus-growing competitors, such as Klabin with an estimated 39.8% margin and VCP with margins of 29.8%. Aracruz's emphasis on eco-efficiency and intensive management have been an integral part of a cost strategy that has had a clear positive impact on the company's financial performance. The results of that strategy can be seen in Figure 4, which compares Aracruz's average price (FOB) of pulp to its production costs per ton from 1990 to 1995.

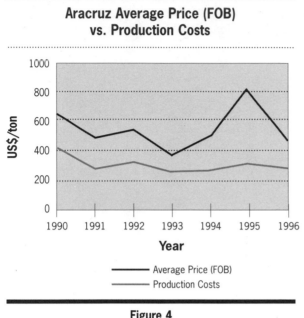

Figure 4

Aracruz is listed on the New York Stock Exchange (ticker: ARA). It routinely takes advantage of opportunities to trade on Brazilian currency fluctuations to bolster its cash flow. The company's good reputation has also allowed it to exchange short-term Brazilian debts for long-term foreign denomination loans. Consequently, its net financing costs per ton of pulp produced have dropped from $105 in 1992 to near zero in 1995. Aracruz's 1996 debt to equity ratio was 33%, down from a high of 86% in 1991. For investment decisions, Aracruz uses a discount rate of 15% above inflation. Appendix B compares Aracruz's financial performance to regional averages. Because of Aracruz's recent large investments, the company's strong financial performance advantage is not as evident by this measure.

In other significant financial developments, 28% of Aracruz's outstanding shares were purchased in 1996 by the South African conglomerate Mondi. Aracruz's technical knowledge and productive land base make it an attractive investment for forest products and other natural resource companies.

A CONTROVERSIAL SOCIAL PERFORMANCE

Perhaps no issue has been as controversial for Aracruz as social responsibility. From indigenous rights issues to community investment, the company's supporters and detractors differ widely in their views of its impact.

Indigenous Peoples' Land Claims

When Aracruz Celulose was founded, Brazil's national government was not receptive to the land claims of indigenous peoples. Consequently, when Aracruz purchased its land from the legally recognized owners it inherited preexisting controversies over land ownership. An indigenous tribe, the Tupinikim, argues that since colonial times they have had rights to 13,000 hectares of Aracruz's land, and have brought a complaint before the National Indian Foundation (FUNAI). Since the 1600s, when these peoples had an estimated population of more than 50,000, the number has dwindled to a few hundred. Those Tupinikim that remain live in poor conditions, according to NGO accounts. These critics report that as many as thirty-two indigenous communities may have been forced to relocate during Aracruz's startup period. The indigenous claims, totaling 43,000 hectares of land throughout Espirito Santo, are not limited to Aracruz land. Yet, the FUNAI complaint argues only for the land surrounding the Aracruz mill, an indication that the company is still a focus of local discontent.

Aracruz argues that when it acquired the land, which consisted mostly of already degraded properties, no indigenous peoples were living there, so relocation was unnecessary. Aracruz also notes that it donated 1,819 hectares of company land, as well as materials and equipment to help establish a permanent reservation for the peoples. Aracruz currently supports a government-NGO-private sector partnership program to provide assistance to indigenous peoples called the Interinstitutional Indigenous Health Nucleus (NISI), giving it $135,000 in 1995-1996. Aracruz management firmly believes that it has never caused any prejudice to the local indigenous communities or to their land. According to an Aracruz position paper on the subject, "If the tribe's current land holding were correctly and efficiently worked, the tribe would be perfectly capable of achieving economic self-sufficiency."

Working Conditions

Another indication of the gap between Aracruz and critics' perceptions is the characterization that Aracruz is unfair to workers. Employees have brought several lawsuits alleging poor working conditions and pressing for greater benefits. Some of these suits have been successful. Critics contend that the company often lays off experienced workers without severance benefits, then replaces them with younger, cheaper laborers. Working conditions at the mill are criticized, with some activists suggesting that the percentage of workers who retire with disabilities is unnecessarily high.

When informed of these criticisms, Aracruz managers expressed surprise and reported that even some of their harshest critics recognize the exceptional working conditions of Aracruz employees. According to the company, it pays the highest salaries in the Brazilian pulp and paper industry, has a variable compensation system for employees that links corporate performance to bonuses, and has a competitive fringe benefits package that includes vacation bonuses, medical and dental assistance, and more. "Furthermore," Aracruz management writes, "the company has always been open to dialogue with labor unions."

As part of continuing cost reduction, Aracruz recently downsized its workforce and outsourced many of its operations, which caused direct employment to drop from 7,440 in 1990 to 3,000 in 1995. During that time, production doubled from 501,000 tons to 1,042,000 tons. Most of the outsourced operations are now performed by local firms, which mitigates some of the economic hurt caused by the downsizing. Aracruz is again expanding its operations—its plans call for an increased capacity of 200,000 tons by year-end 1997—which is generating more local employment. As part of its

outsourcing and expansion, Aracruz now helps local farmers grow eucalyptus for its mill under its Forestry Farmers Program. Without Aracruz's presence, the weak economy of Espirito Santo would certainly be in much worse shape.

Aracruz Celulose has also invested heavily in the local community. The company states that local investments have totaled more than $3 billion, with upwards of $125 million of that plowed into social programs. The company supplies its 3,000 employees and their 8,000 dependents with inexpensive housing, a school, a hospital, and other infrastructure improvements.

Would the communities of Espirito Santo be better off without Aracruz's presence? Probably not, but the extent of the controversies surrounding the company on these issues suggests that Aracruz has not successfully engaged its local communities and should encourage greater dialogue.

CONTENTIOUS ENVIRONMENTAL ISSUES

Aracruz has an impact on the environment in two major areas: in the forest and at the mill. Plantation forestry is becoming more accepted among environmentalists, and Aracruz provides a prime example of a progressive plantation operation. Yet even here Aracruz is not shielded from criticism. The intensity of the company's plantations yields great efficiencies and quality, but also requires intensive resource use.

Issues in the Forest

Aracruz Celulose owns and manages extensive forestlands. Yet the company considers its business agriculture, not forestry. "Eucalyptus is a crop, not a forest," company documents declare. Still, Aracruz does make efforts to preserve many of the services provided by forests.

Aracruz Forestry

These efforts become more evident by following the life cycle of one Aracruz eucalyptus. Aracruz produces seedlings primarily through coppiced

shoot cloning, known as "vegetative propagation."
Eucalyptus lends itself well to this practice. Through
this method, the company selects individual trees
with superior characteristics and makes many identical copies for planting. Then, armed with extensive knowledge of the planting sites' conditions,
Aracruz foresters match individual varieties with
the sites best suited for them, reducing need for
fertilizer, pesticides, or other activities that have a
negative environmental impact. Aracruz uses three
major species of eucalyptus: E. grandis, E. urophylla,
and, most often, a hybrid of the two.

The short growing period of Brazilian eucalyptus—
tree height can reach 100 feet after seven years—has
enabled Aracruz to make great genetic improvements in its eucalyptus population. After years of
research, Aracruz's eucalyptus plantations are now
perhaps the most productive in the world, with an
average yield of 45 cubic meters per hectare each
year, up from 30 cubic meters in 1990 (see Figure
5). That yield compares favorably to a national average yield of 20 cubic meters per hectare annually
in Chile, the closest competing region. Aracruz
maintains orchards for research and hybridization to
protect its lead in the genetic improvement of
eucalyptus.

The cloned seedling is first grown in the company's
nurseries, then moved to a prepared site either on
Aracruz Celulose land or occasionally to land
owned by a local farmer under the company's
Forestry Farmers Program. For the first few years
the seedling is protected from having to compete
for vital nutrients. Years of experimentation has
taught the company important lessons on how to
space trees and produce maximum yield. Until
1994, seedlings were watered once after planting,
and sometimes again in times of drought. Since
then the company has stopped planting seedlings in
the dry season, and stopped irrigation. Pesticides
and herbicides may be applied to the sites to protect the seedlings from undergrowth. When the
seedlings are tall enough, natural undergrowth is
allowed to develop.

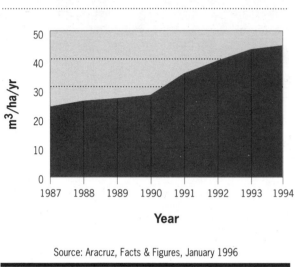

Mean Annual Increment Improvements

Source: Aracruz, Facts & Figures, January 1996

Figure 5

When the seedlings, now trees, are ready for harvest
after seven years, the site is cleared. All trees are
felled and trimmed, and the undergrowth is
knocked down. Recently, as part of its drive for
greater efficiencies, Aracruz adopted the use of
high-tech harvesters to replace the chainsaws formerly used in logging. These new harvesters
improve yields and efficiencies. The logs are sent to
Aracruz's mill, the crowns and smaller branches are
left on site, and larger branches may be sent to charcoal manufacturing operations. After harvesting, the
forest plantation is recovered by natural regeneration
through sprouting (coppicing) and another production cycle is initiated without having to plant.

After the second harvest, the site is prepared for the
next planting. It is no longer burned, which used
to be the practice. Instead, the branches and other
plant material left behind are allowed to decom-
pose to return nutrients to the soil. Fertilizer is also
added. Then a new batch of eucalyptus seedlings
matched with the site is planted, and the cycle
begins anew.

This intensive management system produces sustained fiber yields unmatched in the industry. The cloned trees provide relatively homogenous fiber for the mill, which is of extremely high quality. Aracruz has paid careful attention to its forests to maximize the quality of its fiber: the company is certified under the International Organization for Standardization (ISO) 9000, a global system of quality management standards, which covers the entire forestry cycle, from seed to mill. Critics, however, fault the company for its intensive forestry and argue that its methods deplete the soils, reduce and damage the water supply, and do not preserve diversity.

Biodiversity

The preservation of biodiversity is an important service provided by natural forests. The United Nations Environment Program (UNEP) has concluded that losses in biodiversity can reduce the ability of ecosystems to adjust to environmental changes and disasters; damage their capacity to provide clean soil, water and air; and deny people the future use of species of as yet undiscovered value. A 1995 UNEP assessment of global biodiversity warned of the dangers of altering forests and other ecosystems to maximize yields.

The simplification of ecosystems to obtain higher yields of individual products comes at the cost of the loss of ecosystem stability and of free services such as controlled nutrient delivery and pest control, which thus need to be subsidized by the use of fertilizers and pesticides. Modification of ecosystems by the introduction of alien species, either deliberately or accidentally, has positive and negative ecosystem effects—but too often the latter because of the reduced biotic controls on the invading species.

Although, according to Aracruz and independent reports, the company has not cleared much natural forest, its plantations on previously cleared land come close to monocultures. Although different varieties and species of eucalyptus are used, large tracts of land are maintained by the company with the explicit goal of growing eucalyptus extensively, without the wide variety of plants that grow in natural forests. Moreover, eucalyptus is not native to Brazil. Some environmentalists consider such exotics a threat to preserving the integrity of Brazil's native biodiversity.

Soil Quality

The impressive yields achieved by Aracruz Celulose are also a bone of contention. Plants, through photosynthesis, capture sunlight and nutrients from the air for growth, but they also require nutrients from the soil. Aracruz's trees are not self-sufficient. Although the company leaves much of the trees' nutrients on site after logging, it also adds fertilizer, which critics point out suggests that the net nutrient flow would otherwise be negative.

Water Use

Intensively managed forests are generally big water users. Critics point out that fertilizer, pesticides, and other chemicals applied on intensively managed forests can run into nearby water supplies, damaging the quality, which makes them a source of nonpoint pollution just like agricultural crops. The growing trees also require massive amounts of water. Some research indicates that eucalyptus uses more water than other species. Other studies, however, contradict these findings. They suggest that eucalyptus may actually be marginally more efficient in its water use per unit of wood produced than many other commonly used species.

Either way, Aracruz's goal is to maximize the yield per hectare, and with so much growth, massive amounts of water are required, even if the eucalyptus is an efficient per unit producer.

ARACRUZ'S ENVIRONMENTAL PRESERVATION EFFORTS—THE RESPONSE TO CRITICS

Do these criticisms accurately characterize Aracruz Celulose's impact on the environment? The company's supporters do not think so. The eucalyptus grows on land that was predominantly degraded from agriculture when Aracruz bought it. The land's biodiversity had already been compromised before Aracruz was started. Now, supporters point out, it is covered with trees and harbors, by the company's count, more than 1,500 species of fauna, including 18 that are endangered. The company also maintains 57,000 hectares of natural forest reserves, intermixed with the eucalyptus plantation sites, to preserve biodiversity where once there was none. By the time Aracruz began operations, the Atlantic rain forest, which once covered 90% of Espirito Santo, had been decimated to about one-tenth its original size.

The company takes pains to minimize the environmental impacts of its management. Those efforts include minimizing any disturbance of soil, and using biological, instead of chemical, controls for pests when possible. "In terms of biodiversity," Lorentzen states, "eucalyptus plantations should be regarded as being like any other crop, such as corn, soybeans, wheat, sugar cane, and coffee." When compared to these crops, eucalyptus plantations require less fertilizer and chemicals. Company research begun in 1994 and filed with the state environmental agency shows that there is no contamination of the local watershed's water and soil from chemicals used in the forestry operations. As to water use, Aracruz cites studies which show that the region's annual rainfall of 1,400 mm is more than enough to meet the demands of any crop in the region, including eucalyptus, with a surplus left for ground storage or to replenish springs.

Finally, Aracruz managers point to the growing demand for pulp and their efficiency at meeting that demand. "Fast growing plantations actually help reduce the world's—and Brazil's—wood deficit by alleviating the main pressures on native forests and consequently helping to preserve them," company literature points out. The fiber supplied by Aracruz would have to come from somewhere, and though the company's negative environmental impacts may not be zero, they are still less than many other sources.

Pulping Controversies

Most of Aracruz Celulose's logs are shipped to its pulping facility, which has a 1996 production capacity for 1,025,000 tons a year. The company has relentlessly pursued efficiency in its pulping operations through technology improvements and waste minimization. These actions have not only improved the company's bottom line, they have also reduced the company's impact on the environment. Where environmental protection has conflicted with efficiency, however, controversy again reigns.

Pulp mills typically place serious burdens on the local environment. The logs that enter the plant must be cleaned, chipped, chemically treated to remove lignin, dried, and often bleached. These operations take massive inputs of water and energy, and produce large amounts of chemical and biological waste, which is emitted into the air, water, and soil. Over the years pulp producers have increased their efficiency to lower costs, which also has the environmental benefit of lowering emissions.

Few pulp producers, however, are as efficient as Aracruz. It uses the bark from the logs as fuel for its boilers and dryers. When lignin is extracted through the company's kraft process, the wastes are again used as fuel. This way the company meets 80% of its own energy requirements and recycles 94% of the chemicals used in the digestion phases. Aracruz's recent efforts to genetically select trees with less bark will hamper these efforts to an extent, but will help the company lower costs.

All of Aracruz's pulp is bleached. In the past, most pulp bleaching processes used elemental chlorine as the main active agent, a practice that has been linked to the creation of highly toxic dioxin. Since the discovery of its toxicity, pressure mounted for producers to move away from elemental chlorine and a market has developed, primarily in Europe, for TCF pulp. Aracruz reacted quickly. Currently, about 45-55% of the company's production is ECF, with production of an additional 10% TCF. All elemental chlorine-based production is being phased out, so that soon all pulp produced will be either ECF or TCF.

Aracruz has also acted to eliminate its waste production. Since 1990 effluent levels of biological oxygen demand (BOD), or organic wastes, have plunged 90% and those of toxic absorbable organic halogenated compounds (AOX) even more. Those effluents that remain are piped more than one mile out to sea before release. Aracruz's efforts in the 1990s to cut waste represent a successful eco-efficiency improvement.

Eco-efficiency represents win-win solutions for the company and the environment. Aracruz has been criticized, however, by those who feel that the company has neglected necessary, but costly, air pollution abatement. Local people complain about air quality, and the Pulp Industry Workers Union has accused Aracruz of "incorrect eco-management of both waterborne and airborne emissions." At pulp mills, superior environmental performance may require costly end-of-pipe control solutions that add little, if any, to the bottom line. Aracruz may have successfully focused upon eco-efficiency improvements, but it is unclear how the company has dealt with the more painful trade-offs where improved environmental performance would be "inefficient."

Nevertheless, the company is committed to continuous environmental performance improvement. It has invested more than $2 million in gas treatment equipment and processes. Aracruz has also initiated an innovative community-based program to help detect noxious sulphur fumes, and recently put in place a more effective air emissions control system.

THE ECO-SOCIAL BOTTOM LINE

This study as secondary research cannot draw any conclusions on the net social and environmental effect of the company, but the following statements by two commentators illustrate the widely disparate viewpoints of the company's activities. On the plus side, respected forest analysts credit Aracruz with social and environmental improvements:

> Natural forests in Espirito Santo and Bahia on the Brazilian coast have been heavily exploited by farmers, charcoal makers, loggers, and ranchers for decades. By the late 1960s, many lands had been left eroded or abandoned, and many local people were consequently impoverished and unemployed. Virtually no reforestation had taken place. Aracruz Celulose S.A., with government support, took control of much degraded land within the tattered fragments of natural forest, and has established major eucalyptus plantations. In doing so, it has begun to improve the local environment and social conditions.

—*Caroline Sargent and Stephen Bass, International Institute for Environment and Development, in their book, Plantation Politics.*

Critics, however, fault the company for those same actions:

> [Aracruz] assumes the image of protector of the environment, but its eucalyptus trees have dried streams, destroyed the local fauna, impoverished the soil, impeded the regrowth of native plant species, and drastically reduced the area available for cultivating basic foodstuffs (in a country where many people die of hunger). This is not to mention land concentration and the expulsion of the rural population, which has contributed to increasing the urban population and the degradation of living conditions in the cities. Where is the sustainable development here, we might ask?

— *M.T. Goncalves, et al., "Exploracao florestal no norte do ES e sul da Bahia–impactos e alternativas," FASE/IBASE, Rio de Janeiro, 1994, quoted from The Ecologist.*

WELL POSITIONED FOR THE FUTURE

Aracruz is not only a financial success today, but it is well positioned for continued success. The company has built a competitive advantage based upon a number of factors—geographical, technological, tax relief, product mix, and fiber supply—and is focused upon defending this advantage.

Cost Advantages

As noted above, Aracruz enjoys a significant cost advantage based upon its eco-efficiency programs and the productivity of eucalyptus growing in the triangle of southeastern Brazil, Uruguay and Paraguay. In the region, Aracruz's competitors are currently few in number, although interest in start-up operations in the area by industry leaders from other parts of the world is high. Among these local competitors, such as Klabin's Riocell subsidiary, CENIBRA, Jari (Monte Dourado), and Bahia del Sul, Aracruz produces the superior quality product at much larger volumes (more than double that of the next-largest producer's, Riocell's, capacity) with lower costs.

Other major eucalyptus-growing regions include Australia, Chile, and Portugal, with efforts underway in Indonesia and elsewhere. The productivity of these regions, however, does not match that of Espirito Santo. Aracruz is the low-cost producer in the low-cost eucalyptus-growing region of the world. As recent success has made it cash-rich and the cost of capital is relatively low, the company is looking into the possibility of expanding its forestlands into neighboring states or even Uruguay or Paraguay. Aracruz is also reportedly interested in purchasing the government's controlling interest in CENIBRA when or if it becomes available.

Aracruz is also a world leader in eucalyptus plantation forestry. The company is generations ahead of its competitors in genetic refinement of the eucalyptus varieties it uses, and has developed a thorough knowledge of the best forestry techniques both for matching specific genetic variations to particular sites and for silvicultural activities throughout the growth cycle. Aracruz will continue to improve the genetics of its tree stock and silvicultural methods, thereby ensuring that it remains further up the learning curve than any potential competitor.

The company's significant tax holiday ended in May 1997. The terms gave Aracruz tax exemption for all export revenues, currently over 90% of sales, as long as the company met the following criteria:

1. generated total pulp export revenues of $2 billion between 1987 and 1997;

2. maintained a positive foreign-trade balance annually and attained an accumulated balance of $1.9 billion over the period;

3. made investments equivalent to $1 billion; and

4. maintained its indices for environmental performance within established limits.

By the end of the period the company had met all its conditions. These generous tax concessions have given Aracruz Celulose a tremendous advantage as it expanded into the global market. In the future, analysts estimate that Aracruz will have a 10% effective tax rate.

An Early Mover in Technology

As the world's economy expands, pulp should continue to be in demand. Recognizing that market demand shifts, however, Aracruz has been an early mover into technologies such as TCF and effluent-free production. The company is also taking a serious look into forest certification, either under Forest Stewardship Council (FSC) guidelines, ISO 14000 environmental management guidelines, or possibly both. Thus, if regulatory restrictions or consumer demand mandate a shift in the industry, Aracruz should be well positioned to move in this direction with less disruption than some competitors might experience. The company's emphasis upon quality is another example of this proactive market positioning, as high quality is always in demand even if the commodity nature of the market does not fully reward such efforts today.

Aracruz is also moving into solidwood and engineered board production. The company believes it can parlay its technological expertise into superior eucalyptus solidwood performance, and is interested in Joint Implementation (JI) carbon sequestration projects. In March 1997, Aracruz announced a joint venture, Techflor Industrial S.A., with U.S. based Gutchess International Group to produce solidwood products. Gutchess, a forest products manufacturer and marketer, specializes in U.S. hardwood species. Construction of a sawmill in southern Bahia is to begin in May 1997. By 1997 the mill is projected to produce 75,000 cubic meters per year of structural and decorative millwork, half for Brazilian and half for international markets. The venture will put Aracruz in more direct competition with Amazonian solidwood, which removes pressure from endangered tropical hardwoods. The company plans to supply the mill entirely with wood from its plantation and use all waste either as chips for the pulp operations, or as fuel for the sawmill.

Productive Plantations Are Key

All of these efforts would be for naught if the company's resource base were to disappear. So Aracruz goes to great lengths to sustain the productivity of its plantations. If the predicted global pulp supply shortfalls do occur, Aracruz's sustained yield policies give it an advantage by providing a secure fiber source. Aracruz's efforts to preserve biodiversity by interspersing natural forest set-aside areas in its eucalyptus stands mitigates some of the risk of a catastrophic fire or blight. This emphasis on risk-avoidance may also be at the heart of the company's public relations campaigns to place it in good stead among global decision-makers. Given

the continued controversies surrounding the company and apparently insufficient local stakeholder outreach, these efforts may have only made the company the target of greater scrutiny.

THE LESSONS OF ARACRUZ CELULOSE

Aracruz Celulose is an eco-efficiency success story. Because the company can produce pulp very efficiently, pressure is taken off of natural forests and waste is minimized while the company enjoys financial success. Aracruz's success through a sustained yield, technology-driven, eco-efficiency strategy demonstrates the potential not only for better business and environmental performance throughout the industry, but specifically the environmental and business advantages of plantation forestry for pulp production.

Sustainable forestry, however, requires companies to move further than short-term efficiency in their operations and investment choices. Aracruz management claims that it is, in fact, engaging these challenges in a responsible manner, and the company is looking into forest certification.

Even if that is the case, can any forestry operation fully achieve the moving target that is sustainable forestry? The company is a global leader based out of an economically and environmentally depressed region, and has environmental and social responsibility as explicit goals. The company actively touts its environmental performance and is often held out as an example of the future of fiber forestry. Yet despite this positive attention—or perhaps because of it—Aracruz is often singled out for criticism and protest. The final lesson to be learned from Aracruz's experience is that when it comes to environmental performance, nothing is ever "good enough."

Riocell S.A.–Business Description

Isak Kruglianskas, associate professor at the Universidade de São Paulo, and Tasso Rezende de Azevedo, executive director of Imaflora

Riocell S.A. is a Brazilian pulp producer and exporter affiliated with the forest products conglomerate Industrias Klabin (also known as the IKPC Group). The company obtains all of its fiber supply from tree plantations, primarily of eucalyptus. Riocell annual production capacity is 300,000 tons of short fiber pulp. In 1995 Riocell generated total revenue of US$210.1 million on 212,000 tons produced, which represented approximately 18% of Klabin's annual sales.

Riocell is located in the municipality of Guaíba, within the metropolitan area of Porto Alegre, which is the capital of Rio Grande do Sul, the southernmost state in Brazil. The company, about 30 km from the center of Porto Alegre, is one of the most important in the state of Rio Grande do Sul.

The Guaíba region used to consist mainly of open fields, prairie lands called the "Pampas Gaúchos." Today these prairies have been taken over by agriculture, especially the production of grains such as soya and rice, and cattle grazing

The forests harvested for wood in the state of Rio Grande do Sul have traditionally been spp. acacia, spp. eucalyptus, and spp. pinus. The bark of acacia is used to produce tannin, while the wood is used for the production of paper and pulp by Riocell, as well as a source of energy. Eucalyptus trees account for the second most extensive forest production in the region. Groves of acacias and eucalyptus trees dot the countryside but there are no continuous and closed forests such as those that exist elsewhere in Brazil.

The IKPC Group owns about 208,000 hectares, primarily of Araucaria angustifolia trees, Pinus spp (especially P. elliotti and P. taeda) and eucalyptus trees (especially E. grandis, E. saligna and E. urophylla) along with 100,000 hectares of natural forests that are earmarked for preservation and conservation.

Riocell has concentrated on eucalyptus, aiming to produce pulp for export. The company has a total forest area of 71,717 hectares, 53,216 hectares of which are made up of eucalyptus with the remaining 18,501 hectares (25.7%) set aside as conservation areas.

Table 2 compares Riocell's position to other pulp and short fiber producers on the market. With an annual production of 212,000 tons in 1995, Riocell ranked as the fourth largest producer in the region.

Riocell's ownership has changed hands several times over the past three decades, with periods of national, foreign, and private domestic control. At one point in the 1970s, while it was under Norwegian ownership, the government suspended Riocell's operations in response to the manufacturing facility's poor environmental performance, primarily its levels of air pollution. The government then took over the company. Although Riocell has subsequently been taken over by Klabin and the environmental problems dealt with successfully, this incident continues to influence the outlook of Riocell's management. It is reflected in the company's emphasis on environmental protection and maintaining good relations with the communities surrounding its operations.

Market Pulp Production, 1995 Bleached Short Fiber Production

Company	Thousands of Tons
Aracruz	1,042
Bahia del Sul	331
CENIBRA	364
Riocell	212
Jari	129
Others	253
Brazilian Total	**2,331**
World Total	**13,417**

Source: ABECEL, 1995 AR

Table 2

Those concerns are reflected in Riocell's official mission statement:

> Riocell must grow in Brazil and abroad, producing and marketing pulp, paper and related products of a quality that will satisfy its clients and maximize investment returns, improving the quality of life of its internal and external communities and protecting the environment.

Riocell is oriented to external markets. The world market for pulp is some 34 to 35 million tons a year. The pulp market is growing at an annual average rate of 2.9%. The market for short fiber pulp, however, is growing at an annual rate is 3.9%. Riocell's holds about 1%, or 300,000 tons a year, of the world market in short fiber pulp. However, Riocell's share of pulp made from eucalyptus wood, a 5 million tons a year market, accounts for 7% of world production.

In addition to standard bleached pulp, Riocell also produces some dissolvable pulp, which is used in the production of rayon, limited amounts of ECF bleached pulp, and some fine writing papers. Sales of its main product lines were projected as follows for 1997:

- Sales of dissolvable pulp were expected to increase by 10%.

- Sales of pulp bleached by the ECF process were projected to increase by 40%.

- Sales of pulp bleached by the traditional methods with chlorine were expected to decline slightly. They were expected to drop about 4% to compensate for the increase in the production of pulp bleached by the ECF process, which accounts for 10% of the current production of paper and pulp.

COMPANY OPERATIONS

Riocell's operations are divided into two main units: forestry and industrial production.

Forestry

Riocell owns three basic types of forest areas:

- Fully-owned forests that make up about 83.5% of its land base. On these lands the company not only owns the land, but also the production from it.

- Leased forests that total about 15.6% of its land base from which the company owns all production.

- Jointly-owned forests that comprise about 0.9% of its land base. On these lands Riocell plants trees in partnership with other owners of the land and divides the production with them.

Riocell's history of forest management has evolved through four distinct phases as indicated in Table 3.

As Table 3 shows, pulp production was originally based upon fiber inputs produced by third parties, especially acacia wood, while the company began to plant its own eucalyptus forests. This stage continued until the company began farming out its production system in the late 1980s. In this era, even on its own lands, all the forestry operations were handled by third parties, including inventories, planting, maintenance, harvesting, and transportation.

At present, the company is planning to farm out or subcontract a large portion of the production of its raw material, following the "Scandinavian Model" in which the production of raw materials is carried out in areas owned by third parties. Of course, Riocell will still have large private holdings of its own forests.

Riocell's subcontracting of its forestry activities is more significant than any other Brazilian company's in the paper and pulp sector. Today only seventy of the company's employees work in forestry. Besides activities related directly to forestry, they are also in charge of receiving logs at the factory, and chipping the wood, activities that are usually carried out by the industrial unit in similar companies.

Riocell's Strategic Evolution Phases

Phase 1 (1970s)	Raw materials (forests) in third-party hands and harvesting and transportation are performed by the company itself and by third parties.
Phase 2 (1980s)	Raw materials and transportation are fully owned by the company, period of establishment on own lands.
Phase 3 (1990s)	Raw materials in company hands, transport in third-party hands, company activities are carried out by third parties.
Phase 4 (present goals)	Jointly-owned raw materials (by company and third parties); land, transportation, personnel; resources in third-party hands.

Source: Riocell

Table 3

Although the technology applied by Riocell in forestry production is essentially the same as that used by its competitors, Riocell's procedures are distinct in several areas.

Soil Improvement and Preparation

In 1988, Riocell stopped burning forest debris before preparing the soil for planting. Burning of forest debris makes it easier to prepare the soil because the resulting ash can be readily mixed into the soil using lightweight machinery. However, such practices can have significant environmental impacts, because they release carbon dioxide (CO_2) into the atmosphere, and because they can degrade the soil. Burning also destroys nutrients, especially nitrogen, and microorganisms that contribute to the health of the soil. If such practices are not used, the nutrients remain in the soil and the layer left by the forest waste material acts as a protective cover to guard against erosion and secondary growth, which decreases the need to use herbicides to control the secondary growth during the first six months after planting.

Riocell was one of the first pulp makers to eliminate post-harvest burning, a controversial stance at the time, especially because alternative methods of preparation cost more. To absorb these costs, the company again pioneered a forest practice, this one called minimal cultivation. In this technique, rows of seedlings are planted by hand in soil that is minimally prepared. Herbicides to keep down secondary growth are applied only to the rows, which are planted about three meters apart.

Chemical Use

The company uses only two products for chemical control. Roundup, an herbicide, is used against invasive plants after they emerge. Mirex-s, a chlorophosphorous-free bait, is applied to control ants. The herbicide is applied in targeted areas. However, the insecticide is widely applied because ant nests are spread out over the area.

Bark Removal in the Field

Riocell is the only company that carries out 100% of its bark removal mechanically in the field. It has done so since 1972.

Removing the bark in the field produces a number of benefits. The bark, branches, and leaves of the trees contain 70% of the nutrients, which the plant absorbs from the soil. Thus, bark removal in the field guarantees better conservation of the soil's productive potential. The bark shavings also form a protective carpet for when the machinery moves over the ground, which helps to minimize erosion. Finally, if the bark is stripped off in the field, the volume of wood that must be transported to the pulp facility is lower, which lowers transportation costs as well as fuel consumption. (Other companies, however, can compensate for the higher costs of transporting the bark if they use it as a source of energy in their boilers.)

From its founding, Riocell has considered the soil an asset that must be conserved, and has made protecting the soil one of its principles. To help ensure the productivity of its soil, Riocell runs a Technology Center, which has a laboratory dedicated to analyzing the fertility and the rejuvenation of soils and vegetation.

Industrial Production

The industrial unit is made up of three basic processes: 1) the fiber line, where wood processing by mechanical, chemical, and thermal means takes place; 2) the recovery of chemical products line, whose component units recover the chemical reagents used in the process of wood digestion for subsequent reuse, thus preserving economy of production as well as the environment; and 3) the utilities system, which supports the other two lines, by providing such components for production as treated water, compressed air, steam and electric power, the chemical products used in the whitening process, and the treatment of industrial wastes generated by the production processes.

BARRIERS AND OPPORTUNITIES FROM SUSTAINABLE PRACTICES

Availability of Raw Materials

Riocell produces about 300,000 tons of pulp a year. To produce that pulp it consumes about 1.7 million cubic meters of wood, or about 5,500 cubic meters of wood each day. Taking into account that the average productivity of Riocell's forests is 50 cubic meters per hectare per year, or 350 cubic meters in a seven year forest rotation, the company needs to harvest about 15 hectares of forest lands a day to maintain its production.

At its present production rate, Riocell needs 34,000 hectares of planted forests to produce 1.7 million cubic meters a year, or about 65% of the company's current land base. Even setting aside a series of conditions related to forest productivity

at different stages of planting, among others, that limits production, Riocell clearly owns enough forest lands to maintain its current production.

This is possible because since the 1980s the company had been preparing to double its production of pulp to about 700,000 tons a year, which would have required some 4.06 million cubic meters of wood a year. In order to produce this amount of wood at present productivity Riocell would have needed 81,000 hectares of planted forests.

Riocell, however, never undertook the expansion. In the meantime, it husbanded its forests with an eye to future expansion. The current wood surplus is sold in the form of timber to saw mills for civil construction, power line poles, and as a source of fuel. Furthermore, if additional production capacity is needed, it can be found through efficiency gains at other stages of the production process.

Even with enough raw material to sustain the factory's current production with leeway, Riocell currently buys about 20% of its raw materials from third parties—18% from acacia forests and 2% from jointly owned forests where both the company and the owners of the land are entitled to a share of production, and from development areas. The company does this to keep the timber market alive in the region and to give farmers incentive to plant more forests.

The trees are dispersed among 162 tree farms in 23 municipalities, which averages to 3% of the total land area of each of the municipalities. This is a rather unusual agrarian structure. It is the policy of Riocell not to concentrate the production of raw material in large plantations, which can cause social and environmental disturbances. The combination of surplus and decentralized forest holdings gives Riocell a competitive advantage in terms of social and environmental sustainability. This structure, however, also entails higher transportation costs since the company's raw material sources are spread out geographically.

Marketing Aspects

Riocell's operational structure is dedicated to the pursuit of total quality. As part of its development of total quality management, Riocell was one of the first pulp manufacturers to receive certification under the ISO 9000 standards in 1993 and ISO 14000 in 1996.

As part of this effort, the company tries to match specific customer's needs with specialized product characteristics, using Riocell's modern laboratories to carry out chemical and physical analyses that guarantee the quality and performance of the products. For example, Riocell runs a pilot refining plant with a daily capacity of 100 dry tons per day, which is used to test the effects of slight changes in input mixes and production methods. With this facility, Riocell can help its customers identify the ideal manufacturing conditions for paper production from Riocell pulp.

Riocell has always conducted research to develop new products, discover new uses for pulp and develop innovative techniques that can be adapted to the changing realities of this highly competitive market.

In the case of bleached pulp production using TCF processes, which have less potential to pollute, Riocell faces no technological or economic barriers. According to information released by the company, Riocell has the technology to produce TCF pulp, holding two patents, one of which involves a TCF-type bleaching sequence and an innovative technique to extract lignin from pulp, which further reduces the consumption of chemicals during the bleaching process. TCF technology has not yet been introduced in mass production, however, because the market is not yet ready to pay more for products that use TCF technologies. There is no consensus that TCF is environmentally superior to ECF, which Riocell already uses in part of its pulp production.

Some 99.86% of the solid wastes generated during the industrial process are recycled into fertilizers and correctives for the soil or sold as byproducts, mainly to be added to cement. This is done by Vida Produtos Biológicos, a company owned by

Professor José Lutzenberger. Professor Lutzenberger was one of the major critics of the facility during the 1970s when its operations were suspended, so his involvement and cooperation is a positive sign of Riocell's improved performance. Riocell carries the difference between the production and sales costs of fertilizer as an environmental cost. In addition to this work, Lutzenberger also provides landscaping and gardening services.

In March 1990, Riocell began operating a new oxygen delignification unit, which further reduces its environmental impact by recovering 30% of the pollution from the bleaching process.

Social Conditions

Riocell today is a company that concerns itself with community outreach. Through its employees, the company has set up a network for reaching out to the different sectors of the community: government authorities, politicians, universities, schools, NGOs, service clubs, and community leaders in general.

In 1991 Riocell set up the Educational Center for Training in Environmental Sciences (CECIELL) to train and raise the awareness among local communities on matters related to the environment. Since then, CECIELL has mounted a number of initiatives:

- Environmental science clubs which promote activities to get children involved in solving local problems.

- Publication of the *Regional Environmental Newspaper*, which informs members of the local community about the state of the environment.

- "Children's Voice" program, which meets with authorities to discuss problems related to the environment.

- "Pupa" program, aimed at helping teachers address environmental problems and make environmentally-sound changes in their lifestyles.

CECIELL has also successfully organized a series of seminars in which all of the company's workers can participate, including all "partners" (company suppliers). At these seminars themes related to the environment and sustainable development are introduced and debated, with a goal of making the participants aware of the issues.

FINANCIAL COSTS AND BENEFITS OF SUSTAINABLE ACTIVITIES

Riocell's sustainability efforts can be grouped into four categories:

1. Soil conservation, which was a cultural value inherited from the Norwegians.

2. Community outreach, a concern that arose from the closing of the company for over three months by the public authorities in the 1970s.

3. Technological and environmental development, backed by a research center and a "technological network" of experts.

4. Management strategy.

Riocell's investments in environmental protection, since its establishment, are estimated to be US$135.166 million. Table 4 summarizes the breakdown of these investments, while Table 5 shows Riocell's gross profits for the years 1995 and 1994.

The cost of stripping the bark from wood in the field—compared to wood whose bark is removed at the factory—represents a 5.66% increase in the cost of forestry production. In Riocell's case that leads to an annual additional cost of about US$1.5 million, or an increase of 1.23% in total production cost. Table 6 shows the breakdown of environmental costs in overall forestry and industrial production costs, including fixed and variable costs, as well as depreciation reserves.

The figures shown in Table 6 have taken into consideration not only the financial costs of these activities, but also any quantifiable benefits gained. An example of this eco-efficiency gain would be the reduction in fertilizer use that come from debarking the trees in the forest. These activities, then, actually do result in net costs for the company. Among the various costs that are incurred in the treatment of contaminants, the highest cost is that of treating liquid wastes, which make up about 50% of the total costs incurred by Riocell's efforts to achieve environmentally-responsible production.

Investments in Environmental Protection

Item	Description	Value in US$1000 since establishment
1	Emissions into the air	28,397
2	Effluvia or liquid wastes	85,340
3	Solid industrial wastes	4,987
4	Environmental quality control	1,352
5	Forest areas	14,700
6	Environmental research	390
Total		**135,166**

Source: Riocell

Table 4

Breakdown of Riocell's Business Results

Item	Value in US$1,000* 1995	Value in US$1,000 1994
1. Sales on foreign markets	148,400	116,365
2. Sales on domestic markets	101,810	98,825
3. Taxes	(20,416)	(18,962)
4. Net sales income	229,794	196,228
5. Cost of products sold	(125,516)	(131,510)
6. Gross profits	104,278	64,718

*based on 1995 exchange rates US$1=.973 reals

Source: Riocell, 1995 AR

Table 5

Share of Environmental Protection and Control Costs within the Company's Total Production Cost

Item	Share of Company's Total Production Cost
Forest management	1.2%
Solid wastes	0.7%
Treatment of liquid wastes	4.7%
Emissions into the air	2.7%
Total	**9.3%**

Source: Riocell

Table 6

Impact of the Costs of Environmental Control and Protection on Gross Profits

Item	Value in US$1,000* 1995	Value in US$1,000 1994
1. Net sales income	229,794	196,228
2. Cost of goods sold	(125,516)	(131,510)
3. Gross profit	104,278	64,718
4. Cost of environmental Protection and control (9.3% of 2)	11,673	12,230
5. Gross profit excluding costs of environmental protection and control	115,951	76,948
6. Relative increase in gross profit without environmental protection and control	11.2%	18.9%

based on 1995 exchange rates US$1=.973 reals

Source: Riocell

Table 7

Taking into consideration Riocell's business results in fiscal years 1995 and 1994 as presented in Tables 5 and 6, Table 7 shows that despite any benefits arising for both society and the company's image, the adoption of more environmentally-sound business practices represent a 15% average reduction in gross profits for Riocell. The average between 11.2% and 18.9% corresponds respectively to the estimated value of gross profit reduction in 1995 and 1994.

Aside from the cost of generating environmental protection and control, the company also spends resources on staff education, dissemination of information to the community, environmental technology research, and other initiatives. These costs are not included in the above analysis.

Riocell actively engages in technological research to help it compete in the highlycompetitive pulp market. This research is geared to finding new ways to differentiate its own products and to reduce production costs. An analysis of its unit costs, as submitted in Table 8, shows that the largest cost component of the business is industrial processing, which represents 72% of total production costs.

As the breakdown of production costs indicates, transportation costs are relatively low (3% of the total cost), which indicates that Riocell's "mosaic strategy" of dispersed forest holdings does not necessarily entail significantly higher transportation costs. One reason may be that even though Riocell's forests are spread out, the average distance to the mill at about 70 kilometers, is not great, according to an April 9, 1996, report from Oppenheimer & Co., Inc.

The share of total costs represented by the company's own personnel (manpower) represents 18% of the company's total production cost, with labor costs representing 14% of total costs in the industrial unit and 7% in the forestry unit. It is difficult to estimate the actual value of labor in production costs in the forestry unit, however, because most of this activity is farmed out under contract service. The actual contribution of labor to production costs in the forestry operations may be as high as 75%.

Breakdown of Unit Cost (US$/ton) of Riocell Pulp Production (1995)

Forestry Cost	Transportation Cost	Industrial Cost	Production Cost	Cost US$/ton [1]
24%	3%	73%	100%	320 (estimate)

Source: Riocell
[1] Oppenheimer and Co. Inc.
International Research, April 1996, p. 24.

Table 8

RELEVANT FORESTRY CODE ASPECTS UNDER BRAZILIAN LAW

Current forestry regulations in Brazil are laid down in the Brazilian Forestry Code, Federal Law 4,771 dated Sept. 15, 1965, and therefore existed before the company was founded. A superficial analysis of the company based on the Forestry Code shows up three items in the code that need to be considered in any discussion of Riocell: 1) legal reserve areas; 2) permanent conservation areas; and 3) conversion of native vegetation into plantations.

The permanent conservation areas are defined in the forestry code (articles 2 and 3) as lands along rivers, lakes and dams, steeply inclined lands, hilltops, or any areas higher than 1800 meters.

The legal reserve areas are defined in the same code (article 16, paragraph 2) as 20% of private property areas. These areas, which must be legally registered, may not be used for any purposes other than conservation, and must permanently maintain their forest cover. Riocell has earmarked 22.1% of its areas for conservation and preservation, and no trees are logged in these areas. To ensure compliance with the law, the legal minimal 20% must be observed in all the tree farms owned by the company. It would seem that Riocell is, in general, complying with this important clause of the

forestry legislation, since the law may be interpreted to include the conservation areas among the legal reserve areas. Although debate on this issue is ongoing, this should not in any way become a problem for the company.

Several laws address the controversial issue of conversion, with the emphasis on maintaining natural ecosystems. According to the company's engineers, the company is not allowed to convert even a bush in the designated conservation areas. The rigidity of this position is constantly stressed.

MARKET PRESSURE TOWARD SUSTAINABILITY

Although Riocell's best known product, in terms of both size of production and billing, is market pulp (60% of total sales earnings), other products also account for a large share of its sales. Table 9 shows the main products marketed by the company and their respective share of exports and sales earnings as shown by billing.

Until 1991, dissolvable pulp was sold only on the domestic market. By 1996, about 80% of the sales of this product were destined for foreign markets. In the interim, national textile production had declined because the industry's outdated technology and lack of economies of scale had made it much less competitive in global markets.

In 1995 the international pulp market went through two distinct phases. During the first three quarters of the year, demand was strong and prices bullish. During the last quarter pulp stocks increased rapidly, reversing the above trends, which also happened on the domestic market. Most international analysts consider this a circumstantial phenomenon. Of the amount sold by the company, 67% was exported, of which 30% went to European markets that are generally considered the most demanding environmentally. The remainder went to Asian and Latin American countries and to the United States. Of the 240,000 tons of pulp sold, 181,700 tons were destined to foreign markets, while 58,300 sold in domestic markets.

Riocell's Principal Products

Product	Volume of Exports	Share of Earnings	Observations
Paper pulp	82%	59.5%	10% of production is pulp bleached using the ECF process
Pulp filler	82%	0.5%	waste recovery for use in paperboard manufacturing
Dissolvable pulp	76%	20%	for use in cloth materials, acetate
Paper	5%	20%	printing/writing

Source: Riocell

Table 9

The company benefited from its operational flexibility, producing and marketing 34,200 tons of soluble pulp for industrial purposes. It also sold 37,000 tons of quality paper for printing.

In the opinion of Riocell's executives, the company achieved one of the best performances among Brazilian companies in terms of product prices, especially in the case of bleached pulp. Since pulp in international markets is largely a commodity product, if a client is willing to pay a higher price for the pulp supplied by this company, it indicates that Riocell must have a quality differential. However, it should be pointed out that since Riocell produces different types of pulp (see Table 9) computing the average cost per ton of total pulp sales can lead to incorrect interpretations as to the prices achieved by the different pulp products in the marketplace. Any comparisons of prices between Riocell's and competitors' products need to be made using the same type of products.

Sometimes erroneous analyses are made when the average price of every marketed ton of pulp is obtained by dividing the total earnings by the total number of tons produced based on the aggregate figures submitted by companies, without considering the differences in product mixes between companies. The importance of these logistical market conditioners is indicated, for example, by the fact that in 1996 Riocell was marketing its nonbleached pulp at prices above those obtained for bleached pulp in certain market niches.

The current capacity of the chlorine dioxide producing unit is insufficient for all of Riocell's production to be converted to ECF. Such production is undertaken only when specifically ordered by customers.

RIOCELL'S STAKEHOLDERS

Riocell's stakeholders consist of members of both the external and internal public, the latter being its employees. The most important stakeholder groups include:

- The local community
- The public authorities
- Shareholders
- Suppliers in general
- Service suppliers
- Universities (a considerable part of Riocell's technological research is done in partnerships)
- NGOs
- Pulp and paper consumers (customers)
- Employees

Local Community

The local community concerns cover a wide range of issues. Economic concerns are related to the supply of jobs and opportunities for business initiatives that the company can offer. The community is also concerned over the way the company's forestry activities affect the aesthetics and landscape of the rural environment. Air pollution also remains a concern, which is particularly relevant given Riocell's history and the fact that it is the only pulp and paper producer in Brazil situated close to a large urban area (Riocell is just eight kilometers away from the city of Porto Alegre as the crow flies).

Public Authorities

The public authorities are sensitive to the demands of the local population on environmental matters. After all, it was the public authorities who gave incentives for and sponsored the installation of the original Norwegian company in the 1970s. Those same authorities also shut the company down under public pressure over pollution concerns shortly thereafter. By the same token, the authorities have an interest in the company's welfare since it provides jobs and taxes to the local economy.

Shareholders

One of the main concerns of shareholders, apart from the strategic aspects complementary to other businesses controlled by the shareholders in the paper sector, is their return on investments, or the market value of Riocell's shares. Shareholders view all investments in the context of maximizing the company's profits in the medium and short term, including those related to doubling production capacity, acquiring new environmentally-friendly equipment, or providing technological training through modernization.

Suppliers

Suppliers in general, and service suppliers in particular, have developed strong ties with Riocell in the form of partnerships. Their concerns for the company, however, are similar to those expressed by employees and shareholders, since their continued existence can be strongly affected by any unsatisfac-

tory performance on the part of Riocell. They are also included in the company's ISO 14001 environmental management system.

As a result of its business optimization strategy, any activities which Riocell does not consider essential to core company activities have been farmed out. Riocell's main concerns are the production of pulp, state-of-the-art production technology, and product marketing. A partner is a third party who is more closely connected with and has a greater commitment in the business. To give suppliers incentives to plant forests, the company has used the Forestry Development Programs, which are oriented toward forestry companies and government extension programs. The program is part of Riocell's strategy to keep supply prices of its raw materials attractive while simultaneously reducing competition with other consumers of eucalyptus wood, especially for energy purposes.

Riocell sponsors planting of eucalyptus trees throughout the state, so that wood consumers will have access to a supply of raw material in the microregions, and therefore not compete with Riocell for the purchase of its wood supply.

The price of the wood after bark removal, delivered at the factory, is quite competitive in the region, about US$13.88 per cubic meter, almost the same as the price of the wood produced by the company itself which is about US$13.28 per cubic meter.

Universities

One group of stakeholders that the company is increasingly interested in are universities. Riocell, in an attempt to improve the efficiency of its technological development, has successfully implemented a policy for carrying out joint research projects with certain universities, according to company executives. Riocell has set up a system of technological networks to connect more efficiently with external data bases and with groups knowledgeable in technology that is of interest to Riocell. This gives the universities a stake in Riocell's continued welfare to the extent that the company now represents a source of funding for their research projects as well as a channel for access to different technological know how and the possibility of experimenting in new fields. Riocell also sponsors, in association with the Federal University of Santa Maria, a masters course in pulp and paper technology with an emphasis on environmental issues.

Nongovernmental Organizations

Given the potential environmental impact of pulp and paper producers, NGOs, as well as nongovernmental individuals (NGIs), keep close watch on Riocell. Their concerns center on keeping the negative environmental impacts of pulp and paper production to a minimum.

Customers

Customers are primarily concerned with the quality of the company's products, the prices thereof, and their timely delivery. Riocell pays special attention to these stakeholders and has put at the disposal of its customers a pilot manufacturing facility that tests ways to produce pulp that has the most optimum characteristics to best meet the needs of customers. The possibility of producing paper with differing characteristics is one of the company's comparative competitive advantages. Riocell is an international supplier of world renown in various markets where it has carved out a place for itself among large international paper groups such as Stora, Arjo Wiggins, Kymmene, Courtaulds, and Kimberly-Clark, to mention only a few. At present, it competes in North American, European, Asian and Latin American markets. Riocell, in contrast to the majority of pulp manufacturers, not only stresses meeting the demands of customers, but also the importance of providing post-sales services.

Employees

Lastly, and possibly the group of greatest interest to the company, is that represented by its employees. Their concerns are not limited to ensuring Riocell's com-

petitive edge and, thus, their own jobs. Employees are also concerned about the environmental impacts of Riocell's operations and the environmental and safety conditions in the workplace. Riocell has adopted policies that meet total quality management (TQM) standards, which include rewarding team work, providing incentives for TQM, stressing transparency, and recognizing employee contributions. These policies have helped the company to reach satisfactory levels of quality in both the forestry and the industrial sectors.

Riocell and Aracruz Within a Framework of Sustainability

Robert Day, World Resources Institute

The growing body of experience with sustainable forest management has begun to yield a number of relevant lessons for business. One key precept is that forest management, while unique, is nonetheless a business that has many of the same management challenges and roles found in other industries. The integration of environmental and social sustainability into business strategy is one of those challenges: It is just as vital for forest products companies as it is for businesses in the chemicals, oil, or mining industries.

SOCIOENVIRONMENTAL CONCERNS AS A COST AND AN OPPORTUNITY

In an era when natural resources are becoming a global constraint, and the peoples of lesser developed regions are demanding higher standards of living, no business can afford to ignore its impact on communities or the environment. If a company violates its social contract by breaking the law, for instance, it could well lose its franchise to operate. However, by treating socioenvironmental concerns as merely a cost, and not an opportunity, a business will add inefficiency through higher production costs, greater investments in pollution control, and lower worker productivity—without gaining any

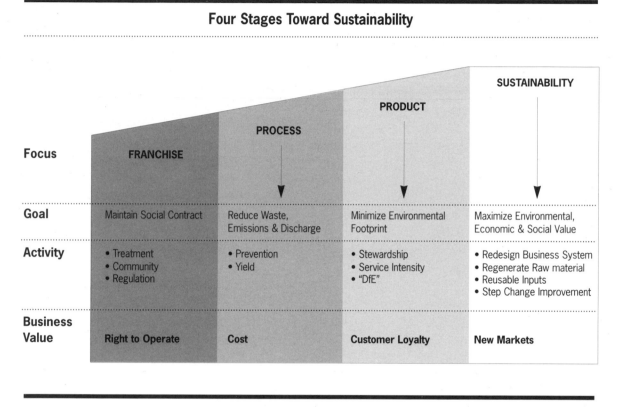

Four Stages Toward Sustainability

Focus	FRANCHISE	PROCESS	PRODUCT	SUSTAINABILITY
Goal	Maintain Social Contract	Reduce Waste, Emissions & Discharge	Minimize Environmental Footprint	Maximize Environmental, Economic & Social Value
Activity	• Treatment • Community • Regulation	• Prevention • Yield	• Stewardship • Service Intensity • "DfE"	• Redesign Business System • Regenerate Raw material • Reusable Inputs • Step Change Improvement
Business Value	Right to Operate	Cost	Customer Loyalty	New Markets

Figure 6

compensatory benefits. Indeed, mounting evidence shows that those companies that integrate socioenvironmental concerns into a comprehensive business strategy can create opportunities to enhance efficiency and the quality of production, and even develop sustainable competitive advantages. This is especially so for raw-material-intensive industries such as forest products, where escalating environmental and social pressures are daily constraints.

The full integration of these issues into business strategy requires major changes in thinking by management and stakeholders. The process can be described by the four stages outlined in Figure 6.

Franchise Protection

In this first stage of thinking a company seeks to prevent environmental and social issues from creating problems that could eventually cause the company to lose its social contract to operate in the economy. Sustainable activities in this stage tend to focus on compliance with environmental regulations and end-of-pipe pollution control, activities that present the company with added costs, but little benefit. On the social side, investments tend to be donations for infrastructure improvements and improvement of minimum standards of living that are needed to ensure the cooperation of surrounding communities. A company can be mired in this first stage while operating at standards above the legal requirements—the key is that the company does the minimum necessary to insure that it can continue to operate.

Process Improvement

In this second stage, the company tries to reduce waste by streamlining its production, which enables it to find the classic win-win solution. Here, the benefits of activities can easily outweigh the costs. This can be especially true in commodity industries, which forest products manufacturing tends to be, where the low-cost producer can hold a significant market advantage. Nevertheless, creating sustainable competitive advantage at this stage is difficult because the company continues to make the same product for the same customers without differentiating itself from competitors.

Product Stewardship

At this point, the company examines the entire life-cycle of its products to identify further opportunities for efficiency gains. Raw materials are reconsidered with an eye toward ensuring consistent quality and lower costs. Use of the product is examined in hopes of identifying new services or products for which there is latent demand in the market. The company that has integrated product stewardship into its business strategy tries to use environmentally-sound, low-cost raw materials to produce superior products efficiently. The benefits from such thinking are often not as easy to identify as those gained from simple process efficiency improvements, but they are no less real and are potentially much greater.

Sustainability

In the first three stages, companies try to minimize their environmental and social footprint—to reduce the "harm" their activities cause. In the final stage, sustainability, the company looks to maximize its environmental and social benefit, providing superior returns to financial shareholders and external and internal stakeholders alike. This frame of think-

ing is most difficult to define and identify, because so few examples exist. Sustainability requires that a company attempt to make the environment better while doing business, instead of being satisfied with merely avoiding environmental damage. In the process, management needs to recognize four areas where the value of sustainability can be maximized—financial, social (external stakeholders), personal (internal stakeholders), and ecological. This stage is potentially the most powerful: over time environmental and social constraints promise to get worse. In the future the economy will increasingly reward the company that produces eco-social benefits. The forest products industry is uniquely suited to adopt this point of view, given the long growing periods and renewable nature of its raw material. Yet, few companies are considering this frame of thinking, much less adopting it.

In practice, of course, these stages are not nearly so clear cut. Many companies have adopted business practices that reflect the contemporary thinking on eco-efficiency, for instance, while many other activities in these same companies remain mired in compliance-focused franchise protection. Just as product stewardship thinking encourages synergy throughout the production process—from design to disposal—sustainability must be integrated throughout the business to enhance opportunities and benefits.

This generic framework can be applied to any industry. However, the natural and human resource-intensive industrialized manufacturing processes make them the most suitable to find examples to draw from: it is no coincidence that some of the most obvious example are in the chemical and forest products industries. In the forest products industry, this thinking is most evident in the production of pulp from tree plantations. A comparison of two companies that epitomize plantation forestry, Brazilian pulp producers Riocell S.A. and Aracruz Celulose S.A. can best illustrate these stages in action.

THE COMPANIES

Riocell

Riocell S.A. is a subsidiary of the giant Brazilian paper products conglomerate Klabin, the largest integrated pulp and paper producer in Latin America. The fairly independent Riocell, however, is only loosely linked in production to the other subsidiaries. Riocell produces eucalyptus and acacia pulp from 71,500 hectares in the Rio Grande do Sul state of southeastern Brazil.

Riocell started in 1972 as the pulp mill of a Norwegian company, Borregaard. During its first few years the company became infamous in the region for high levels of air and water emissions. After the mill was finally shut down by the government, the company changed hands from Borregaard to the Brazilian government and then to Klabin. Riocell's management has not forgotten its past as a public enemy, nor will local environmental activists let it.

In 1995, Riocell produced 240,000 tons of pulp. Of this, 34,200 tons were dissolvable pulp used in textile manufacturing. Riocell also sold another 37,000 tons of fine writing paper. By volume, however, bleached pulp is the company's biggest product. Two-thirds of the company's product is exported, 30% of which goes to Europe, the rest to Asia, Latin America, and the United States. Net sales in 1995 totaled $230.5 million.

Aracruz Celulose

Aracruz Celulose is the world's single biggest producer of bleached eucalyptus pulp. Not only does it produce this pulp in high quantities—capacity is over 1 billion tons per year—but the company's pulp is of the highest quality and is used in the production of fine writing papers, tissue, and other paper products. The company does not manufacture any of these products itself, and exports almost all of the pulp that it produces. Net sales in 1995 totaled $767 million.

The company was started in 1967 as a joint venture between domestic investors, foreign investors, and the Brazilian government. Aracruz grows and uses only eucalyptus for its fiber, and has historically invested heavily in research. Aracruz was one of the pioneers in the cloning, or vegetative propagation, of the species, and has focused its genetic research upon improving fiber yield per hectare.

Aracruz is based in the southeastern state of Espirito Santo, where it has a pulp mill and a dedicated port facility, but is headquartered in Rio de Janeiro. The company has forest holdings of 203,000 hectares.

COMMON ADVANTAGES

Market pulp is a global commodity, characterized by a general lack of product differentiation, which gives the low cost producers an advantage in the market. As the globalization of trade allows for easier transportation between markets, Brazil's pulp and paper industry enjoys a global low-cost advantage, principally due to low energy costs and the high productivity of the soil and climate in the southeastern states. In addition, the introduction of fast-growing exotic species, specifically eucalyptus from Australia, has allowed the country's pulp and paper industry to make the most use of its advantages. The cost per ton of bleached hardwood pulp in Sweden is 185% higher than in Brazil, and over 15% higher in the southern U.S. than in Brazil. The growth of similar plantations in southeast Asia is challenging this advantage, but so far only to a limited extent.

To the extent that plantation forestry can take pressure off natural forests, it has the potential to be good for the global environment. Although Aracruz's plantations do not substitute directly for Brazilian tropical hardwood products, they do provide fiber that would otherwise be required from other pulp-fiber producing regions of the world, such as the temperate forests of the United States and Canada. As fiber suppliers, plantations can be highly effective: an intensively managed plantation three-quarters the size of Texas could theoretically supply all the world's demand for timber and fiber forever at current rates of consumption, according to one producer. Because plantation forestry also typically involves ownership of the forestland, owners usually take measures to sustain the land's productivity. For these reasons plantation forestry can potentially be both profitable and environmentally friendly.

Nevertheless, the intensive nature of plantation forestry makes it a target for critics. Tree plantations can require the clearing of existing forestland, the intensive use of chemicals and nutrients, monoculture systems, and the disruption of local communities. The WBCSD International Institute for Environment and Development report, the *Sustainable Paper Cycle,* which generally supports plantation forestry, states that, "Forest management and plantations reduce species diversity within forests, especially monoculture, short rotation systems." Viewed from another perspective, however, plantation forestry represents a transition in forest management from a mining orientation of extracting resources to an agricultural one, which considers timber-based fiber a crop.

On the face of it, both Riocell and Aracruz stand to benefit from common advantages in the current market. Both are Brazilian pulp producers that depend primarily upon eucalyptus fiber supply. Both were started in the late 1960s and early 1970s, during a period of Brazilian government incentives and economic growth. Both had Scandinavian backing, which brought in expertise from that region's well-established forest products industry. Both companies export to markets around the world. Both strive for environmental and social responsibility, and feel, with some justification, that they are contributing to sustainable development.

STRATEGIC AND FINANCIAL DIFFERENCES

Yet the two companies view the environment and surrounding community, in light of their business strategies, quite differently. Riocell considers the environment as a cost for its manufacturing process. Company documents primarily address the costs incurred by the company through its efforts to comply with environmental regulations. They give little attention to the potential cost savings of preventing waste. Indeed, the company states that adoption of environmentally-friendly business practices cost it 15% of its profits annually (see Table 7 in Riocell case study). Riocell has focused primarily on reducing end-of-pipe emissions rather than pollution prevention, which has contributed to apparent inefficiency at the mill.

Aracruz, by contrast, views the environment as an opportunity for both efficiency and new markets. The company has emphasized eco-efficiency in its manufacturing operations, seeking to drive down costs by minimizing waste. It has extended this effort to the plantations, by working to ensure that its eucalyptus fiber has consistent yield and quality. The company has achieved these goals through aggressive research to match superior genetic strains of a particular eucalyptus hybrid with advanced silvicultural methods. As a result, Aracruz is now the world's low-cost bleached eucalyptus pulp producer. The company has also sought to take advantage of the shift in some markets, notably Germany, toward environmentally-friendly products by investing in TCF technology, and presenting itself as a sustainable producer by supporting efforts like the International Institute for Environment and Development (IIED) study on sustainable paper cycles and the WBCSD.

Given the franchise frame of thinking that appears to categorize Riocell's approach to the environment, and the product stewardship frame that Aracruz appears to have adopted, it is not surprising that the companies demonstrate differences in financial performance. In a commodity industry, Aracruz achieves consistently positive returns. Even in the cyclical-low year of 1996, when pulp prices dropped from nearly $800/ton in the first half of the previous year to less than $500/ton, and with significant periods of maintenance shutdown at the mill, Aracruz still managed to produce a net income in the first half of the year of $50 million (down from $191.5 million for first half 1995). One financial analysis of the company concludes that it has established, through its genetic improvements and land holdings, a competitive advantage that should be sustainable for at least twenty years. This performance has attracted investors such as the conglomerate Mondi, which acquired a significant stake of the company's ownership in 1996. Compared to a three year average operating margin for the Latin American pulp and paper industry of 11%, Aracruz's 1993-95 average operating margin of 17% was impressive to these investors.

Riocell, however, managed to earn only $20 million in net income for the entire cyclical-high year of 1995, which yielded an operating margin of under 10%. Inefficiency at the mill is apparently hurting the company, as the equipment is reportedly outdated and waste is high. Cost is a significant driver in this industry; Riocell's costs per ton produced have been estimated at $20 more per ton than Aracruz's ($320/ton vs. $300/ton).

Next Steps

Yet Riocell is taking promising steps. Although Riocell has not integrated eco-efficiency or product stewardship throughout its business, it is proactive in some activities. For instance, the company places great value on the productivity of its soil and tries to preserve it by removing bark from logs in the forest rather than at the mill, which returns nutrients to the soil and reduces fertilizer use. Riocell has also considered the needs of local communities by contracting out as many operations and

services as possible—the entire forestry unit consists only of seventy employees. This represents socioefficiency because it lowers the company's costs while providing local entrepreneurs and farmers with more opportunities. While some may debate the value of outsourcing as a socially responsible practice, local communities appear to welcome Riocell's "mosaic approach" of small, outsourced forest holdings. The mosaic approach and dispersed debarking appear to be inefficient, yet Riocell's cost per ton of raw fiber at the mill is significantly lower than Aracruz's—about $86/ton vs. $100/ton for Aracruz. If Riocell were to better integrate environmental and social sustainability throughout its forestry and manufacturing, it might be able to preserve this value better and capitalize on these investments.

While Aracruz has been proactive on environmental issues, on social issues its frame of thinking appears to be in line with protecting its franchise. The company acts beyond its legal requirements to provide healthcare for local communities, housing for its employees, and free education, but without these investments the company could not operate. Even though the company contributes heavily to the development of the regional economy and to strengthen the local infrastructure and social system, outsiders occasionally criticize Aracruz, arguing that it is insensitive to the needs of local stakeholders. These significant social investments do not provide commensurate returns for Aracruz. By treating social sustainability as an opportunity for strategic advantage as Riocell does, rather than as a franchise protection, or "social obligation" issue, Aracruz might be better able to capitalize on the greater relative prosperity it has brought to the region, avoid controversy, and reap the benefits of better relations with neighbors and a more positive image with the public.

Conclusion

The lessons from Riocell and Aracruz are relevant to all forestry operations, even those that are not plantation based. A company locked in a compliance-focused, franchise-protection frame of mind will derive few benefits to outweigh the costs of environmental protection. But environmental and social sustainability, when integrated fully into the business, can open up opportunities for superior business performance. If these considerations, however, are only partially integrated into a company's strategy, opportunities will be lost. In this way sustainable forest management is comparable to total quality management. It needs to be applied throughout a company's business operations, or the potential performance gains will not be fully realized.

Appendix A

Aracruz Consolidated Statement of Operations

Year ended December 31...	1995 US$thousands	1994 US$thousands	1993 US$thousands
Operating Revenues - Sales of Pulp:			
Domestic:			
Related Party	27,638	17,411	11,393
Others	46,423	29,980	29,029
Total	74,061	47,391	40,422
Export:			
Related Party	34,129	25,293	17,919
Others	688,295	456,590	315,341
Total	722,424	481,883	333,260
Total Sales	796,485	529,274	373,682
Value-added tax and other sales deductions	29,172	23,757	25,010
Net operating revenues	**767,313**	**505,517**	**348,672**
Operating Costs and Expenses:			
Cost of sales	295,534	279,939	277,458
Selling:			
Related party	6,762	9,555	8,522
Others	26,442	21,124	21,460
Administrative	51,298	55,847	42,603
Loss and provisions for loss on sale of operating assets	2,228	7,257	18,392
Other, net	36,121	(5,201)	7,304
Total Operating Costs and Expenses	**418,385**	**368,521**	**375,739**
Operating Income (Loss)	**348,928**	**136,996**	**(27,067)**

Source: 1995 AR

Appendix B

Aracruz Financial Performance vs. Regional Averages*

■ Operating Margin Avg. 1993-1995
■ Return on Equity Avg. 1993-1995

*averages are for pulp and paper industry,
some data unavailable for Latin America

Source: Morgan Stanley Research

Collins Pine:
Lessons from
a Pioneer

CASE STUDY

PREPARED BY:
ERIC HANSEN
JOHN PUNCHES

A Case Study from "The Business of Sustainable Forestry"
A Project of The Sustainable Forestry Working Group

The Sustainable Forestry Working Group

Individuals from the following institutions participated in the preparation of this report.

Environmental Advantage, Inc.

Forest Stewardship Council

The John D. and Catherine T. MacArthur Foundation

Management Institute for Environment and Business

Mater Engineering, Ltd.

Oregon State University
Colleges of Business and Forestry

Pennsylvania State University
School of Forest Resources

University of California at Berkeley
College of Natural Resources

University of Michigan
Corporate Environmental Management Program

Weyerhaeuser Company

The World Bank
Environment Department

World Resources Institute

CCC 1-55963-621-1/98/page 6-1 through page 6-16
© 1998 Oregon State University
All Rights Reserved

Collins Pine:
Lessons from
a Pioneer

CASE STUDY

PREPARED BY:
ERIC HANSEN
JOHN PUNCHES

A Case Study from "The Business of Sustainable Forestry"
A Project of The Sustainable Forestry Working Group

Contents

Collins Pine: Lessons from a Pioneer

Collins Pine Company, headquartered in Portland, Oregon, produces a variety of lumber products for industrial and construction markets. The privately held company, with revenues of $220 million in 1996, runs forestry and manufacturing operations in California, Oregon, and Pennsylvania, operates three retail stores under the name of Builders' Supply in California, and markets products internationally through Collins Resources International Ltd. In 1996, Collins Pine expanded into the production of plywood, hardboard, and particleboard through the acquisition of Weyerhaeuser Company's Klamath Falls operation.

Collins has a long history in sustainable forest management. The company that eventually became Collins Pine started in 1855, when Truman D. Collins bought forestry and milling operations in Pennsylvania. Truman W. Collins, the founder's grandson, adopted sustained yield forest management on company lands near Chester, California, in 1940. The Collins Pine management system, based on U.S. Forest Service models under research at the time, emphasized selective cutting, a practice that creates stands of uneven-aged trees similar to those found in some natural forests. The Forest Service later switched to techniques that foster even-aged stands of trees, but Collins Pine retained and improved its uneven-aged management. This management style remains the foundation of the company's western operations today.

Throughout the company's 142-year history, the Collins family has maintained its ownership and an active interest. Maribeth Collins, widow of Truman W. Collins, is the current board chairman. The company's values and philosophy reflect those of family members, past and present. Stewardship is the cornerstone of that corporate philosophy, which the company defines as commitment "to the long-term management of our forest resources and to the responsible utilization of these and other resources to produce the finest-quality finished products."

As part of this commitment, in 1993 Collins Pine became one of the first companies in the world to have an independent organization certify that some of its timberlands are well managed. As an early adopter of sustainable forest management and certification, Collins Pine has confronted a variety of challenges in its pioneering efforts to practice sustainable forestry and to market certified wood markets. Although the company has had difficulty finding premium-paying markets for its certified products, certification has enabled Collins Pine to gain access to new markets and to improve its business practices.

Company Operations

Over the years, the Collins Pine land base has changed in response to company needs. Some of its original lands were sold to pay inheritance taxes, while others were purchased as opportunities arose. Today, the company manages timber holdings in conjunction with manufacturing operations through Collins Pine in Chester, California; through Kane Hardwoods in Kane, Pennsylvania; and through an affiliate, Ostrander Resources's Fremont Sawmill in Lakeview, Oregon. Collins Pine also manages the recently acquired plywood, hardboard, and particleboard facilities in Klamath Falls, Oregon (see Table 1). Collins Resources International Ltd. (CRI), which acts as a wholesaler and the international sales force for the company, operates out of Portland. Traditionally, CRI's marketing efforts were concentrated on western Europe, but the company has recently identified the Pacific Rim as an area for expansion.

In 1993, Scientific Certification Systems of Oakland, California, certified Collins Pine's Almanor Forest in Chester as a "State-of-the-Art Well-Managed Forest," one of the first in the U.S. In 1994, Scientific Certification Systems awarded the Kane Hardwoods forest its "Well-Managed Forest" designation. A Scientific Certification Systems pre-audit of the Fremont Sawmill was completed in July, 1997. Collins Pine executives expect the mill and its timberlands to be certified in the spring of 1998.

The Collins Pine Operations

Almanor Forest, Chester, Califor.

Acreage: 94,000 (softwoods).

Facilities: Sawmill with a 75-million-board-foot (MMBF) annual capacity; dry kilns; planer; remanufacturing plant.

Products: Specialty and dimension products; construction and industrial lumber products; fuelwood or wood chips from logging.

Kane Hardwoods, Kane, Pa.

Acreage: 122,000 (hardwoods).

Facilities: Band-mill with a 20 MMBF annual capacity; dimension plant with an annual capacity of 4 MMBF; dry kiln with a 14 MMBF annual capacity.

Products: Logs to domestic and export markets; dimension blanks and squares; glued panels, flooring, and moldings; material for shipping pallets; bark; small logs for pulp; wood residue to pulp facilities.

Fremont Sawmill, Lakeview, Oreg.

Acreage: Over 80,000 (softwoods).

Facilities: Softwood lumber mill with an annual capacity of 40 MMBF; remanufacturing plant; dry kilns; and planer capacity.

Products: Industrial and construction lumber.

Collins Products, Klamath Falls, Oreg.

Facilities: Plywood manufacturing (160 million square feet capacity annually based on a ⅜-in. measurement); particleboard mill (120 million square feet annually based on a ¾-in. measurement); hardboard manufacturing (130 million square feet annually based on a 7/16-in. measurement). Facility uses sawdust, wood chips, and small-diameter logs from other Collins Pine locations.

Products: Structural underlayment, sheathing, and tongue-and-groove plywood; particleboard ½-in. to 1¼-in. thicknesses; ½-in. and 7/16-in. hardboard siding.

Table 1

Business Definition and Strategic Intent

SUSTAINABLE FORESTRY ACTIVITY

Company-owned timberlands at Collins Pine's three manufacturing operations supply about 50 percent of each location's raw materials. Each site then buys logs on the open market to supplement its own log production and maintain manufacturing levels. Log buyers at each location reported that some of the material they buy comes from sustainably managed forest lands, although the sustainability of these other operations has not been verified. The combination of its own certified production and that purchased from the outside makes it likely that more than 50 percent of the total raw material volume used by Collins Pine's three solid-wood manufacturing locations comes from sustainable forestry operations.

Products made from wood that comes from certified timberlands can be marketed as certified and may carry the Forest Stewardship Council logo. The Forest Stewardship Council (FSC) is an international accreditor of independent product certifiers. Companies certified by an FSC-accredited certifier may incorporate the FSC logo into their marketing materials. Scientific Certification Systems is accredited by FSC, so by virtue of its SCS certification, Collins Pine has access to the marketing logos of both SCS and FSC. Collins initially utilized the SCS logo, but switched to the FSC version when it became clear that it was gaining recognition.

Both of Collins Pine's certified locations (Kane and Chester) carefully segregate material from their forest land that can be sold as "certified" and track it through manufacturing and shipping, so that it will not get mixed with products coming from non-certified sources. Neither location, however, markets more than five percent of its total production as certified, even though at least 50 percent qualifies. The limited market demand for certified wood accounts for the discrepancy between certified production and sales.

BUSINESS GOALS AND OBJECTIVES

Collins Pine's ability to survive in the competitive forest products industry is indicative of a sound business strategy based on the foundation of sustainable forestry. The obligation of mill and forest managers to provide an economic return to the owners is tempered by the company's policy of cutting no more timber from forests than can be sustained over the long term. Harvest levels are determined not by mill requirements but by forest growth. This long-term outlook has its advantages. Mill managers are able to anticipate timber production from company lands well in advance and plan accordingly to supplement their own supply with outside purchases.

Collins Pine manages its timberlands with multiple objectives. Forest management is designed to maintain and enhance diversity in the forest (among species and sizes of trees), improve forest health, and increase the production of high-quality timber to feed the company's production facilities. Broader goals of maintaining the forests' functions as watersheds and habitats for wildlife also are included in forest management planning. The overriding objective, however, is that Collins Pine's forest management personnel conduct their activities while keeping management options open for future generations. The willingness of the owners to forgo potential short-term profits in favor of long-term sustainability makes it possible to carry out these objectives.

The company also has other goals related to sustainable forestry and its corporate values. Producing high-quality products and developing markets for certified products are priorities. Collins Pine strives for public recognition as a socially and ecologically responsible company. As part of that objective, it makes long-term commitments to the communities in which it operates. Collins Pine actively promotes certification to help foster public support for commercial forestry and to help regain access to public forests in the Pacific Northwest that have been off-limits to logging in recent years.

THE COLLINS PINE BRAND OF SUSTAINABLE FORESTRY

Collins Pine managers in the mill or in forest operations share a distinct, cohesive vision of sustainable forestry. The company's timberlands are considered a resource base from which growth can be removed, but the net growing stock cannot be depleted. Managers commonly refer to company timber as "principal" and the growth as "interest," indicating that they are free to draw from the interest but that the principal must remain stable.

The activities of forest management personnel reflect that philosophy. In interviews, Collins Pine managers confirm that they consciously act in ways that will retain options for future managers, promote forest diversity, allow the forests to regenerate naturally whenever possible, and protect wildlife habitat and watershed functions. Forest managers tailor their management practices to each site using a variety of silvicultural techniques dictated by tree species, age, and other characteristics. Collins Pine uses both even-aged and uneven-aged practices to mimic the natural processes that create diversified tree stands and promote natural regeneration of trees.

Ecological and Social Effects

HEALTHIER FORESTS, SATISFIED EMPLOYEES, AND GOOD RELATIONS WITH NEIGHBORS

The effect of Collins Pine's land management practices is visible in the forest. Typically, foresters log sites at 12-year to 20-year intervals. They leave standing and downed dead wood in significant quantities. On lands under active management, harvests are light. Good road maintenance, relatively moderate terrain and climate, and careful timing of logging minimize damage to soil and water quality. In its audit, Scientific Certification Systems documented that the company's management practices cooperate with nature and that the company's

This massive ponderosa pine has been retained in the Collins Almanor Forest for its grand stature and the diversity it lends to the stand's structure and appearance.

foresters have a "commendable level of recognition for all forest resources, including wildlife, water quality, natural biodiversity, and visual aesthetics." A walk through Collins Pine's managed stands finds stands of trees that are often greener and visibly healthier than forests on neighboring public and privately held lands—even when those nearby forests are less intensively logged.

Collins Pine goes to great lengths to support the communities in which it operates. The company gives the public liberal access to its forests. It directly and indirectly supports local land-use consensus-building groups made up of all types of individuals, including preservationists. Collins Pine encourages

research and educational projects on its forest land, supports schools and hospitals, and, through its long-term commitment to stable employment, is recognized as a contributor to community economic stability.

The philosophy of the Collins family seems to have a profound influence on employees. Numerous interviews indicate that relationships among mill personnel often approach those associated with an extended family rather than a corporate employer, and that turnover is lower than the industry norm. Employees are often eager to speak at length about their relationship with the company and the respect they have for the Collins family. In one typical remark, an employee commented: "This is a wonderful company. The Collins family is ...just great to work for and that's why you don't see very high turnover. They are concerned for their employees and the environment."

The corporate culture encourages Collins Pine employees to carry the company's philosophy beyond the workplace. At both the Lakeview and Chester operations, employees are active in consensus groups organized to help communities resolve the competing demands on land so prevalent in the Pacific Northwest. As Collins Pine's ambassadors to these groups, employees educate and communicate with neighbors who do not always favor logging, and are able to stay "in tune" with the needs and concerns of their communities.

As part of its outreach, Collins Pine allows local schools to establish research plots and projects on its land. In Lakeview, for example, high school students have installed plots and monitored how well various techniques regenerate trees. The activity has accomplished more than simply educating young people in forestry. Collins Pine managers think that it also has given students an opportunity to see for themselves that managed forest lands are dynamic systems that can be harvested without sacrificing scenic beauty, wildlife, or sustainability.

Through these actions, forest certification, and the willingness of employees to spend time in the woods explaining their practices to the public, Collins Pine has earned sufficient credibility to work cooperatively with environmentalists on forest management issues. At the Chester operation, for example, this credibility has enabled Collins Pine to become an active member of the Quincy Library Group, a local consensus group that includes stalwart environmentalists.

On occasion, Collins Pine managers have made sacrifices to maintain the company's credibility. In 1995, Collins Pine was ready to bid on a salvage sale of burned timber on nearby federal land. Since the most feasible access to the salvage timber was through Collins Pine land, the company had an obvious competitive advantage to win the bid. Members of the Quincy Library Group, however, opposed the logging. Collins Pine withdrew from the bidding even though the sale made good business sense. In this case, the company preferred maintaining its relationship with the consensus group to proceeding with a deal of relative short-term importance.

CONSTRAINTS TO SUSTAINABLE FOREST MANAGEMENT

Private ownership undoubtedly plays a key role in Collins Pine's ability to employ the conservative forest management style that helped it gain certification. Under the company's forest management objectives, the land cannot be pushed to its maximum production level. Harvest levels may not exceed growth and often are below this level. Rotation lengths are significantly longer than those used by competitors, and management costs are higher on a per-unit basis. To achieve those objectives, Collins Pine's owners consistently place less emphasis on maximizing short-term profit than do most publicly held companies.

Sustainability can increase production costs or diminish profits in a number of ways. Harvest plans tied to the status of the forests may hinder the company's ability to respond to fluctuations in market demand and/or price. Allowing trees to grow longer, leaving larger and more trees standing, and protecting non-timber resources often requires lower harvest levels and the use of more expensive harvesting methods. Finally, the company's need for comprehensive timber stand information and significant control over harvesting operations makes forest management labor-intensive.

It is difficult to fully quantify the cost of sustainable forest management for Collins Pine. A company adopting a Collins Pine style of forest management would likely recognize distinct increases in costs and perhaps decreases in profitability. Collins Pine, however, has operated under these constraints for years; any additional costs are an accepted price of its corporate philosophy. In recent years, Collins Pine has invested from $16 to $36 per thousand board feet of logs to cover forestry costs. These include the costs of marking trees for sale, overseeing harvest contractors, measuring forest growth, and maintaining forest roads. Company managers acknowledge that their costs are higher in some areas than those experienced by many other industrial forest land owners. They counter that the long-term stability of their wood supply afforded by conservative forest management practices compensates for any sacrifice in short-term profits.

Collins Pine's annual allowable cut gives the company an average of 316 board feet of logs per acre per year on its Almanor Forest. Its timberlands supporting the Fremont Sawmill in eastern Oregon produce an average of 125 board feet per acre. Because the Almanor Forest receives considerably more rain and has been under active management for much longer than the Fremont operation, this difference in productivity is not surprising.

By way of comparison, other private industrial forest lands in eastern Oregon averaged 268 board feet of logs per acre in 1995. Statistics specific to industrial forest lands were not available for northern

California, but the average for all private timberlands in the area, including Collins Pine's Almanor forest, was about 230 board feet per acre per year in 1994. This indicates that Collins Pine's Lakeview forests are producing well below industry averages, while its more-established Almanor Forest produces at a level that actually exceeds the area's average.

Eastern hardwood forests typically produce at a much slower rate than western softwood forests. Collins Pine's Kane hardwood forests are no exception. Annual allowable cuts there average 57 board feet of logs per acre per year, but recent measurements of the forests' growth indicate that cut could be considerably higher—perhaps even doubled. The average harvest for all of Pennsylvania's timberland was 102 board feet per acre in 1989. At that time, state estimates indicated that growth exceeded harvesting by 2.6 times.

These harvest estimates support Collins Pine's contention that it harvests at rates lower than the industry average. However, since many of Collins Pine's trees are allowed to reach greater age before being harvested, the overall quality of the trees and

their resulting value should be higher. Logs originating from Collins Pine's Almanor Forest do, in fact, tend to be of higher grades and larger sizes that those the company buys from outside sources. However, Collins Pine may not be purchasing a uniformly representative sample of logs produced on lands other than their own. If an area company were aggressively seeking quality logs on the open market, the material obtained by Collins Pine might be skewed toward smaller, lower-quality logs. What is clear is that Collins' forest management practices provide it with higher quality raw material than what it is able to purchase on the open market.

The costs of Collins Pine's certification activities can be quantified. Each certified location had a preaudit and a full certification audit. Each is charged a yearly fee to maintain its certification and will be re-audited five years from the initial certification date. Initial fees for the two certified locations totaled $60,000 to $80,000, and yearly fees for each location will cost as much as $7,200. The company will pay additional costs as its Lakeview and Klamath Falls locations go through the process. Collins Pine estimates that capital improvements made as a result of certification may cost as much as $250,000 per year for the next three years. These include setting up new systems to measure and document timber volume and growth in its Pennsylvania forests. Forest management costs have risen since certification—they have roughly doubled in the Almanor Forest—as forest managers responded to suggestions made by the certification team. In addition, the increased materials handling costs associated with tracking certified wood from the forest through manufacturing may reach $150,000 per year. While these costs may seem impressive, they represent only about

This stand is characteristic of the components of the Collins Almanor Forest not yet brought under active management. It is predominated by white fir.

one percent of Collins Pine's total sales. Company executives consider this cost modest. They are quick to point out that many of these costs paid for improvements that were needed regardless of certification, and that those investments will return dividends through increased efficiency.

Market Analysis

THE ELUSIVE MARKET FOR CERTIFIED PRODUCTS

Collins has been relatively unsuccessful in marketing its wood products as certified. As stated previously, less than five percent of the lumber and other products produced at the Kane and Chester operations are sold as certified, even though at least 50 percent could qualify. At one point, the Chester operation sold more than 15 percent of its production as certified, but the level dropped after two separate arrangements dissolved. Although management has invested considerable time and energy, including 35 percent of the vice president of marketing's time for the last three years, no significant markets for certified product have materialized. Failure to develop these markets, while frustrating for salespeople, is not necessarily surprising given Collins Pine's early entry into the market. The company's pioneering efforts have, however, proved instrumental in bringing the issues of sustainability and certification to public attention and may provide a foundation for companies that enter the market later.

Collins Pine has identified specific geographic and demographic market segments that are receptive to certified products. Receptive consumers tend to be highly educated and have significant levels of disposable income. These geographic markets include Austin, Texas; Sante Fe, New Mexico; the San Francisco Bay Area in California; and Vail and Aspen, Colorado, as well as the United Kingdom. In the U.S., the company has found that areas with harsher climates often harbor more "green" consumers.

The failure of the company's certified products to meet expectations in Portland, Oregon, is an indication, according to Collins Pine managers, that consumers are often more inclined to talk "green" than to act "green." Company personnel referred to a California Forest Products Commission study that investigated the nature of consumer perceptions of certification. Focus groups conducted for that study indicated that most consumers would not pay a premium for certified products. In fact, these consumers did not think certification should be necessary. They felt that companies should already be following stringent regulations and that sustainable forestry should be a given.

Collins Pine's evaluations of consumer demand come from the company's experience in dealing with their markets rather than from primary research. Salespeople often field calls from people interested in buying certified wood products, but those calls come mostly for consumer products, for which Collins Pine can provide only the raw material. Salespeople at corporate headquarters also get similar calls. This may happen partly because the company's 800 number is printed on the sticker that accompanies its certified products, and because it has received extensive press coverage for becoming certified. In any case, Collins Pine has become a source of information for consumers trying to find certified products.

MARKET BARRIERS TO CERTIFIED PRODUCTS

Collins Pine has encountered a number of barriers to marketing its certified products. These barriers fall into five general categories:

Limited market demand. The actual demand for certified or otherwise sustainably-produced wood products is currently limited and segmented. As a market pioneer, Collins Pine has struggled to identify and serve these small niches efficiently.

Unfavorable consumer perceptions. Collins Pine sales and marketing personnel have discovered that their customers often harbor the misconception that certified wood must be inferior to wood produced through "standard" industry practices. These individuals think that companies sacrifice quality to reduce environmental impacts. This belief was evident even when marketing to another environmentally oriented firm, The Home Depot. In Collins Pine's case, however, the opposite is actually true. Trees are allowed to grow longer than on comparable industry forests. These older trees tend to have a higher proportion of clear, defect-free wood. Collins Pine personnel have had to educate potential certified product customers by demonstrating the relationship between their forest management practices and the quality of the products they produce from that wood.

Limited distribution channel development. Existing wood products distribution channels are reluctant to carry certified wood products. These products must be tracked from the forest floor to retailers' shelves, which requires sophisticated systems unless certified product remains segregated during storage and transport. In this way, certification adds complexity and cost to the distribution process.

Difficulties in meeting specific market demands. If there are markets with a significant demand for certified products, as in the U.K., other problems surface. CRI personnel receive phone calls almost every day from companies interested in certified products. But those potential buyers have precise demands. They typically require the highest-grade lumber of a specific species and thickness. More often than not, the volume requested in the specific grade, species, and thickness exceeds what Collins Pine can produce or CRI can get through other sources.

Limited product availability. Certified wood products are currently available only in extremely limited volumes, which has a number of implications.

Most wood products producers have neither sought nor obtained certification, which makes distributors hesitant to carry the certified products available. It is difficult for distributors to find enough product volume to justify allocating floor space, storage, and other distribution resources to certified products. In turn, the dearth of readily available sources of certified materials makes product specifiers, such as architects and engineers, reluctant to use these products in their designs.

This combination of market inhibitors has created a dilemma for certified wood product producers—one that resembles the old chicken-or-egg story. Product volume will not grow until distribution channels are developed. Distribution channels will not develop unless sufficient quantities of product are available and consumers demand it. Yet consumer demand appears to be stifled by a lack of product. Collins Pine has yet to identify which component of the market is a precursor to the development of the others. In the meantime, those consumers who now want certified wood products often cannot get them, even though certified raw material is available from Collins Pine and other forestry operations.

Strategies and Experiences

BUSINESS STRATEGIES

Collins Pine operates under six strategic priorities:

Quality. Collins Pine management and employees recognize product quality as the company's paramount competitive advantage. As Lawrence Potts, general manager in Chester, commented, "When consumers are walking down the alleys in The Home Depot looking for lumber, they are not looking for a sticker that says certification. They are looking for a board of quality." The company's forest management strategy facilitates quality because it produces larger, higher-quality logs; its manufacturing operations follow through by maintaining high levels of technical sophistication.

Collins Pine pays close attention to customer concerns, which undoubtedly contributes to the quality of its products. The Chester operation's general manager explained that they regularly bring customers to the mill and ask them to evaluate how effectively Collins supplies quality products. Customers are encouraged to examine lumber piece by piece and share their likes and dislikes. Collins Pine also surveys customers quarterly to determine their satisfaction with the products. To improve quality, managers at the Chester operation indicated that they may compete for the Malcolm Baldridge Award for Quality, a process that can help companies improve their quality management.

Price. Collins has adopted a long-term strategy to develop markets for certified products. It tries to establish relationships with customers at market prices, and has an informal agreement with one customer to share profits when a premium is realized. Although the company would like to realize price premiums for certified products, it has not yet required a price premium as a prerequisite for offering certified products.

Employees commented that they think that certified products can justify a price premium. When asked whether they felt that the price of certified products "should" be higher than non-certified but otherwise equivalent products, Collins Pine personnel nearly always answered in the affirmative. They said that since certification has associated costs, Collins Pine should recover these costs. Employees also commented that since sustainable forest management requires less intensive logging, resource owners should be entitled to a premium to offset the lower harvests. Employees who considered the question from a consumer perspective responded differently. They said that consumers may have a right to demand sustainability from the forest products

industry, and that companies should not necessarily expect a price premium for certified products.

Distribution. Traditionally, Collins Pine has sold to commodity markets. In the early 1990s, the company shifted its efforts from commodity markets toward higher-margin markets, such as furniture and specialty shelving, that would be more likely to pay for high-quality products. Collins Pine has since streamlined its distribution channels by selling more products directly instead of through brokers or wholesalers. The shorter channels allow Collins to deal more effectively with niche markets, help offset the added costs of selling in smaller volumes, enhance its ability to communicate with customers, and facilitate quality improvements.

Company Image. Collins Pine works diligently to maintain a respectable corporate image and tell its story to anyone who is interested. Many of the foresters have become "expert" public relations people because they spend so much time talking with the public and giving forest tours. One forester claimed to spend ten times more time on public relations for Collins Pine than he spent in a similar position for his previous forest industry employer. The Chester operation's chief forester claimed that virtually no one who wants to visit is denied.

Certification has generated numerous positive articles in newspapers, magazines, and forest industry and environmental publications. The company also was recognized for its efforts by the President's Council on Sustainable Development, which awarded Collins Pine the President's Sustainable Development Award in 1996. This type of promotion, company executives pointed out, cannot be purchased at any price.

At the same time, R. Wade Mosby, Collins Pine's vice president of marketing, contends that the company's brand name, "CollinsWood®, The First Name in Certified Wood Products," is gaining recognition. Customers are beginning to recognize CollinsWood® even though they may not recognize the Collins Pine name. A well-recognized brand name may prove valuable in the future to build demand for Collins Pine certified wood products.

Competition. Defying conventional corporate logic, Collins encourages competition in certified products to overcome the limited availability of products and the poorly developed distribution channels that now inhibit the market. Collins Pine would like to see larger companies enter the certified arena. One company executive estimated that to make a market for certified products work efficiently about ten percent of the wood consumed should be certified. At present just one-half of one percent comes from certified production.

Low consumer demand contributes to Collins Pine's difficulty in establishing significant market share for its certified products. Most consumers are not only unaware of sustainable forestry issues but also do not understand what certified products are. If more certified products are stocked on retailers' shelves, consumer awareness may rise. The existence of sufficiently large volumes of certified products to make an impression on the average consumer may be the most important catalyst for demand. That, at least, is the hope of Collins Pine managers.

Strategic alliances. In the face of lackluster consumer demand, Collins is considering alliances with other companies to market certified products. One suggestion on the table entails teaming up with several suppliers of certified products for home construction. Collins Pine envisions creating a package of certified products that could be marketed to the professional builder or final consumer building a home. Another concept involves forging an alliance with an industrial customer to produce a final consumer product. Through an association with a consumer goods producer, Collins Pine could begin educating the final consumer—where demand needs to be generated, according to company managers—and start to build demand for its brand name certified products.

A CHECKERED SUCCESS IN MARKETING

Collins has mounted five significant certified product initiatives. Two of them, pine shelving and white fir lumber for furniture, are now defunct. Of the other initiatives, veneer logs are still sold to a slicing operation in Kentucky, white fir construction lumber is sold in Austin, Texas, and low-grade hardwood lumber is sold to a flooring manufacturer. Other, smaller efforts have involved a small wholesaler, a builder in California, and a builder in Sun Valley, Idaho.

Pine shelving. Collins Pine developed and sold pine shelving to The Home Depot of Atlanta, Georgia. The product was stocked in six stores in the San Francisco Bay area. By selling directly to the retailer, Collins realized 15 percent more profit on the product than it would have through normal distribution channels. Concurrently, The Home Depot was able to lower its retail price and maintain profit margins. The shelving sold as CollinsWood® appearance grade, a proprietary grade designed to meet customer preferences and optimize the value of the raw material.

Even though the pine shelving product sold well and store managers liked it, The Home Depot dropped the product in late 1996 for reasons that remain unclear. Collins Pine managers attribute the action to the difficulties of warehousing the shelving. The Home Depot warehouse in Stockton, California, had to store the product separately to meet the chain-of-custody requirements, since Collins could not supply enough shelving to meet the demand generated by more than a few of The Home Depot's many stores.

White fir furniture stock. White fir lumber sold to Lexington Furniture, part of the furniture maker Masco, for a line of designer furniture was a great success from Collins Pine's perspective: The company realized 40 percent more for the wood than if it had sold it as construction lumber. The "Keep America Beautiful" furniture line was featured on cable television's "The Furniture Show," which included footage of the Collins Pine mill in Chester along with interviews with the chief forester and general manager. The line also was covered in the December 1994 issue of *Furniture Design & Manufacturing*.

The line did not, however, fare well with consumers for a number of reasons. The furniture was bulky and would overpower rooms in an average single-family home. More than 100 different pieces were available, but individual pieces were priced fairly high, and no suite prices were offered. Customers, who were more accustomed to hardwood furniture, did not take readily to white fir. The pieces were often damaged during shipping (if dropped, white fir tends to split) and as a result the packaging had to be redesigned, which caused frustration at Lexington. During its first year, the line sold over $5 million. That sales level might have been considered a success with a smaller company, but the cash flow was insufficient for Lexington and the line was discontinued.

The pine shelving and fir furniture initiatives demonstrate that certification is only one of the many product attributes evaluated by consumers. Certification cannot serve as a crutch for an ill-conceived or poorly marketed product, nor will it as yet spark enough consumer interest to pull difficult-to-handle products through distribution channels.

Veneer logs. The Freeman Corporation purchases high-quality, veneer-grade logs from the Kane and Chester operations. Freeman, which operates a veneer slicing operation in Kentucky, markets the veneer as certified. Freeman has agreed to share with Collins Pine any profits above a certain level that it realizes on sales of certified veneer. Although profit sharing has not yet reached significant levels, Collins Pine benefits from a stable buying arrangement and alliance with a company that helps promote certification.

White Fir construction lumber. Collins Pine has found a ready market for construction grades of white fir lumber in Austin, Texas. Sales are directly related to the Austin Green Builder Program, which encourages the use of "sustainable" building materials. The white fir competes with southern pine in this market and in general costs less in the larger dimensions (2X8 and 2X10). Collins Pine, however, so far has not realized a consistent premium for the certified wood. Some months produce a premium as high as two percent , while in others the product is actually sold at a slight loss. The level of the premium is related to the fluctuating price of southern pine. While certification gave Collins Pine entry into the market, it is unclear whether any premiums can be attributed to certification or simply to the availability of larger-dimension lumber.

Hardwood flooring. The Kane division sells about one truckload of low-grade cherry lumber each month to a company that produces flooring. Demand for the cherry outstrips what Collins Pine can supply. Traditionally, this low-grade material was sold as pallet stock. The rustic-looking lumber, however, appeals to certain segments of the building market. When sold into these niches, Collins Pine can sell the wood for almost twice what it gets as pallet stock.

Is there a "green" premium? Collins Pine uses certification as one component in the marketing of its total product offerings. It has had little success certifying an existing product line and recognizing a market premium, which makes it difficult to attribute any premium directly to certification. What the examples clearly show, however, is that certification has opened up new markets for Collins Pine. In several instances, the profits from the company's products in these new markets exceed those that the raw material would otherwise generate if sold into its traditional markets. Certification can positively influence market success if it is properly exploited, although—at least for this company—it is difficult to attribute a price premium to the certification itself.

FOREST MANAGEMENT STRATEGY

At Collins Pine, forest management revolves around its system of "principal and interest," which essentially means cutting only the growth of the forest while keeping the overall stands of trees stable. The success of the company's forest management depends on accurate inventory and growth estimates. Different methods are used in each of its three major forests to gather the information. The Almanor Forest near Chester, California, has a long history of forest inventory. Permanent growth plots were established in the 1940s and now number over 550. Timber in these plots is managed identically to surrounding timber and the plots are remeasured every ten years, which gives an accurate estimate of timber growth that is then used to determine harvest levels.

The other two locations do not have the same amount of stand information on which to base harvest decisions. At these operations, foresters have relied on their personal knowledge of the condition in stands to set harvest levels. Critics have faulted these methods for their potential lack of accuracy. The company has responded by investing in the development of stand inventory data and systems to gather and manage this information. Although Scientific Certification Systems initially cited Collins Pine's lack of stand inventory data as a significant concern in its certification audit of Kane Hardwood, in the end it agreed with the company foresters that the growth of the stands exceeded harvest levels.

Silviculture in Oregon and California.

Each of Collins Pine's timber-growing operations uses forest management techniques tailored to local species, sites, and climate. The 94,000 acres at Chester and the 80,000 acres at the Lakeview location feature rela-tively moderate terrain dominated by ponderosa pine/Jeffrey pine and white fir. The Lakeview operation produces about 35 percent white fir and 55 percent ponderosa pine/Jeffrey pine, with the remainder a combination of lodgepole pine and incense cedar. Trees grow slowly in this region, and as a result produce high-quality wood. In both places, the company uses predominantly uneven-aged management and natural regeneration. Units are logged in 12-year to 20-year intervals. Trees are selectively cut, particularly those that have begun to decline in vigor. In recent years, the foresters have concentrated on removing white fir from stands, an action recommended by Scientific Certification Systems to improve sustainability. Collins Pine's practices of selective logging and suppressing natural fires on its western lands had created stands that were overstocked with white fir, a situation that increased the risk of fire and discouraged regeneration of the more desirable ponderosa pine/Jeffrey pine trees.

Foresters mark sections of forests to be logged, indicating which trees are to be cut and which are to be left standing. In general, those trees with the

Collins Almanor Forest, Chester, California. In these uneven aged stands a wide range of tree sizes and species is present.

poorest health and/or form are removed, with the exception of those left for wildlife habitat. Foresters sometimes leave particularly old trees standing out of reverence for their age and stature. These management activities create healthy stands that contain trees of a variety of sizes and species. Foresters rarely use herbicides because the partially shaded, managed stands tend to keep the density of undergrowth moderate. Chemicals are used primarily in disturbed forest areas, such as those that need planting or that have been damaged by wildfire.

Collins Pine hardwood forestland, Kane, Pennsylvania. In these even-aged stands, although several species are present, the stem diameters are relatively uniform.

Silviculture in Pennsylvania. The 122,000 acres of Kane woodlands in Pennsylvania's Allegheny Plateau are quite different from Collins Pine's West Coast land holdings. These forests are dominated by black cherry and other hardwoods that have a limited tolerance for shade, so they are managed in essentially even-aged stands. Rotation lengths are relatively long, about 100 years.

To encourage the trees to regenerate naturally, foresters use a shelterwood method of management. They mark the trees in a given area that are to be left standing and those that will be used as seed trees, then cut all the others. This drastically reduces the density of the stand, allowing sunlight to reach the forest floor. Herbicides are used, when necessary, to reduce competing undergrowth. The shelterwood trees are left standing until the seedlings have regenerated to the desired level. The overstory trees, except those designated as wildlife habitat or left for aesthetic reasons, are then removed and full sunlight is allowed to reach the forest floor. The process promotes rapid growth among the hardwoods. If the overstory is not removed at the proper time, competing vegetation will outgrow the young trees and it will take many more years for the area to ade-

quately regenerate. Even then, the density of trees may be too low or the trees poorly formed.

In its Pennsylvania forests, the company logs small (approximately five- to ten-acre) plots that are distributed over a broad area to maintain diversity in tree ages, forest structure, and wildlife habitat in the larger landscape. Harvest levels in the Kane forests are conservative; foresters estimate that they remove only one-third of the annual growth. They base their growth estimates not on plots, but on their own deep knowledge of the land base. The staff, however, is installing a geographic information system to keep records as Scientific Certification Systems recommends. It plans to "cruise" the timber holdings to establish necessary baseline data such as the number of trees and their size, age, and condition. Periodic remeasurements to document growth rates and forest conditions will serve as the basis for future management decisions.

Remedial actions. Much of the land purchased by Collins Pine has a legacy of overcutting, poor regeneration, or simple lack of attention. Some of these lands may require many years of growth and remedial actions to create the diverse tree stands that company managers want. On the western lands, remedial actions involve cutting down unhealthy trees, reducing the number of white fir, and thinning to promote the desired age and size distribution of trees. In the company's eastern holdings, past practices have left foresters with a number of overstocked stands dominated by small-diameter trees. As these stands are thinned and otherwise managed, the diversity of the stands, average tree size, and overall forest health will increase.

Fire and grazing. Fire and grazing are not used as forest management tools to any great extent in the Chester or Kane locations. However, the Chester forestry staff is interested in experimenting with these tools to control vegetation after wildfire and to encourage the regeneration of ponderosa pine/Jeffrey pine. In no case is it company policy to allow grazing near streams, although the company may try grazing cattle in other areas to help keep down the vegetation that prevents the growth of desired tree species.

Business Performance

Collins Pine's performance has declined over the last several years, reflecting market conditions. Table 2 shows the company's financial performance based on Dun and Bradstreet's Business Information Report and includes the average Random Lengths Framing Lumber Composite price for the same period for softwoods only.

Comparing lumber prices with the company's financial performance reveals that Collins Pine's lower financial performance corresponds to dips in lumber prices in those years. In addition to experiencing low prices for lumber, log prices were high during the same period. At the Chester operation, the average value received for lumber dropped by over 15 percent between the 1994 and 1996 fiscal

years, while the cost of buying logs from outside sources increased by 4.5 percent. The situation at the Lakeview operation would have been similar. Wood chip prices, which are extremely cyclical, also were at the lower end of the cycle during the 1995 and 1996 fiscal years. Profits from the Chester operation's co-generation facility dropped from nearly $500,000 in 1994 to break-even in 1996. Hardwood markets experienced a similar cycle to that of softwoods, with low prices realized in the 1995 and 1996 fiscal years.

By a number of measures, the company operates efficiently and competitively when compared with similar operations. A recent study by The Beck Group in Portland, Oregon, compared the Chester operation to 16 other western softwood mills on a variety of performance indicators. The total conversion costs at Chester were lower than average and the production volume per man hour was higher. Sales average for ponderosa pine was far above, while sales averages for Douglas fir, fir/larch, white fir, and hemlock-fir were higher than for any other mill in the study. The evidence indicates that the forestry approach taken by Collins Pine has not negatively impacted its business efficiency or performance.

Sustainable Forestry Outcomes

LESSONS LEARNED

- Foresters have tremendous authority under Collins Pine's land management regime. They dictate to the mills what harvest levels will be, rather than the other way around. This is not the industry norm. It takes individuals committed both to the company's land and to its owners to strike a balance that allows the company to remain profitable without impinging on long-term land management goals.

- Sustained-yield forestry reduces the uncertainty of supply that mill managers typically face, to the extent that company-owned land supplies its own raw material needs. It also encourages long-range planning and long-term investment.

- Collins Pine's commitment to sustainable forestry and sustained communities appears to engender significant goodwill from community members, employees (who may stay with the company for longer than the industry average), and persons and organizations that do not typically support the forest products industry.

UNINTENDED CONSEQUENCES

- Collins Pine was not required to alter the management of its forests in any significant way to achieve certification. The company's long-term commitment to sustainable forestry allowed it to work through the certification process with relative ease. The certifying organization did, however, make a number of reccommendations for potential improvements. Collins Pine responded by increasing supervision of logging crews, increasing documentation of its forest management plans, increasing road maintenance, and making significant investments in forest inventory measurement and tracking systems.

- It could be argued that the company's commitment to sustainable forestry has allowed it

to stay in business when others have failed. Many mills in the Northwest have gone out of business in recent years because they depended on harvests from public lands. As more public forest lands were taken out of production for various reasons, those operations had their supplies disrupted. Through sustainable forestry, Collins Pine is ensured that at least part of its supply will remain stable over the long-run.

Financial Information and Lumber Prices for Collins Pine Softwood Lumber Operations

	Consolidated 1994	Consolidated 1995	Consolidated 1996
Ostrander Resources Co.			
Current Assets	$18. 86m	$20.19m	$19.72m
Current Liabilities	$ 1.94m	$ 2.19m	$ 1.37m
Current Ratio	9.70	9.20	14.4
Working Capital	$16.92m	$18.00m	$18.35m
Other Assets	$20.89m	$21.03m	$23.02m
Sales	$26.12m	$27.03m	$26.9m
Long-Term Liabilities	N/A	$ .011m	$ 1.23m
Net Profit (Loss)	$ 2.81m	$ 1.21m	($0.21m)
Profit as % of Sales	10.7%	4.5%	0.0%
Collins Pine Company			
Current Assets	$32.09m	$39.79m	$36.91m
Current Liabilities	$10.98m	$17.92m	$20.53m
Current Ratio	2.92	N/A	1.79
Working Capital	$21.11m	$21.87m	$16.38m
Other Assets	$38.10m	$39.09m	$78.07m
Sales	$76.87m	$82.13m	$78.10m
Long-Term Liabilities	$ 2.82m	0	0
Net Profit (Loss)	$6.02m	$ 087.m	($2.13m)
Profit as % of Sales	7.8%	1.1%	0.0%
Random Lengths Framing Lumber Composite Price ($/thousand board feet)	$402	$382	$329

Sources: Business Information Reports, Dun & Bradstreet Information Services, and Random Lengths Yardstick, Ramdom Lengths Publications Inc., Eugene, Oregon.

Table 2

- Some Collins Pine personnel claim that by increasing their understanding of the forestry and manufacturing operations, the process of certification has made them better managers. The inventory control requirements to ensure the chain-of-custody for certified products call for precise tracking of volumes, species, and points of origin of wood, which has made inventory systems more efficient and reliable. The process of marketing certified products has helped the company shift from a commodity market orientation toward higher-value specialty products. If generally accepted marketing principles hold, that in itself should enhance Collins Pine's financial performance.

- The industry does not unanimously support Collins Pine's position on certification and sustainable forestry. The company's stance, in some instances, has caused other forest products companies to view it with distrust. They feel that Collins Pine has broken rank with the industry to curry favor with environmentalists.

What Can Be Learned from Collins Pine?

- Market demand for certified products is currently limited and characterized by demographic and geographic segmentation. While future levels of demand cannot be predicted, it is clear that a variety of factors—such as unfavorable consumer perceptions, limited distribution channel development, and limited product availability—dampen that demand. (*Author's note: Conversations with Collins Pine executives during final editing of this manuscript suggest that general demand for certified products is increasing, albeit slowly.*)

- Many different forest management regimes may qualify for some level of sustainability certification. Collins operates even-aged and uneven-aged management schemes and uses herbicides, yet it falls within the definition of a well-managed forest as defined by Scientific Certification Systems.

- Certification provides a number of non-market, image-enhancing benefits that are difficult to account for monetarily. These include public goodwill, credibility with environmental organizations, and interest from the news media.

- Certification may provide competitive advantage as one characteristic of the overall product. In addition, certification can open up new markets and opportunities. Certification will not, however, compensate for a low-quality or poorly marketed product.

- The ownership structure of a company can have a significant influence on sustainability. Publicly owned companies are typically driven by short-term profit. As a privately held company, Collins Pine can more readily forgo short-term returns in favor of long-term objectives. But private forest ownership can also hinder the ability to operate sustainably. The U.S. estate tax structure imposes inheritance taxes that can sometimes be paid only by selling timber lands.

Colonial Craft:
A Rich Niche

CASE STUDY

PREPARED BY:
CATHERINE M. MATER

A Case Study from "The Business of Sustainable Forestry"
A Project of The Sustainable Forestry Working Group

The Sustainable Forestry Working Group

Individuals from the following institutions participated in the preparation of this report.

Environmental Advantage, Inc.

Forest Stewardship Council

The John D. and Catherine T. MacArthur Foundation

Management Institute for Environment and Business

Mater Engineering, Ltd.

Oregon State University
Colleges of Business and Forestry

Pennsylvania State University
School of Forest Resources

University of California at Berkeley
College of Natural Resources

University of Michigan
Corporate Environmental Management Program

Weyerhaeuser Company

The World Bank
Environment Department

World Resources Institute

Colonial Craft:
A Rich Niche

CASE STUDY

PREPARED BY:
CATHERINE M. MATER

A Case Study from "The Business of Sustainable Forestry"
A Project of The Sustainable Forestry Working Group

Contents

Colonial Craft: A Rich Niche

Introduction

The discussion of the certification of forest systems has, until recently, revolved largely around the forests and those landowners who elect to *invest* in certification. However, the response of wood products manufacturers to certification efforts and their willingness to work with certified wood is as important to the acceptance of certification as timber producers' willingness to adopt it. If certification is, as many argue, incentive-based and market-driven, then a system must be in place *beyond the forest* that tracks certified wood flow through to finished products for consumers. Between the forest and the consumer stands the wood product *manufacturer.*

Wood product manufacturers have their own set of criteria for deciding if and when to invest in certification. Some argue that in the present environment investment in certification is premature, since many questions about its economic viability and performance remain unanswered. They ask, for instance:

- Is there documented demand *of sufficient size* for certified wood products in the marketplace to warrant manufacturers to change their traditional business practices?

- Can a wood product manufacturer capture a premium off the sale of certified wood products?

- Is there added market and business advantage to offering certified wood products that is demonstrated in either increased product market share and/or increased company visibility?

- Can a manufacturer be cost competitive in product development if required to separate certified and noncertified wood supply and finished product at the production facility?

- Can certified wood production make a positive difference to the business bottom line?

The business case surrounding Colonial Craft provides some surprising answers.

Company Background

Within the United States, it is difficult to discuss the merits and advantages of independent third-party certified forestry and wood product development without the name Colonial Craft entering the conversation. Since 1992, the company has been a leading thinker and innovator in this arena, and has helped to shape both the industry and the direction of sustainable forest management (SFM) and development of certified wood products throughout the world.

Established in 1965, and headquartered in St. Paul, Minnesota, Colonial Craft is a leading producer of hardwood moulding and millwork products. Today, under the Forest Stewardship Council-approved (FSC) SmartWood certification program offered through the environmental organization Rainforest Alliance, Colonial Craft remains the only moulding and millwork operation to hold chain-of-custody (COC) certification throughout Canada and the United States. The company is the nation's largest producer of window and door grilles and small mouldings used to provide pane designs for all shapes and sizes of windows and glass-exposed areas on all types of the doors. Along with the window and door grilles, the company is also North America's only certified picture frame moulding producer and certified wood products producer serving the barbecue grill manufacturing industry.

TESTING THE TALK—IS IT REAL OR SIMPLY GOOD PR?

Colonial Craft's efforts to advance certification of forestry and forest products have been viewed with not-so-subtle skepticism by some in the industry who believe that the company's actions are based more on achieving noteworthy public relations than good business. They question whether the efforts have actually produced bottom-line results for the company and speculate that without those bottom-line results, Colonial Craft's efforts to market certified wood products are destined to be short-lived.

Colonials Craft's stated corporate values and mission are similar to those used by most companies in the wood products industry today. Environmental stewardship, employee empowerment, community support, workplace safety, service, and quality are standard stated industry values. Certainly no one could suggest that Colonial Craft's mission to be a profitable wood product company by continuing to improve process, product, and service strays far from the standard industry business mold.

Even so, differences in Colonial Craft's business presence and practices are evident. From the front cover of the company brochure *Sustaining Our Living Legacy* quoting an old Hopi Indian legend, "Earth gives life and seeks those who gently walk upon it," to the statement by Colonial Craft President Eric Bloomquist, "We cannot maintain a healthy economy if we neglect the health of our environment," it is clear that the company means to be held accountable to the business of wood products *and* the environment.

The company's commitment to sustaining resources, economies, and communities has spanned at least two decades. In its efforts to sustain economies, the company has generated continued full-time, family-wage employment for citizens of Minnesota and Wisconsin for several decades. In 1984 the company had just 15-20 employees. By the early 1990s, it employed 100 people and by 1996, more than 200.

Colonial Craft has also helped facilitate sustainable communities by providing jobs for physically- and mentally-challenged individuals, and through the part-time employment today of more than seventy autistic individuals.

In addition the company has established a track record in sustaining forest resources through efforts that have included:

- Converting wood waste into beneficial products such as animal bedding for local farmers.

- Allocating employee manpower and company funds to facilitate the planting of more than 30,000 tree seedlings in Wisconsin and Minnesota.

- Participating in environmental education curriculums for K-12 to teach children the importance of wood products and the need to sustain tree growth.

But it was Colonial Craft's venture into wood products certification that illustrated its willingness to assume a high-risk pioneering role in furthering SFM practices through "green" certification at the manufacturer level.

EARLY INTEREST IN CERTIFICATION

In 1992, Colonial Craft became interested in independent third-party certification efforts as a result of listening to a presentation from Dr. Jim Bowyer, head of the forest products department at the University of Minnesota and president of the International Forest Products Society. Dr. Bowyer provided data that projected that the demand for wood and wood products would likely double over the next fifty years, as would the world's population. Dr. Bowyer's comments, along with concerns over the health of U.S. national forest systems, prompted Colonial Craft president Eric Bloomquist to help facilitate the establishment and operation of the international FSC. Today, the FSC serves as the certifier of the four key independent third-party certification programs in the world: The Rainforest Alliance SmartWood program in Vermont and New York, Scientific Certification Systems (SCS) in California, and the Soils Association and SGS both located in the United Kingdom.

Since Colonial Craft does not own its own forestlands and must rely on others to supply its production facilities with lumber, the company needed to document that its vendors were adhering to certification criteria to achieve SmartWood certification. To accomplish this, Colonial Craft conducted an intensive survey of its lumber suppliers. Suppliers were asked to verify where their logs came from, who harvested the logs, how the logs were harvested, and whether the forests from where the logs were harvested were sustainably managed. In 1994, based on the survey results and its capability to separate certified wood from noncertified wood in its operations, Colonial Craft became the first moulding and millwork producer in North America to receive green certification.

In May 1995, the company received the first load of certified hard maple at its production site. This was processed into and sold as certified wood product in June 1995. Colonial Craft has made a commitment to work only with certified wood by the year 1999.

Currently, with more demand in some product areas than can be supplied due to a lack of certified forests in the United States, Colonial Craft offers a full line of 100% certified picture frames mouldings (called Warm Woods), and over 17% of its window and door grille volume annually is produced from certified wood.

Recognizing the importance of the widespread adoption of the International Organization for Standardization (ISO) environmental standards and SmartWood's environmental performance requirements in the international marketplace, Colonial Craft began the ISO 9000 registration process in 1994. That process is expected to be completed by fall 1997 and ISO 9000 certification issued in early 1998.

A History of Smart Growth

Like all wood products companies Colonial Craft must continually deal with market fluctuations, production efficiencies, and strategies to offset increased competition. In this environment the company's history of growth in sustainable forest product development stems from five general attributes:

- Leadership at the top.

- Corporate and company attitude.

- Willingness to engage in intelligent risk.

- Consistent quality performance in business.

- Long-term vision development and follow-through.

STRATEGY BRINGS SUCCESS

The specific factors that have been significant to the growth of Colonial Craft are discussed below.

1) Strategic connection with high-performance partners that have a presence in moving traditional and environmentally-innovative products into the domestic and international marketplace.

Although currently the nation's leading supplier of wood window and door grilles, Colonial Craft did not occupy that position until after 1982, when it formed a business relationship with Andersen Corporation, makers of well-known Andersen Windows and Andersen Doors. Colonial Craft's growth exploded when this relationship was forged. Grille sales in 1982 prior to the Andersen partnership were less than $400,000 annually. By 1988, annual sales were running about $2.5 million; by 1990, they exceeded the $5 million mark, and by 1997 had reached $14 million.

From a strategic standpoint, the affiliation with Andersen has been of great benefit to Colonial Craft in several areas. Andersen Corporation was, and continues to be, a major player in the production and sales of wood windows and doors for U.S. and Canadian markets. *Professional Builder* magazine's 1996 Brand Use study documented Andersen's position as a top brand product used by a substantial percentage of U.S. builders. According to survey results, over 50% of builders install Andersen windows in their projects.

That same study also documented Andersen's good

performance in the category of "exclusive brand uses" by builders. Survey results showed that nearly 12% of U.S. builders interviewed reported using the Andersen line of Frenchwood hinged patio doors exclusively. Results also found that 14.3% stated exclusive Andersen product use for Frenchwood and Perma-Shield Gliding patio doors; 18.8% in Andersen wood windows; and 35.1% in Andersen clad windows.

The combination of Colonial Craft and Andersen proved logical in another arena—concern for producing environmentally-appropriate wood products. A producer of over 100 million pounds of sawdust a year, Andersen had been researching waste disposal options since the 1970s. And with the production of vinyl products, the company had recycled some of its vinyl waste into nailing-flanges, but still had large amounts left over. During the 1970s, Andersen discovered a way to combine wood fibers and PVC into a material that achieved similar structural properties as wood. The company made its first prototype composite window of wood and PVC waste in 1972 and installed it in test homes in Minnesota. But wood seemed inexpensive and plentiful at that time, so the project was shelved. Twenty years later, in 1991, when the wood products industry was experiencing serious problems in both wood supply and quality, Andersen revisited those test windows to see how they held up.

They were impressed with what they saw, so in 1991 Andersen introduced a new composite material called Fibrex, which combined 40% wood weight with 60% PVC weight. New independent testing to evaluate Fibrex's performance in durability, resistance to decay, and other structural properties showed that the material's performance greatly exceeded Andersen's expectations. The research showed that Fibrex provided stiffness, resistance to heat, thermal expansion superior to vinyl, and insulating properties that put it on par with pine and vinyl. It also found that Fibrex was impervious to decay.

Anderson produced a new window product line called Renewal made from Fibrex in 1992, the same year that Colonial Craft initiated the SmartWood certification process. By 1994, Colonial Craft was supplying certified wood grilles to Andersen's Renewal line, but not by request from Andersen. Independently, Bloomquist made the commitment that all Colonial Craft wood grilles going into the Renewal line would be made from certified wood.

During 1995, Andersen Corporation received independent, third-party certification through the environmental organization Green Seal for its product performance in energy efficiency. In granting the certification to Andersen, Green Seal stated that the door and window products produced by the company "are among the most energy-efficient doors and windows available, offering considerable savings in heating and cooling costs, and reducing air and water pollution associated with extracting and burning fossils fuels."

Although not active in the promotion of certified grilles for its Renewal windows, Andersen does include as part of its product display the following statement:

> Andersen window and patio door grilles are green certified by the SmartWood program to provide you with valid assurance that you are purchasing products made from wood that has been harvested and processed in a sustainable manner. Thus, the green certification process helps to ensure we have healthy forests for future generations.

2) Management style, employee awareness, and commitment to company's vision, direction, and goals.

The willingness of Colonial Craft to engage in intelligent risk is most clearly evidenced in its management. Employees are empowered to help shape the future of the operation, as well as the day-to-day functioning of the company. Employees from top to bottom appear to share the company's commitment toward producing environmentally-appropriate products, and the ongoing efforts to produce a safe working place for all employees. The following actions illustrate this commitment.

- Company employees volunteer as teachers in local school systems using a curriculum designed to teach K-12 students the importance of wood products in their lives and the need to manage and use these wood resources sustainably.

- Company employees develop safety teams in the production operations. These teams conduct monthly inspections to ensure that safety conditions are being met throughout the operation. This includes the president's office, which reportedly received employee safety team mandates to remove a multioutlet electrical plug from which a mass of extension cords were strung, and relocate the office phone so that it did not require cord extension to a table across the room. The president complied.

- The willingness to involve employees begins with the attitude of top management and appears to be producing positive results for Colonial Craft's environment.

3) Visibility as a pioneer in the environmental and industry communities.

Colonial Crafts' efforts in sustainable development and resources are visible in both industry and environmental camps, which normally separate themselves as much as possible. Within the last three years, the company has been asked to present information on sustainably-managed forest resources and certified wood product development to traditional wood products industry organizations including the American Wood Moulding and Millwork Producers Association, the National Hardwood Lumber Association, the American Hardwood Manufacturers Association, Custom Woodworking Business magazine, and the International Forestry Division of the USDA Forest Service.

The company has played an equally visible role on notable environmental platforms including the First North American Conference on Trade and Sustainable Products sponsored by the Smithsonian Institute in Washington, D.C., the FSC Founding Assembly, the Forest Trust, and the Forest Policy Center in Washington, D.C.

4) Production, in part, is based on flexible manufacturing.

One of the benefits Colonial Craft appears to bring to its customers is an ability to quickly adapt to varying design modifications in product development. This has been particularly profitable for the company in its relationship with Andersen Corporation. Andersen has sometimes requested product design changes that need final production results within a week. The ability to cost-effectively and efficiently conduct flexible manufacturing in a processing plant takes careful thought and attention to product quality and cost control.

5) Production is focused on a few products.

This decision in product manufacturing can prove to be both an asset and a liability. The asset accrues from being able to focus production and service to a few products and product buyer types, and in adapting production lines for flexible manufacturing options. The liabilities arise from increased competition in the marketplace that can leave a focused product producer scrambling to identify and develop new products and nurture new markets. Additional risk lies in assuming that existing customers will maintain market share. Colonial Craft's decision to focus a large portion of the company's product development to Andersen Corporation is beneficial so long as Andersen retains its market share and is able to supply business to Colonial Craft.

One way that Colonial Craft has lowered the risks of its product focus on Andersen is to identify product opportunities that solve processing bottlenecks while addressing sustainability concerns. The picture frame moulding line developed at Colonial Craft resulted from trying to figure out what to do with the thin longer wood waste pieces left after lumber ripsawing. Picture frame moulding was the answer.

Financial Performance

Colonial Craft has demonstrated strong financial growth during the last decade. As Figure 1 illustrates, a review of the last five years of financial performance shows that Colonial Craft has continued to experience double digit growth in annual sales, with the exception of a 2.3% annual growth in sales experienced for 1995.

Colonial Craft Financial Performance

Year	Annual Sales $	% Increase from Previous Year
1991	$12,641,000	18.3%
1992	$16,641,000	31.6%
1993	$20,219,000	21.5%
1994	$23,492,000	16.2%
1995	$24,030,000	2.3%

Source: Colonial Craft

Figure 1

Wood and Wood Products magazine, in reporting the results of its annual survey of the top 100 value-added wood product producers in North America, has listed Colonial Craft in the elite ranking for the last five years. Approximately 10% of the total 100 companies making the list for 1995 were companies that also manufacture moulding and millwork products.

INTERPRETING SURVEY RESULTS

Comparing financial and production performance of Colonial Craft to the other similar product producers making the list leads to the following conclusions.

- Colonial Craft was the second largest moulding and millwork operation in both annual sales and total number of employees making the top 100 list of performers.

- Between 1994 and 1995, the company experienced the smallest percentage growth rate in annual sales of all ten product competitors making the list. The averaged percentage increase in annual sales during that period of time for the nine remaining companies was 30.8% *Wood and Wood Products* notes the following in its survey results introduction for 1995:

  > The combined sales volume for those making the list cracked $1 billion—nearly 35% greater than the amount the top 100 companies listed for 1994 amassed. However, the 14.1% average sales growth posted for 1995's group was less than half of what 1994's group accomplished, and 10% below the previous record low set in 1990.

- Colonial Craft appears competitive in recognizing industry standards of practice which are traditionally viewed as good business strategy. Colonial Craft attributes its success in production to new equipment purchases, improving its quality of production, and listening and implementing employee suggestions in their management and facility operations. Colonial Craft competitors ranked their reasons for success as follows:

  New equipment purchase......................................80%

  Increase internal and external communications......40%

  Increased plant size...40%

  New product development...................................20%

  Broadened geographic market area........................20%

  Dedicated largest percentage of product production to a specific customer..........................20%

• Competitors that indicated equipment purchasing as a success indicator identified optimizing systems and new moulder systems as the manufacturer's top new investment choices. Colonial Craft invested in optimizing systems for its operations.

All of this points to a company that understands the business of successfully producing wood products, certified or not. And Bloomquist is often the first to underscore the fact that certified status of a product will not substitute for consistent product quality, customer service, intelligent product design, marketing acumen, and good overall business management.

CERTIFICATION A GROWING IMPORTANCE TO FINANCIAL PERFORMANCE

Even though Colonial Craft has had the ability to process certified wood products since 1995, certified wood products do not yet make up a substantial portion of its product offerings. The ongoing lack of certified wood supply in the species and grades required to meet Colonial's production needs is largely responsible for the lack of certified products.

In 1996, about 8% of the total annual volume of the 10 million board feet of wood processed by Colonial Craft came from certified forests. The company expected to increase that amount to 13% of total annual volume by the end of 1997. Currently, Colonial Craft pays a slight premium on the majority of that wood supplied to its operations. Andersen's Renewal product line made from Fibrex which uses certified wood grilles supplied by Colonial Craft was expected to expand during 1997 by opening seven new retail stores.

Of Colonial Craft's total annual sales of over $24 million for 1995, less than 1% resulted from certified wood sales that had premiums attached. Colonial Craft, for example, received a premium for certified wood sales on the sale of approximately 182,000 board feet of certified basswood product used in the manufacture BBQ grill parts. The completed BBQ units were then sold into the U.K.

market. Certified wood sales in this product category resulted in about $200,000 in sales for Colonial Craft during 1996.

The BBQ manufacturer purchasing this product identified Colonial Craft by the environmental and industry visibility the company had received in certified wood products. Originally attempting, without success, to sell noncertified BBQ units in the European markets, the BBQ manufacturer was essentially forced into locating and obtaining certified wood components for its BBQ units. Colonial Craft was able to secure a 15% premium on these certified BBQ component parts.

As a result of supplying grill parts for certified line offered by the BBQ manufacturer, in 1997 Colonial Craft was negotiating a contract to produce the noncertified product line for the same company. Although Colonial Craft produces a quality product in required volumes on time, it is clear that this new business opportunity came as a direct result of Colonial Craft's visibility as a company that produces certified product. The added business is expected to increase annual sales well over $1 million, reflecting more than a 2 million board feet added annual production output.

The architectural moulding line manufactured by Colonial Craft was introduced on the shelves of Home Depot in 1996. Although difficult to verify, Colonial Craft's entry into the Home Depot market may have, in part, been due to the visibility of its certification efforts. Home Depot has been a leader in moving certified wood products into the North American do-it-yourself (DIY) markets.

Other certified wood products such as Colonial Crafts's window and door grilles and the picture frame mouldings are sold as certified product but have no premium attached to the sales.

The company maintains separate inventory space in its operations for certified and noncertified wood. When producing for customers requiring certified wood, separate production of "menu" runs is conducted, ensuring that no noncertified wood source gets mixed in with certified wood source. The added cost to the sales of the certified product as a result of this inventory and process separation is negligible.

FUTURE STRATEGIES

Colonial Craft continues to look for new growth opportunities that will increase business and the quality of the environment. Five key opportunity areas were identified during the development of this case which reflect Colonial Craft's continued commitment to increased production efficiencies, new product offerings, identification of new markets and new product distribution channels, as well as a commitment to certified products.

Increased Production Efficiency

Colonial Craft has invested in color optimization systems that help it to more efficiently and effectively separate material based on grade/appearance. The optimization system was installed in midsummer of 1996, with the first generated readings on yield produced for the month of August. The computer-generated results are summarized in Figure 2.

Colonial Craft Production at Luck Plant for August, 1996

Species	Volume (lineal feet)	Average % Yield	Volume Waste (lineal feet)
Certified Basswood	78,984	93%	5,528
Hard Maple	390,670	90%	39,067
Red Oak	337,837	97%	10,135
Total	**807,491**	**93%**	**54,730**

Source: Mater Engineering

Figure 2

From the data and on-site observation, several conclusions are clear:

- The loss factor for the three key species run during that month was approximately 7%—an exceptionally low loss factor for many value-added wood processing operations, and not necessarily representative of the company's average yield.

- For the month of August, approximately 55,000 lineal feet of solid wood material went to the waste bins destined for the chipper or set aside for giveaway wood scraps to local residents.

- For this case, waste bins were spot checked to evaluate piece size and shape. From visual review, it appeared as if about 80% of the waste material had color variation as its only defect and was six inches or longer in length.

If Colonial Craft were to invest in scrap recovery equipment that can process wood scraps down to six inches in length, good quality fingerjoint blocks or cabinet parts could be produced. With fingerjoint blocks selling at approximately $500 per million board feet, Colonial Craft might realize an annual increase in gross sales of up to ten times the gross sales that would be generated through selling chips of the same volume.

In further evaluating options to convert wood waste into value-added product, Colonial Craft estimates that about 50% of the total annual volume of raw wood that comes into its operations for value-added production ends up as waste material. Currently, the company generates approximately 8,000 tons of sawdust annually from its two processing operations. Evaluating opportunities to convert that wood waste into more valuable wood component parts might be an immediate goal for the company. Aside from the environmental benefits to gaining as much value-added product from the wood resources made available for production (certified or not), as noted in the increase yield discussion above, the bottom-line benefits could be significant.

With the high level of configured wood and wood that might be converted to fingerjointed material that Colonial Craft has to work with, looking for outsourcing options to produced high-value products should be evaluated. Examples might include outsourcing fingerjointed material for veneer overlay application or lamination. Outsourcing of fingerjoint blocks into fingerjointed lumber might be another consideration for the company.

Commitment to New Product Offerings

New product offerings that capture the use of custom grade material and increase new products and product lines—especially those that do not rely on the cyclical housing industry and take advantage of development opportunities in nontraditional products—are also on Colonial Craft's radar screen.

According to Colonial Craft representatives, about 30-40% of the hard maple it processes has color variation rather than the preferred clear look desired by the company's existing customers. Research conducted by Mater Engineering during 1996 to evaluate market demand for certified custom grade wood found that "brown-stain" maple has excellent market potential for certified wood flooring in European markets. Understanding these unique opportunities for developing custom grades may prove to be a profitable option for the company and illustrate a higher and better use of the resources harvested from sustainable forests.

Colonial Craft has already moved into some new product offerings that do not rely on the cyclical housing industry with the development of its component parts for BBQ grills and picture frame moulding. With the introduction of Fibrex products, Colonial Craft will opt to evaluate further opportunities in product development using this nontraditional material to manufacture window and door grilles.

Identifying New Markets

While Colonial Craft sells the value-added wood products to companies that have fairly extensive market distribution reach, the company itself provides manufactured product to local and regional customers. Identifying well-researched opportunities to expand its market base into other geographic areas would be an activity worth pursuing for Colonial Craft.

With the success of the Andersen Corporation partnership, it is understandable why Colonial Craft's future efforts to expand its customer base might be focused on finding a few good large customers rather than pursuing many small customers. Both strategies have advantages and disadvantages, but Colonial Craft has an excellent track record of expanding its customer base and seems well positioned to carry out this effort.

Evaluating New Product Distribution Channels

Wood product producers working with certified wood in an era of still limited supply face a set of challenges that can dramatically affect the movement of certified wood flow. Lumber brokers and wholesalers who often serve as the middlemen in the selling and purchasing of lumber between the forest landowner selling the logs, to the primary sawmill, and from the sawmill to the value-added producer may try to "lock-up" certified wood supply. Suppliers may be reluctant to sell too large a percentage of their certified resource to one buyer and may do so only they if are able to increase the base price plus any premium they may charge for certified wood. Added to the middleman costs the buyer must assume when purchasing the lumber, the price could then become noncompetitive and the wood flow would stop. Colonial Craft is well aware of these negotiated cost factors and distribution channel issues that need constant attention while certified wood supply is limited.

Customers that rely on maintained and/or increased levels of new building construction have product needs that rise and fall with the number of new building permits issued. As a result, suppliers to these customers follow the same path. Colonial Craft has identified diversification away from the new housing construction industry into other product distribution channels as an intelligent way to even the flow of product demand and supply over time.

Commitment to Producing Certified Products

Colonial Craft intends to continue to be at the forefront of increasing the supply and use of certified wood products throughout the world. The company has stated its intent to purchase only certified wood for its own production by 1999.

These ongoing efforts will maintain its high visibility to customers and consumers alike through printed brochures and videos created by Colonial Craft that highlight the benefits of certified wood and wood products.

Social Implications

Colonial Craft's pioneering efforts in sustainable forestry and forest products development have produced measurable positive impacts for the company on several levels. The company's leadership in economic, environmental, and community efforts has been recognized by the wood products industry as well as the public. Honors and recognition received from such divergent groups as the United Church of Christ's Corporate Social Responsibility Committee, *Wood and Wood Products* magazine, the Wisconsin Environmental Work Group, and the Wisconsin Manufacturers and Commerce for its efforts in SFM and certification are an illustration of the impact Colonial Craft is having in this new arena for the wood products industry.

What may be more important for the company, however, is the influence it has had in helping to bring sustainable forestry to the management of publicly-owned forests in the United States. Colonial's efforts to highlight the advantages and opportunities in certification of forests, coupled with the need for additional certified wood to meet its own customer demand, was the catalyst for the creation of a groundbreaking partnership in January 1997. Minnesota State foresters and forest managers for Aitkin County in Minnesota joined forces to participate in the nation's first pilot project involving an FSC-certification assessment of state and county forestlands, which encompassed more than 600,000 acres of state and county forestland. The independent evaluation conducted by SmartWood was completed in 1997. It recommended that the lands be officially certified. State and county forestry officials have indicated they hope the end result to be official certification of the forestlands included in the pilot project assessment.

Lessons Learned

The experience of Colonial Craft provides four important guideposts for wood product manufacturers evaluating the business of green certification.

1) Forestry and forest products certification is not yet well established, but it is clearly growing. As such, the financial success of a manufacturing operation is not predicated on offering certified wood products to consumer markets. But, as Colonial Craft demonstrates, offering certified products can strengthen the bottom line.

Offering certification as a product differential does not substitute for meeting the other requirements for obtaining, maintaining, and growing a solid customer base. Substituting consistency in quality production, customer service, on-time delivery and competitive-range pricing for certification status is not a viable option. If, however, all other solid business attributes are in place, certification can serve as a clear product differentiation for the wood product producer. Simply put, certification may be one of many tools used in the practice of SFM.

2) Although some products currently appear to command premiums in the market as certified product offerings, the long-term expectation of capturing these premiums may prove secondary to the more important long-term issue of achieving consistent cost-plus-normal-profit recovery matched with increased market share for any products produced. Further, as Colonial Craft's experience illustrates, companies that decide to invest in certification may find that the increased general business exposure resulting from offering certified products can produce new business development in even noncertified product arenas. This is especially important when a limited supply of certified resource constrains the ability to move certified wood products into the market.

3) In contrast to conventional wisdom, the cost of working with certified wood appears less a constraint to the wood product manufacturer than might be expected. It is more important for the producer to understand that the demand for targeted wood products, especially in the European markets, currently exceeds supply.

4) Finally, as Colonial Craft has demonstrated, one company can have significant influence on the development of sustainable forestry and certification. Colonial Craft's leadership in the private sector has dramatically increased the awareness of and opportunity for the evaluation of SFM practices for use on public forestlands. Public forestlands certification pilot projects could significantly help to increase the future growth of independent, third-party certification of forests and forest products.

J Sainsbury plc and The Home Depot: Retailers' Impact on Sustainability

PREPARED BY:

JAMES MCALEXANDER

ERIC HANSEN

A Case Study from "The Business of Sustainable Forestry"
A Project of The Sustainable Forestry Working Group

The Sustainable Forestry Working Group

Individuals from the following institutions participated in the preparation of this report.

Environmental Advantage, Inc.

Forest Stewardship Council

The John D. and Catherine T. MacArthur Foundation

Management Institute for Environment and Business

Mater Engineering, Ltd.

Oregon State University
Colleges of Business and Forestry

Pennsylvania State University
School of Forest Resources

University of California at Berkeley
College of Natural Resources

University of Michigan
Corporate Environmental Management Program

Weyerhaeuser Company

The World Bank
Environment Department

World Resources Institute

J Sainsbury plc and The Home Depot: Retailers' Impact on Sustainability

CASE STUDY

PREPARED BY:

JAMES MCALEXANDER

ERIC HANSEN

A Case Study from "The Business of Sustainable Forestry"
A Project of The Sustainable Forestry Working Group

Contents

J Sainsbury plc and The Home Depot: Retailers' Impact on Sustainability

Introduction

Sustainable forest practices have become a pivotal issue within the forest products industry for a variety of reasons ranging from a broad sense of environmental awareness and responsibility to a more self-interested concern for maintaining the economic productivity of forests. Whether the forest products industry widely adopts sustainable practices, however, depends on their long-term economic viability. The development of broad demand and markets for sustainably produced wood products will be a key component of that economic viability.

The efforts of retailers J Sainsbury plc (JS) in the United Kingdom and The Home Depot (HD) in the United States to stock their shelves with products drawn from well-managed forests place them at the forefront of this global issue. These large, respected retailers are uniquely positioned to merchandise sustainable forest products to the mass market and by so doing, lend credibility to these products and demonstrate the importance of the issue to the industry and the public. The buying power of these two companies is of such a magnitude that their purchasing practices can exert a strong influence on the forest products' industry worldwide. However, the use of such buying power as the panacea for ensuring responsible forestry greatly oversimplifies the complex issues involved in changing the management techniques of the timber industry.

The initial programs of these two retailers and that of the 1995+ Group, a group of major wood products buyers in the United Kingdom, indicate that retailers and large wood products buyers will be instrumental in cultivating consumer awareness of certified products, as well as pulling suppliers toward certification and sustainable forest practices. A comparison of the activities of the two companies, which operate in different competitive, cultural, and political environments, identifies a variety of salient issues that will influence whether or not their initial efforts to market certified products are successful. The ability of these retailers to obtain and merchandise sustainable forest products is a barometer for the future direction of sustainable forestry.

The material presented is drawn from a number of different sources and research methods. In-depth interviews with senior executives, wood products buyers, marketers, environmental managers, store managers, and retail employees from both companies were the primary sources of data. These interviews were balanced by discussions with the 1995+ Group, competing firms, and suppliers, visits to stores of both companies in different regions while posing as consumers, and supplemented with a review of published materials.

Environmental Management

Both HD and JS are widely recognized as socially responsible companies. Both firms consider this reputation and associated goodwill to be instrumental to their long-term strategic positions. JS's management prides itself on a long tradition of providing customers with products of high quality and good repute, and they work hard to maintain this reputation (see box). Management speaks of having cultivated the trust of its customers, and that shoppers at JS believe that the company will "do the right thing." HD has an ethic similar to that of JS, but not the long history with its customers. In 1996 HD managers began to revise the corporate mission statement to reflect that ethic by including the notion of being the "most trusted retailer."

J Sainsbury plc: Service for the Affluent Homeowner

With sales in 1995 approaching $22 billion, JS is among Europe's largest retailers. The company holds the Sainsbury supermarkets in the United Kingdom with about 12% of the grocery market; Savacenter, a hypermarket store group; and Homebase, a DIY home improvement outlet. The Homebase is similar to HD in that it claims a 10% share of the U.K. market, a proportion close to that of HD, and nearly $2 billion in sales.

A typical Homebase store carries more than 20,000 items and has about 3,500 square meters of floor space. The affluent DIY customers is its principal target customer. Unlike HD's warehouse stores, Homebase outlets, consistent with Sainsbury's supermarkets, are upscale stores that are well appointed. Management is especially attentive to merchandising concerns, such as product facing and displays. JS broadly integrates its family of stores by managing with similar principles and strategic commitments. Overlap among the products carried by the JS subsidiaries gives the company purchase economies and clout with suppliers. Since Homebase is so well integrated in the JS family, it reflects the efforts of the parent company.

JS's decision making is highly centralized, with some 200 centrally located buyers purchasing for the supermarket group and 50 buying for the DIY outlets. To generate customer loyalty, JS is committed to marketing its own brands. Unlike the private label strategies used in the United States, however, JS claims that its own brand name merchandise is of equal or superior quality than manufacturers' brands. The extensive use of its own label gives JS additional leverage with suppliers. At the same time, though, it exposes the company to additional responsibility and liability for the safety and quality of these products. The significant staff and resources assigned to the Scientific Services Divisions (SSD) reflect that added responsibility.

Issues of social responsibility are wide reaching for both retailers. The JS family foundations, for example, are active in numerous philanthropic activities. The HD has provided over $1.5 million to nonprofit environmental organizations over the last five years. The Home Depot also participates in many local and national programs such as the charitable organization Habitat for Humanity, which facilitates home ownership for families of limited financial means. In recent years, as environmental issues have climbed higher on the public agenda, these companies have more actively incorporated "environmental responsibility" into their management decision making and corporate actions.

J SAINSBURY PLC

The primary responsibility for JS's environmental programs is housed in the Scientific Services Division (SSD). The SSD has a significant staff that is responsible for evaluating and agreeing upon standards for the quality, safety, legality, and performance of products sold by JS. Since more than 60% of the products sold carry its brand name, issues of responsibility and liability receive a high level of scrutiny and scientific evaluation. Because of its evaluative role, the SSD is linked closely to the company's centralized buyers, who look to SSD to evaluate products and approve suppliers. The

SSD has the power to refuse or deselect suppliers, a sharp contrast to the advisory role of HD's Director of Environmental Marketing.

The SSD has mounted a number of environmental management initiatives: It has, for examples, ferreted out products that have been developed using animal testing, investigated fishing practices, and examined the practices of its suppliers of wood products. JS recently published its first annual environmental report which outlines the company's policy on the environment and the management of environmental issues. This 1995-96 report highlights such environmental achievements as creating a Best Practice Guide on Ground & Surface Water Protection, reducing energy consumption, and

gaining a commitment from 90% of its wood products suppliers to use responsibly-managed wood sources. The report also lists future actions and targets, which include eliminating chlorofluorocarbon (CFC) refrigerants in new and existing stores, and stocking five hundred wood products that have qualified for third-party certification by 1998.

THE HOME DEPOT

As the market leader in the $130+ billion U.S. home improvement market, HD is uniquely positioned to increase the awareness of customers, suppliers, and competitors about environmental and sustainability issues (see box). HD developed a formal environmental program that began after Earth

---

The Home Depot: A Warehouse for the Do-it-Yourself Consumer

HD, the world's largest home improvement retailer, has over 580 stores and over 118,000 employees. Established in 1978 in Atlanta, Georgia, the company has experienced phenomenal growth, reaching sales of $19.5 billion in fiscal 1996. Management expects to have 1100 stores operating by the year 2000. Currently, HD operates in the United States and Canada, but will soon enter Chile. Entry into other international markets is likely.

The typical HD store provides a "warehouse" environment with spartan facilities, wide aisles, merchandise that is often displayed on steel shelves, and the occasional motorized forklift working its way through the aisle. The average store stocks about 45,000 items within a floor space of about 9,750 square meters inside, and 2,300 square meters of outside garden space. The HD customer is primarily the do-it-

yourself (DIY) homeowner, but professional contractors are becoming a more important customer segment. In 1995, HD held about 12% of the $134 billion U.S. home improvement market. In 1996 its share climbed to 14%.

HD practices decentralized management, especially for its buying function. Regional buyers and merchandisers from six regions make product mix and inventory decisions. Consequently, while product categories are similar from store to store, suppliers can differ. The sheer sales volume of HD stores enables them to exert considerable influence on suppliers, even dictating such important aspects of the buyer-suppliers relationship as delivery terms and product labeling. HD's power over suppliers, though, is tempered by the need to maintain adequate inventory in so large a distribution channel.

Day 1990, about the same time it hired Mark Eisen* as Director of Environmental Marketing. Soon thereafter came the publication of the company's environmental principles in 1991. The later adoption of substantially the same principles by the National Retail Hardware Association and Home Center Institute is an indication of the influence HD wields in the industry.

Eisen's perspective is reflected in a comment attributed to him in HD's 1994 Social Responsibility Report, "Our concern for the environment is part of our ongoing, everyday business and planning. Improving the environment is the ultimate home improvement we can make."

The Environmental Greenprint® (Appendix A), first published in 1992, is another early component of the environmental program. It documents methods for consumers to make their homes more environmentally friendly through such things as reduced energy consumption, and improved indoor air quality. In 1993, HD placed permanent banners on the front wall of each retail outlet documenting a commitment to continually improve environmental programs (see Figure 1). Dennis Ryan, senior vice president of merchandising, described the degree to which a philosophy of social responsibility is integrated into the corporate culture:

Our founding fathers...care about their fellow associates, they care about the communities... you find that it rubs off on the company. We know that the right thing to do is the right thing to do... whether it's a business decision for shareholders, the care of our associates, for our suppliers, or for our communities...

As Director of Environmental Marketing, Eisen manages a broad range of environmental issues for HD. His duties include auditing the environmental claims of the suppliers in the decentralized supply chain, educating and sensitizing management and corporate buyers about environmental issues, and cultivating support and visibility for environmental issues among HD's external stakeholders. Eisen is attached to the company's community affairs division, however, rather than his original appointment in marketing. This affiliation suggests that HD views success in the environmental arena as dependent upon building broad support among business and the public.

Eisen has proposed that HD implement a formal Environmental Management System (EMS) to further institutionalize environmental management. An EMS would more broadly and formally integrate environmental issues into managerial decision making, especially purchasing decisions. As of late 1996, HD had approved hiring a consultant to make the business case for implementing an EMS. The consultant's recommendation will then be brought before the company's senior executives for assessment.

Figure 1

* Eisen left the company in 1997 and is now retained by HD as an independent consultant. In June of 1997, Kimberly Woodbury was hired to continue the development of HDs Environmental Program.

Sustainable Forestry and Retailing

Buyers for both JS and HD indicate that they perceive an inherent difficulty in personally assessing forest management practices. They generally do not have access to adequate information and are not qualified to assess the sustainability of a particular producer's practices. Nor do they have the time to complete a careful on-site investigation. Moreover, buyers and other managers rarely agree on the meaning of "responsible supply." To some buyers, responsible forestry requires avoiding those products made from wood harvested through clearcuts. Others view sustainable practices as good crop management with emphasis upon sustaining a yield for future harvests. A product manager at HD explained this view of lumber as a marketable crop:

> Because of human demands...if you don't manage the production of lumber, then you're going to lose it. Sustainability is just the management of the crop. There's a lot of poor management. They have to manage the ground and they have to manage everything that affects the growth of that crop... so you don't end up with situations like what we had in the dust bowl.

Additionally, the fact that there is still much debate within the scientific and environmental communities about the meaning of sustainable forestry makes it even more difficult for the retailers.

RETAILERS SUPPORT THE FOREST STEWARDSHIP COUNCIL

Since the retailers find it difficult to assess the practices of suppliers themselves, both HD and JS have turned to certifying organizations such as those accredited by the Forest Stewardship Council (FSC). The nonprofit FSC approves organizations to certify forests as meeting sustainability standards. Accordingly, FSC-sanctioned certification indicates that a particular product comes from a forest that meets environmental, social, and economic standards. For both companies, certification simplifies the buying process by removing from them the responsibility for evaluating forests and forest practices. Neither company claims expertise in forestry or a desire to develop it. As William Martin of Sainsbury's, the senior manager responsible for coordinating certification, explained, certification has great value: "If FSC certification tells me, that it is a well managed forest, I am happy to honor that. I know if there are any questions I can refer them to the FSC or to the certifier."

Sainsbury is also heavily involved with the U.K.-based 1995+ Group. This buyers group, formed by the World Wide Fund for Nature (WWF) in 1991, operates under an agenda to move members towards the exclusive purchase of third-party certified wood products. The five main requirements for membership include supporting independent systems of forest certification, buying a substantial and increasing volume of wood fiber from certified sources, and naming a senior manager responsible for implementing the program (see Figure 2). Group members, now numbering eighty, have publicly committed to purchasing wood and wood-based products from "well-managed" forests. Prominent companies in the group include B&Q plc, Boots the Chemists, WH Smith Retail Ltd, and Tesco plc.

Many of these companies, including JS, are committed to buying 100% certified wood products by the year 2000. Since the Group accounts for between 11% and 25% of the total U.K. consumption of wood and wood-based products (according to Delphi International Ltd., a market research firm, and the World Wildlife Fund), the importance of this commitment to the wood products industry cannot be overemphasized. The U.K. market consumed approximately 59 million cubic meters of wood and wood-based products in 1995, about 15% of total European consumption. Delphi International estimates that demand by the Group will grow around 15% to 20% per year.

WWF 1995+ Group Membership Requirements

Commitments

1. Members are committed to supporting internationally applicable, independent systems of forest certification, based on standards which take full account of environmental, ecological, biodiversity, social and economic needs: such as those promoted by the pioneering work of the Forest Stewardship Council (FSC).

2. Members are committed to phasing-in the purchase of forest products from well-managed forests as verified by independent certifiers accredited to certify systems as described in 1 (above), for example Scientific Certification Systems (SCS), Rainforest Alliance, SGS, Soil Association or other independent certifiers when appropriate.

Target

3. The long-term target is for Group members to purchase a substantial and increasing volume of their requirements for forest products from certified sources as described in 1 (above). Members are free to set their own timetables and to make this information public.

Obligations

4. A named senior manager will have responsibility for implementing the above commitments and target.

5. Members will be required to demonstrate progress toward the target through a system of six-monthly progress reports. Reports must include a database of forest products used by quantity, type, and forest source. Sources must be categorized as (a) forest meeting the performance standards set out in 1, (b) known well-managed forest, or (c) unknown and/or not well-managed forest.

All information supplied by Group members shall be kept confidential with the exception of the following information that will be extracted for use in publicly disclosed reports: the identity of Group members, the broad commitments made by Group members, the aggregated status of members' achievements, and the total forest products purchasing power of the Group. No proprietary or individual company information will be disclosed without that company's consent.

Figure 2

HD and JS have publicly encouraged industry-wide adoption of third-party certification. Fear of a public relations calamity appears to have been the initial motive behind support for certification. In the past, environmental advocacy groups have targeted European retailers with actions that have embarrassed and hurt them. Several years ago, environmentalists demonstrated at storefronts and retail lumber yards in the U.K. with large inflatable chain saws to emphasize their claims that the retailers were aiding the destruction of rain forests by carrying products made from tropical wood. Although environmental groups have not actively targeted U.S. retailers, managers are keenly aware of the potential damage that unfavorable publicity may inflict upon their reputation and bottom-line.

Even though senior managers for both retailers acknowledge that their environmental efforts are, in part, a response to the perceived risk associated with bad publicity, they also affirm that their continued commitment to acquire certified wood products reflect their corporate missions. This level of commitment is essential because efforts to provide certified products are in their infancy, and, as a result, neither HD nor JS have had extensive experience in buying and selling these products.

Visitors, posing as consumers, to a number of stores revealed the embryonic nature of these programs. Retail stores in both the United States and the United Kingdom carried a number of products that made environmental claims—"for every tree harvested 4-6 seedlings are planted or 10-20 seeds sown," for instance—but the stores stocked few certified wood products. The environmental claims that were made are clearly a source of contention between the retailers and suppliers. On a couple of occasions, managers, when asked about these claims, were visibly embarrassed and even frustrated that the claims were printed on the packages. Both retailers are working to rid their stores of any specious claims.

The salespeople at these stores were generally knowledgeable and helpful. Although employing salespeople knowledgeable about home improvement is important to the competitive positioning of these stores, the salespeople interviewed seemed to be unaware of certification and related issues. It was apparent that the salespeople were unaccustomed to receiving customer inquiries on this topic.

In examining the progress each company has made in its commitment to support sustainable forest practices and market certified wood products, it would be easy to conclude that their commitment is insufficient for the job or that their strategies are ineffective. However, the effect of the companies' commitment on the overall issue of forest sustainability goes well beyond the mere physical presence of certified products on store shelves. Behind the scenes and headlines are successes not evident to the casual observer. Over time, the results of the continued investments made by these companies will probably become evident through growing numbers of certified wood products on store shelves.

J SAINSBURY PLC—PROGRESS TOWARD SUSTAINABILITY

JS is an important player in the 1995+ Group. As part of its commitment to the Group, JS has invested considerable resources in documenting the sources of its wood supply. A computer database, aptly named TimberTracker, is used with comprehensive surveying of suppliers to monitor "forest of origin" for all wood-based products sold by JS Group companies in the United Kingdom (see Appendix B). JS tracks not only sawnwood, but also pulp and paper products, which represent a far larger portion of its sales of wood fiber than does sawnwood. The tracking system assigns suppliers a letter grade. Grade A indicates that the supplier is fully FSC certified, grade G that the supplier has been delisted.

The formal TimberTracker program implemented by JS provides benefits to the retailer and the certification movement beyond mere data collection. As the TimberTracker questionnaires work their way up the supply channel, they alert suppliers who may be two or three steps removed from JS that the company is an important customer. JS's size and reputation give it leverage when it approaches timber producers to inform them about its commitment to buy FSC certified wood. As TimberTracker and other similar questionnaires from other important 1995+ Group customers, such as B&Q plc, the largest do-it-yourself (DIY) retailer in Europe, accumulate on the desks of their common suppliers, the collective importance of these customers is likely to impress upon them the significance of obtaining FSC certification and the consequences of resisting.

Although JS has delisted suppliers if they refused to participate or had poor performance, it tends to stick with the stated philosophy of the 1995+ Group, which attempts to move suppliers towards rising levels of sustainable practices. As one member of JS's SSD pointed out, "It's easier to work with suppliers who we know than it is to get new ones, because you don't know what other problems they bring in as well." Martin and many other representatives of JS have visited suppliers worldwide to inform them about JS's commitment to FSC certified products. As part of its overall program, suppliers are asked to create action plans that will take them toward obtaining their raw materials from certified sources. Lobbying by JS and 1995+ Group members was apparently influential in moving large pulp and paper suppliers such as STORA in Sweden toward certification.

JS also has formal internal documentation of its policies toward wood and wood-based products. Three documents are used: a one-page issues and position statement and two Environment Best Practice Guides (Paper; and Timber and Forest Products). Both documents were produced by the environmental management department within the SSD. The Best Practice Guides educate company employees about environmental issues and provides alternatives and strategies for finding responsibly-managed supplies of wood and wood-based products. The documents specify such things as the rationale behind the firm's commitment, affected products, and offers a general evaluation of supply sources. The Timber and Forest Products Best Practices Guide, for example, identifies as the "least contentious sources" the United Kingdom, European Community (EC), and Scandinavian countries; as "questionable sources" such countries as Malaysia, Mexico, and the United States; and as "sources requiring special attention" such places as former Soviet Union countries, most Latin American states, and China.

JS also tries to inform its customers. Although retail employees were not knowledgeable about certification, educational brochures were prominently displayed near the certified rubberwood doors at Homebase stores (see Appendix C). The brochures explain the importance of FSC certification. JS has also aggressively tried to generate media publicity for its environmental programs.

Although a measure of self-promotion is inherent in these efforts, JS management is communicating its involvement with, and the importance of, this environmental issue to the public.

As of November 1996, JS had been more successful than HD in stocking FSC certified products. JS and its subsidiaries had a limited assortment of products with the FSC label: cutting boards, charcoal briquettes, and rubberwood doors. However, certified cutting boards were not found at all outlets and the packaging for the briquettes had not been updated to include the FSC logo and certification claims. JS plans to redesign the packaging for the next barbecue season.

By 1998, JS expects to have 500 FSC endorsed products on the shelves, although many of those will come from one door manufacturer. JS is counting on Swedish companies such as STORA and Assi Domän to complete large-scale FSC certification of their forests. If this certification does not happen, it will be a major setback to JS's current efforts and its goal of full FSC certification by the year 2000 for all the 13,000 wood and paper products it sells. Achieving full certification is an extremely complex, challenging proposition. Paper accounts for about 80% of the wood fiber that JS uses. The sources of fiber that contribute to the production of such diverse products as napkins, diapers, paper plates, newspaper, and magazines are scattered around the world, making it difficult to ensure that the raw materials used in the products come from responsibly-managed forests.

HD is committed to providing its customers the opportunity to choose certified wood products; JS intends to sell certified wood products exclusively. One member of the JS board of directors, pointed out that, even if the market does not currently demand certified wood products, "good marketers will anticipate the need." William Martin offered an analogy to explain JS's position:

If I turn to a customer, and say, "We have two eggs and we'll offer you one guaranteed to be salmonella free, but it will cost ten cents more than the salmonella iffy one" the customer is going to say Sainsbury shouldn't be selling the questionable one anyway.

THE HOME DEPOT—PROGRESS TOWARD SUSTAINABILITY

Although HD does not yet benefit from the actions and power that come with membership in a buying group as does JS, its overall commitment to the sustainable forest products issue is similar. HD managers articulated a goal of providing certified wood products to their customers; and it is also the company's policy not to do business with suppliers that are "not doing the right things." Currently, managers are working with suppliers to increase awareness of the issues and to motivate them to improve the sustainability of their forest practices. They emphasize screening out any new, truly bad suppliers, and improving the practices of current suppliers. Managers hope that these efforts eventually lead suppliers to adopt third-party certification.

HD uses several related tactics to approach the challenge of obtaining products from sustainably-managed forests: avoiding acquisitions from obviously problematic sources, species, and suppliers; carefully investigating questionable suppliers; and using existing and developing relationships to cultivate improved practices and interest in certification. Eisen provides periodic reports and consultative support to buyers as they consider product acquisitions. With access to Scientific Certification Systems (SCS), FSC, and relevant publications, Eisen can provide buyers with informed assessments of product sources that help to make them aware o f environmental trouble spots. As Eisen explained:

If [a buyer] were in China and a supplier said we'll get this wood from Burma, the red light would go off because I sent some articles to [the buyer] about the illegal logging in Burma and the fact that they are going to have to stop [logging] or those people aren't going to be able to cook their food.

As part of an evaluation of new suppliers and to manage current relationships, HD buyers and Eisen visit suppliers' facilities. One manager in the lumber and building materials unit reported spending over a week with a Chilean supplier investigating its reforestation and harvest practices. HD buyers also work with existing suppliers to educate them about third-party certification and motivate them to participate. One buyer explained that his strategy is to present forest sustainability as a marketing issue, telling suppliers that if they want to "market their products, there will come a day in the U.S. when you must be certified." Once a business relationship has been established, HD managers said they are better position to exert leverage on suppliers and motivate changes.

In 1994, HD became the first home center in the world to offer tropical and temperate region wood products from certified forests—shelving from Collins Pine Co. and doors from Portico S.A. For these and related efforts that advanced eco-labeling, in 1996, Collins Pine and HD received the President's Sustainable Development Award. At the time, the President's Council on Sustainable Development praised HD for educating consumers and building support for sustainable product design and production, and helping to open new markets for sustainable forestry products through eco-labeling.

As of late 1996, however, HD carried no forest products with a certification logo. The company had discontinued its Collins Pine certified shelving, although a line of interior doors is certified for its doorskin substrate, which is made from eucalyptus (Premwood® by Premdor). One of its products, doors made by Portico S.A. carried in about 25% of HD stores, melds materials from certified and noncertified forests and, therefore, cannot use the SCS certification. However, Portico uses the logo in its brochures and point-of-purchase materials.

HD is committed to the exclusive use of the SCS Green Cross logo, which is accredited by FSC, principally because the SCS logo has widespread cachet.

Six San Francisco Bay area stores had previously sold certified pine shelving from Collins Pine Co. with an SCS "green cross and globe" logo. Through a unique arrangement and direct distribution, Collins Pine received a premium for the product while HD was able to cut the retail price and maintain profit margins. In late 1996, HD quit stocking the pine shelving. HD managers suggested that the company discontinued the product because of a variety of business reasons unrelated to certification. Collins Pine interpretation is that the decision stemmed from problems—and added costs—associated with maintaining dual inventories for certified and noncertified products.

Even though this ended HD's symbolic first effort in certified products The Home Depot has continued to experiment with other certified product lines. If consumers really want certified products, Eisen said that it is likely that one of HD's competitors will begin selling the Collins Pine shelving, which might cause HD's buyers to reconsider their decision to drop the Collins shelving. While HD executive vice president Larry Mercer stated that, everything else being equal, HD would always choose certified products, the action on the pine shelving indicates that many considerations beyond certification dictate the viability of a product. Concerning the future of certified products at HD, Eisen commented that "...What limits the success of the program is the fact that it's not seen as something that is really ringing the register."

HD managers clearly feel that they do not have the resources or ability to unilaterally develop the market for certified products. Ryan expressed the difficulty that Home Depot retail managers perceive in educating consumers about such complex issues:

It really takes an infomercial where you can explain it for an hour on television...why it's the greatest thing that ever existed. But if you don't have that opportunity, you just put it in the store, it sits on the shelf and it collects dust.

In the context of HD's business environment, managers look for outside groups to share the burden of educating consumers, environmentalists, and regulators. According to Eisen, if environmental groups started promoting certified products as a way for consumers to be sure they weren't harming the forests, "we would have consumers coming into the stores, and our buyers would be going back to companies and the buyers would say get certified tomorrow." Despite the company's desire for an industry-wide effort, managers do not yet find supportive voices resonating with that of HD.

HD's product-line decisions, though guided by environmental and social considerations, are ultimately driven by consumer purchasing decisions. The company intends to purchase sustainably- produced forest products and to expose them to the market. Whether or not consumers buy them, however, will ultimately determine the degree to which these efforts succeed and are broadened at HD.

Different Business Contexts

The ability of these two companies to accomplish their shared goal of buying and merchandising wood products from well-managed forests depends, in no small part, upon the larger business environment in which they operate, and how central the efforts are to corporate strategy.

MACROENVIRONMENTAL ISSUES

HD and JS operate in substantively different macroenvironments that have a dramatic impact on their business strategies. The most striking difference between the two business environments is the relative size of their markets. The U.S. home improvement market is more than six times larger in sales volume than that of the United Kingdom and, of course, the geographic size difference is even larger. While JS is able to use centralized management and purchasing that gives it a high degree of control and uniformity in product specification and acquisition, the size and diversity of the U.S. market make that impractical for HD, which covers more area with one regional division than JS does with its entire operation. This difference, along with the sheer volume of products sold, exerts an undeniable effect on the relative abilities of the two companies to implement an integrated system for purchasing certified wood products.

The United States and the United Kingdom also differ in their politics, legislation, sources of wood supply, and business norms. It is unclear whether a group similar to the 1995+ Group would have the influence in the United States that it has in the United Kingdom. Membership in the 1995+ Group has helped JS learn from other companies' experience how to implement a system to track supply (JS's TimberTracker). The coordinated efforts of the 1995+ Group have increased JS's ability to get access to certified products and lowered its risk of stocking them. Although members do not purchase cooperatively, the combined buying power of the Group speaks much louder to suppliers than a lone company. Since the largest U.K. retailers are joined in marketing certified wood products, it eases the burden on each company of developing primary demand for the products and reduces the risk associated with the adoption curve. Many U.K. consumers may not have, or recognize alternative product sources. HD, however, currently has no other companies in the U.S. industry, retail home improvement or otherwise, to help it move suppliers towards forest certification and to cultivate consumer awareness and preference.

Environmental advocacy groups also appear to be more accepted and wield more influence in the United Kingdom than in the United States. The cooperation of the WWF and business in the 1995+ Group is but one example. This credibility has influenced the corporate culture of many companies, and managers are acutely aware of the need to address the concerns of environmental groups. It could be argued that JS has become a partner with environmental groups through the company's participation in the 1995+ Group, although JS personnel would not characterize it as such.

While WWF recruited JS to participate in the 1995+ Group, a similar situation is only now developing in North America. So far, HD managers have been reluctant to align themselves so obviously with environmental groups or certification schemes for fear of alienating important suppliers and, most likely, some consumer groups. HD prefers an "arms-length" relationship with groups on all sides of the sustainability issue.

HD operates in a fragmented retail sector, compared to JS's markets. HD's current market share is about 14%, while the next largest competitor is about half that size. By contrast, in both the home improvement and grocery sector, JS operates with a market share slightly less than its nearest competitor (12% share in grocery and 10% in home improvement). This reality has an obvious impact on the strategies these companies choose.

STRATEGIC ISSUES

As Larry Mercer, HD Executive Vice President, described it, what is at stake for both companies is the ability to be on the cutting edge of environmental concerns and market sustainably—produced forest products, while still remaining profitable. "The two have to balance," Mercer states. Both firms are responsible to stockholders, and cannot put their companies at a strategic disadvantage. Consequently, fundamental strategic differences between HD and JS influence how each implements its environmental program.

HD is strategically positioned as a discount DIY retailer and strives to provide "excellent customer service and floor-to-ceiling products at day-in, day-out low prices," according to the 1995 Annual Report. The pursuit of overall efficiency, economies of scale, preferential and efficient access to crucial supplies, and low overhead are crucial for business success. The need for efficiency for HD dictates a lean organizational structure. This strategic requirement likely influences the administrative resources that management feels can be applied to programs or developing markets for the limited number of certified wood products that it has been able to purchase.

HD's strategy depends on maintaining high sales volumes and the related economies of scale that brings, which clearly affects its effort to acquire certified wood products. HD management is very sensitive to the sales productivity of the items it carries in inventory. Continued financial success for HD also depends upon maintaining a reliable and efficient flow of supply into its stores. Out-of-stocks, high inventory costs, and other supply inefficiencies make it vulnerable to competitors. Under those circumstances the uncertain and limited supply of certified wood products introduces a greater degree of risk by developing too strong a reliance upon limited sources of supply.

J Sainsbury has a 128-year reputation in England as a well run company that has an almost paternalistic attitude towards its customers. The company's strategy consists of offering quality, value, and superior service to its customers, who are drawn evenly from all socioeconomic groups that live near each store.

JS's management feels that an important component of customers' goodwill comes from its reputation for reliability and responsiveness to customers' concerns. Justin Stead, a consultant for the WWF and manager of the 1995+ Group, characterized the expectations of JS customers regarding environmental issues: "Their customers want to buy every-thing at Sainsbury's and they expect Sainsbury's to address these issues on their behalf." The central role JS's reputation plays is seen in the formal, aggressive, and comprehensive approach it has taken to acquire FSC certified wood products. The seriousness of this commitment is also manifest in the pressure some buyers feel to obtain certified products. As one paper products buyer observed:

> The suppliers see it as a good opportunity to be the first FSC product on the shelf. They know that Sainsbury's will back it. If someone was to come to me tomorrow with an FSC approved product, I would have to take it. They could theoretically charge what they wanted for the product, because we have in some ways tied our hands behind our backs, in that we want FSC products on the shelves.

The comment, while not be taken literally, expresses the pressures placed on buyers. Strategically, JS seems to have more latitude than HD to incorporate programs that depend upon administrative resources—the SSD division, for instance—but the company is still sensitive to obtaining reliable suppliers of quality goods.

Conclusion

Absent consumers demanding certified wood products, both HD and JS have taken actions that are reactive, but also driven by principle. They are reactive because these efforts were established in part to avert the threat of embarrassing publicity. They are driven by principle to the extent that the companies have, within their unique contexts, approached the challenges of retailing certified wood products.

In comparing and contrasting the two companies, three related issues are key to gauging their current and future impact: the development of the distribution channel (gaining supply and cultivating demand), the influence of buyers groups, and the influence of individual environmental entrepreneurs.

Questions of supply and demand are much like the old philosophical query, "what came first the chicken or the egg?" Retailers face a considerable challenge in acquiring a sufficient and economical supply of certified products to make a credible and visible offering to the market. Suppliers respond that there is no profit in certification because the market is not demanding the product, although suppliers clearly are a component of the market.

ACQUIRING SUPPLY

Both retailers bemoan the lack of certified wood products. Their publicized commitment to selling certified products has generated pressure to find supply, yet few vendors are able to supply product in the quantities these retailers need. Since certification also has to be coupled with other critical product attributes as quality, consistency of supply, and price, the difficulties increase. Yet these varied criteria seem to be inherent to good suppliers. George White, of JS's SSD, expressed an opinion echoed throughout the interviews at both companies:

The companies that don't deliver environmentally, don't deliver in other areas as well—according to product delivery, they don't meet specification, they don't deliver on time, and might have price fluctuations...it seems that the important companies also deliver on environmental issues.

As mentioned earlier, JS depends on large companies in Sweden to move towards FSC certification in the near future, which would move the company and other members of the 1995+ Group well towards their goal of 100% certified products. About 40% of the sawn softwood consumed in the United Kingdom comes from Scandinavia, according to Delphi International Ltd. JS management thinks that if Sweden embraces FSC certification on a large scale, Finland will follow quickly, so as to not place its companies at a competitive disadvantage to their most significant competitor.

CULTIVATING DEMAND

Consumer demand, while potentially the most powerful force in this equation, currently appears to be inconsequential. Managers from both companies agree that consumers generally have limited knowledge or even awareness of issues related to sustainable forest practices. Managers also recognize that consumers are not enthusiastic about certified wood products, nor are they likely to pay a premium price for them. This conviction contradicts published consumer surveys that consistently find that a significant proportion of people is concerned about environmental issues and how their behaviors impact upon them. As one HD manager commented, "consumers who way they would buy an environmental product over a non-environmental one rarely do if the price and quality are not equal."

Both HD and JS have stressed the importance of eco-labels (SCS and FSC, respectively) in stimulating consumer awareness and demand for certified wood products. However, it may be that the mass market just expects the products that they buy will not destroy the environment in which they live. Consumers are unwilling to pay a premium, but, in the case of forest practices, may expect that sustainable forest practices are the price of admission for a company that wants to sell to them. Future research might examine this issue.

Consumers simply may expect products drawn from responsibly managed forests, so short-term setbacks should not deter further development by these retailers. The JS and HD managers expressed some frustration over the lack of consumer interest and demand for environmentally responsible products. At the same time, when Eisen remarked, "We are always trying to go ahead of the curve and do it before the customers would demand it," he voiced a forward-looking perspective that is shared among the managers of both firms.

INFLUENCE OF BUYERS GROUPS

JS's participation in the 1995+ Group represents a key difference in the business environments of the two companies. As a member of the Group, JS has support from other retailers, suppliers, and environmental groups in informing consumers. When all 1995+ Group members carry the same certified wood products, it will build awareness of, and preference for, the FSC label specifically and certified wood products generally. The buyers group offers the additional benefit of effectively leveling the competitive playing field. Collectively, the group's members represent huge market share. Since they intend to carry certified products exclusively, their customers will have no alternatives to choose from, so all the competitors would receive any price differential for certified wood. In effect, this would allow them to pass any price premiums on to customers. That puts JS in a better position than HD to push certified wood products because there is no competitive price disadvantage to the company, and the investment of building initial consumer demand is shared.

How global timber suppliers react to the demands of the 1995+ Group companies will be a determining factor in their success. Suppliers have a variety of options. They might certify all of their forest land, a portion of their forest land, or simply abandon U.K. customers that demand certified products. The first two alternatives would probably move the issue forward by familiarizing suppliers with certification and eliminating many of the unknowns that currently exist. They would also increase supplies of certified product to other companies. Abandoning the U.K. market, however, would set a precedent that could discourage other country-based buying groups from such a rigid commitment to certified wood products as that of the 1995+ Group. Such a tactic would likely lead to uncertain supplies and higher prices in the U.K. market.

Success of the 1995+ Group, however, may add power and credibility to buying groups currently organizing in other parts of the world. HD has participated in the discussions regarding the creation of a North American based buyers group, the Certified Forest Products Council, and has financially supported its formation. The Home Depot's participation would be mutually advantageous. In return for lending its credibility and buying power to the group, HD would be able to share the costs of developing consumer awareness and the buying leverage provided by other members. It is unclear, though, whether the Certified Forest Products Council will wield the same influence as the U.K. group. The U.K. group was able to agree upon using the FSC certification and an associated label. A North American group might be divided among differing loyalties to available labels as the FSC, SCS (preferred by HD), or others. Division over such an issue might hinder the group's ability to focus the buying issue and diminish its overall impact.

INFLUENCE OF ENVIRONMENTAL ENTREPRENEURS

A common element for both companies is the important part played by environmental entrepreneurs. At HD, Mark Eisen's efforts have been instrumental in educating senior managers and buyers to the importance of acquiring products from responsible sources. At JS, Alison Austin, William Martin, George White, and Jayne Gilbert, all have been visible and effective advocates for certified wood products. Interviews with individuals in other parts of both organizations, indicated that the efforts to improve the sources of supply for wood products would not have survived without the advocacy of these people. As other companies consider such programs, they will need to choose a manager to spearhead the program carefully. The effort requires someone with the appropriate leadership skills and the qualifications to champion the issue internally. Someone who can command the respect needed to motivate change.

LESSONS LEARNED

It is too early to gauge the success of the efforts by JS and HD to market certified wood products. It is apparent, however, that both are influencing the global forest products industry, and that their long-term success would be a significant force moving the industry toward more sustainable forest practices. Supplies of certified products, of course, would have to increase many times over to meet the needs of these retailers. Success would demonstrate to environmental groups, other firms, and to consumers that an effort by retailers to move to 100% certified products is possible despite seemingly monumental difficulties.

Literature Cited

Delphi International, Ltd. 1996. "The Demand for Certified Wood Products." Unpublished working manuscript. London, England.

World Wide Fund for Nature 1996. "The WWF 1995 Group, The Full Story." WWF-UK, Panda House, Weyside Park, Catteshall Lane, Godalming, Surrey GU7 1XR.

VI. Appendix A: Home Depot's "Environmental Greenprint"

VII. Appendix B: The Sainsbury Group's TimberTracker

TimberTracker Supplier Questionnaire   **The Sainsbury Group**

PART A: SUPPLIER COMPANY DETAILS

A 1   Sainsbury's Supplier Code:

Full Company Trading Name:

A 2   Company Full Postal Address:

Postal / Zip Code:
Country:

A 3   Name of Managing Director:

A 4   Environmental Contact:

Company Telephone Number:
Company Fax Number:
A 5   Country Dialling Code:

About the TimberTracker Questionnaire
Contents:

Part A: Supplier Company Details
Part B: Supplier's Timber Policy
Part C: Forest Sources
Part D: Supplier's Timber Inventory
Part E: Products Supplied to the Sainsbury Group

(forms for parts C, D and E
may be photocopied as
required to match the volume
of data you have).

How to complete the TimberTracker questionnaire.   (notes for each Part).

Supporting Documents:

Brochure "The Sainsbury Group TimberTracker Programme".
FSC and WWF leaflets.
Other publications and guidance on request.

TimberTracker has been developed for the Sainsbury Group,
and sponsored by J Sainsbury, Scientific Services Division.

TimberTracker Supplier Questionnaire   **The Sainsbury Group**

HOW TO COMPLETE THE TIMBERTRACKER QUESTIONNAIRE

GUIDANCE NOTES TO PART A

A 1   (This is our code for your company; it is inserted automatically). If the company trading name shown contains any errors, please correct it.

A 2   Take care to include the full postal address, including any postal or zip codes. If we have pre-printed your address, please check it for accuracy and make any necessary changes.

A 3   Please enter the full name and title by which your Managing Director wishes to be described on any correspondence relating to timber certification.

A 4   Insert here the name of the manager or staff member responsible for day-to-day implementation of your environmental policy. It is this person whom we might wish to contact directly to clarify any points in your response to this questionnaire.

A 5   Please supply your country's international dialling prefix (e.g. UK = 44).

GUIDANCE NOTES TO PART B

B 1   We support the general principles of the FSC and the specific objectives of the WWF'95 *plus* Group, as set out in our explanatory booklet "The Sainsbury Group TimberTracker Programme". We need to have confidence that our Suppliers are also committed to phasing out the purchase of wood and wood products that do not come from well-managed sources (as defined in the booklet). An essential beginning is the issue by you of a formal, written policy on the sourcing of timber, with a commitment to the phasing-out of non-certifiable or non-traceable sources. Answer Yes or No to the question. If "No", but you are preparing such a policy, enter the date by which we can receive a copy.

B 2   The policy should include a target date at least for the first phase (i.e. ceasing to use supply chains whose sources are not traceable, and phasing out forest sources which plainly do not and cannot meet the "well managed and sustainable" criteria.

B 3   Please do include a current copy of your policy. We cannot assess you, the Supplier, without it. If unavailable, state when we can have a copy.

TimberTracker Supplier Questionnaire   **The Sainsbury Group**

PART B: SUPPLIER'S TIMBER POLICY

Your Company Policy:

B 1   Has your company an official written policy, for timber-based products, for using only timber from well-managed sources? (YES/NO):

B 2   If yes, does your policy specify a deadline date for sourcing only from well-managed forests? (YES/NO):

If yes, what is that date? (dd mmm yy)

Is your policy reviewed and updated at least annually? (YES/NO)

B 3   Is a copy of your policy enclosed with your response? (YES/NO)

Membership of World Wildlife Fund 1995 plus Group:

B 4   Has your company formally joined the WWF'95 plus group? (YES/NO):

B 5   If yes, are you a 1, 2 or 3 Star Member? (1 / 2 / 3):

B 6   **Membership of Timber Certification Schemes:**

(e.g. Scientific Certification Systems (SCS), Rainforest Alliance's Smart Wood Program, Soil Association (SA) Responsible Forestry Programme, SGS Forestry. .)
Please give details of your company's membership: Name of scheme, your company's status, effective date. Enclose a copy of the scheme's objectives:

_____

_____

_____

_____

B 7   **Other Relevant Standards:**

_____

_____

_____

_____

B 8   **Management of Your Supply Chain:**

_____

_____

_____

_____

_____

_____

_____

_____

_____

_____

_____

B 9   On a separate sheet, please provide a diagram of your supply chains.

TimberTracker Supplier Questionnaire

The Sainsbury Group

B 4   Please say if you have joined the WWF'95*plus* Group. This, and your "star rating", will make our evaluation task much simpler.

B 5   The WWF will not release other members' star ratings to third parties. However, **you can tell us.** Unless you say otherwise, we shall assume you are a 1-Star member.

B 6   There is a growing number of Timber Certification Schemes. Membership would provide clear evidence of your commitment. Please state the name of any such scheme, your company's status within it, and the date that status was achieved. We have listed some of the better-known schemes. If your's is not one of them, please enclose a copy of your scheme's charter and objectives.

B 7   You may also demonstrate commitment if your company has been certified under recognised international quality and / or environmental standards. State the name of the standard, and the level and date of your certification.

B 8   As a supplier committed to using well-managed forest sources, you will need to check and manage your supply chain to ensure that every source is known and assessed. This will include not only Forest Sources, but also intermediate suppliers, importers, manufacturers, and converters (such as sawmills and pulp / paper mills). Please give brief details of what measures you have taken, and will take, to verify and maintain the consistent integrity of your supply chain.

B 9   If possible, please draw, on a separate sheet, a **diagram of your timber supply chains** (as illustrated in our TimberTracker booklet). This will help us to appreciate the scope of your supply chain management task, and to verify that all sources have been covered in the questionnaire.

GUIDANCE NOTES TO PART C

C 1   A Forest Source is a combination of the geographical area of forest or plantation, together with the company which exercises forest management over these timber resources. One geographical forest may contain several such sources. We need you to complete a separate copy of part C for each forest source. To avoid repetition of detail, we suggest you give each source a unique reference. (For example, Forest 1, Companies 1, 2 and 3 could be F1/1, F1/2 and F1/3).

C 2   Forests may go under several national or local names. We need a recognised name that can be found on general regional maps, and in contracts or deeds.

TimberTracker Supplier Questionnaire   **The Sainsbury Group**

PART C: FOREST SOURCES

C 1 Reference Number of Source :   /

C 2 Name of Forest / Plantation or
Woodland:

Country:
Location / Region:

C 3 Type of Ownership / Operation:

C 4 Details of Logging Concession / Licensee Holder / Primary Mill

Name of Company:
Company's Full Postal Address:

Name of Managing Director:
Phone number:
Fax. number:
Country's International Dialling Prefix:

Concession Holder (CH) / Licensee / Primary Mill's Management Policy:

C 5 Has CH / Mill a forest management plan ? (Y/N/NK)
Have you read the plan ? (Y/N)
C 6 Is the CH / Mill already certified ? (Y/N/NK)
If certified, or if certification is in progress please give details of scheme:

_____
_____

C 7 Please give details of measures you have taken to check whether this source is
well-managed and sustainable?

_____
_____

C 8 Given your knowledge of the source and local conditions, how do you rate the
possibility of the source being certified?

_____
_____

TimberTracker Supplier Questionnaire   # The Sainsbury Group

C 3   State whether the forest is owned by the state, privately, or a combination of both. Please also describe the type of operation:

(1) Community - based , small-scale, forestry project.
(2) Selective cutting of natural forest.
(3) Plantation.
(4) Other (please give details).

C 4   We are interested in the smallest trading unit that **can and does pursue its own forest management policy.** In most cases, such a source is likely to be a primary converter (i.e. a sawmill or pulp mill). The contact details we need are those which would enable direct contact to be made with that source (not those of some larger company or group).

C 5   You should know from your own contact whether the source has its own written forest management policy. Enter Yes, No, or NK for "not known". An "NK" will prompt you to find out. If there is a policy, please state whether you (or a senior member of your company staff) has actually read it.

C 6   The source may already be certified or undergoing certification. If so , please give the name of the scheme, the source's status within it, and the (approximate) date when that status was achieved. If you simply don't know, then enter "NK".

C 7   We are interested in your approach to managing your sources. Give **brief** details of the means you have used to verify **this** source's approach to forest management? (e.g. Site visits, questionnaires, documents **and contracts,** recommendations from agents, etc.).

C 8   **Most importantly**, from your personal knowledge of this source, we ask you to make your own assessment of whether it is likely now or in the future to satisfy the FSC criteria for certification. **Please give brief reasons for your assessment rating.** We suggest these categories:

(0) Insufficient data available now to make such an assessment.
(1) Very unlikely to be able to qualify for certification.
(2) Certification may be possible (with some local or external encouragement).
(3) Certification probable (with some local or external encouragement).
(4) Undergoing Certification
(5) Already Certified

Note: If you have to use category 0, we would expect you to seek further information about the source, and to send us details separately within **8 weeks** of completing this form.

TimberTracker Supplier Questionnaire

The Sainsbury Group

PART D: SUPPLIER'S TIMBER INVENTORY

Group	Timber (Popular Name)	Species (Latin) name(s)	Hard or Soft	Ref. of Forest Source (from Part C)	% of timber type bought from this source for all purposes.	Quantity of this timber type bought for Sainsbury's.	Units in which quantity is given.
(D 1)	(note D 2)	(note D 3)	(D 4)	(note D 5)	(note D 6)	(note D 7)	(note D 8)

TimberTracker Supplier Questionnaire   **The Sainsbury Group**

GUIDANCE NOTES TO PART D

D 1   In Part D we ask you to group together , regardless of source, the different types of timber you buy. We are not concerned here with the form in which the timber is purchased. Where the same basic timber type comes from several forest sources, you will need to complete a separate line for each source. In the first column you can allocate a reference for each timber group, (or simply bracket together all the lines concerning the same basic type).

D 2   Please list in this column whatever local or popular name you use for that timber when trading with that source.

D 3   Different sources may have different local names to describe what is essentially the same type of timber. In the third column we ask you to give the formal Latin species and sub-species names for each timber type, so as to define it precisely.

D 4   State whether this timber is classed as hardwood (H) or softwood (S).

D 5   Note D1 asked you to complete a separate line entry for each source from which you obtain a given timber type. In this column please indicate that source by using the same forest source reference you provided for it in Part C.

D 6   For each of the basic **groups** of timber you marked in column 1, please state in column 6 what **percentage** of your total purchase of this group comes from each source. By total purchase we mean the timber bought for all purposes - not merely that destined for Sainsbury Group products.(It does not matter whether you measure total purchase by weight, volume, or value, provided you use the same measure in each case).

D 7   We need to know **how much timber** (in a recorded or predicted annual period) has to be removed from the forest sources in order eventually to **produce the products you supply to the Sainsbury Group.** Please give the figure for each source, if you know it, or alternatively bracket together the lines for a whole timber group and give one total for the group.

D 8   Give the quantity in whatever units of weight or volume you normally use (cubic metres, metric tonnes, etc), provided those units are clearly stated in column 8.

TimberTracker Supplier Questionnaire

The Sainsbury Group

PART E: DETAILS OF PRODUCTS SUPPLIED TO SAINSBURY'S

Product Code:	Product Description:	Details of Product Timber Content	
		Name of Timber used in product	Form of Timber in finished Product
(note E 1)	(note E 2)	(note E 3)	(note E 4)

TimberTracker Supplier Questionnaire

The Sainsbury Group

GUIDANCE NOTES TO PART E

E 1 We shall provide with this questionnaire a list of Product Codes and Descriptions for the products you currently supply to Sainsbury's. It may not be necessary for you to repeat them all on our form. We use different codes for different versions of the same basic product (e.g. for different colours, or different numbers of the item per retail pack). Where a group of related codes / descriptions shares the same timber content, you need make only a single entry - but indicate in column 1 the **range** of codes you have covered.

E 2 (As for Note E 1)

E 3 In column 3 please give the name of the timber contained in the product. You may use either the popular name or the full Latin names (as in Part D). Note that where a product contains more than one type of timber you will need to enter a separate line for each type.

E 4 Please state in what form the timber appears **in the final product** (i.e. not the form in which you obtained it from the source). Categories are:

(1) Solid wood
(2) Plywood
(3) Veneer
(4) Chipboard
(5) Blockboard
(6) Fibreboards
(7) Paper
(8) Card
(9) Chippings
(10) Charcoal
(11) Moulded composite (e.g. Papier Maché).
(12) Fluff
(13) Viscose, Cellulose
(14) Other (please specify).

VIII. Appendix C: Sainsbury's point-of-purchase brochure

sustaining

the forests

FSC

Sainsbury's. Where environment matters

*Sainsbury's,
Homebase & Savacentre*

a commitment
to saving the world's forests

Most of us are concerned about the fate of the world's forests. We're alarmed by reports that large swathes of virgin forests are still being destroyed.

Some are cleared for agriculture or development. But many are being cut for their timber.

This goes to make the vast quantities of wood and paper products, used in a variety of everyday items which we all buy.

Some of this timber comes from sources which are well managed, in a way that sustains both the timber harvest and the natural ecology which has developed within the forest. But not all forests are managed in this way.

the Sainsbury's commitment

Sainsbury's shares our customers concerns for the future of the world's forests. And we stock over 13,000 products containing wood or paper. These include everything from cocktail sticks, wooden mouldings and chopping boards to nappies, tissues, bath sponges and stationery, as well as newspapers and magazines.

And so we have made the following commitment:

Sainsbury's will only buy paper and wood products which we can be sure come from well managed forests.

In support of this, we are working with independent environmentalists, and our suppliers, to set up a system which will track all our wood and paper purchases. In time, this will ensure that we know exactly which forests all our purchases come from – and how these forests are managed.

We want our customers to know that when they buy products containing wood or paper from Sainsbury's, they are not contributing to the loss of the world's forests.

FSC Trademark © 1996 Forest Stewardship Council A.C.

Most of ... buy w... unwitti... natural...

Man... environmental claims. You've probably seen a few yourself. "This card is printed on environmentally friendly paper ... this paper comes from sustainably managed forests .... three trees are replanted for every one cut ..."

But the sheer number and variety of such claims is confusing and counter productive. And without any independent verification, it can be hard to know just what these claims mean and whether they have any meaningful validity.

Now at last there's a chance of seeing the wood for the trees. From February, shoppers will see a new symbol appearing on a range of timber and paper products.

seeing the wood for

the trees

FSC

An outline of a tree combined with a 'tick mark', above the letters "FSC". FSC stands for the Forest Stewardship Council – a major alliance of environmentalists, industry, foresters and indigenous peoples' groups, who have set world wide criteria for the good management of forests.

Products bearing the FSC Trademark are made from wood which has been independently certified as coming from "well managed" forests. (Obviously this only applies to products made from 'virgin' timber, not recycled paper).

Sainsbury's strongly supports the FSC

initiative. Together with other retailers and also timber importers, we have joined the World-Wide Fund for Nature's *"1995 Plus Group"*. This is a group of companies which is committed to phasing out all wood products which do not come from well managed forests, and moving to those which have received the FSC stamp of approval.

It's early days yet for the FSC. But in preparation, Sainsbury's has set up its own "TimberTracker" system, to find out for ourselves the exact source of all our wood and paper products.

This will enable us to find better sources for all those products which do not come from well managed forests. Longer term, our intention is for all our forest products to have FSC approval.

the environment

so *what exactly is a* well managed *forest?*

Phrases like "well managed forests" and "sustainable forestry" are bandied about freely. But they can mean different things to different people.

After all, "forest" just means an area largely covered in trees; so it can be anything from an untouched jewel of ancient woodland, rich in wildlife, to a uniform plantation of conifers.

In the narrowest sense, managed forestry simply means replacing trees felled, so as to ensure a continuing supply of timber.

This is fine as far as it goes. But in itself, this definition of "well managed" forestry does little to preserve a rich environment. Some such "forests" are really just tree crops, sometimes planted on land that isn't "naturally" forest at all. In some of these,

non-commercial species are weeded out or poisoned, and when the trees have been harvested – sometimes by "clear cutting" a whole patch of forest – the land is replanted with only the commercial species.

Today, there's a growing consensus that a "well managed forest" should be more than just a tree farm. It should not only yield a sustainable timber crop, but also show some of the characteristics of a "natural forest". It should, in short, have a whole range of uses and meanings, environmental and social, as well as commercial.

And it's this wider definition of "well managed" forestry which the Forest Stewardship Council has adopted – and which Sainsbury's supports.

Much of Britain's paper and pulp, for example, is imported from Scandinavia. Here, forestry practices have been continuously evolving, always looking for improvements. These include allowing a range of different native tree species, of varying age, abandoning large "clear cuts" in favour of selective felling; letting the forest "re-seed" itself rather than ploughing up the ground and replanting; letting some dead and decaying trees remain in place, as a vital habitat for birds, insects and fungi; and protecting watercources and wetlands. There are always changes to be made though, and Sainsbury's is firmly encouraging its suppliers to ensure that their timber meets the highest standards.

who decides?

So who decides which forests make the grade?

This will be the job of independent certifying organisations, mostly commercial organisations, approved by the FSC but paid for by the timber companies whose operations they are checking. In essence, it is a little like businesses paying to have their accounts checked by an auditor who, in turn, has been approved by a professional body.

Exactly what constitutes well managed forests will vary from country to country, even region to region. In some forests, only careful, selective cutting will be allowed. Some parts of such forests may even be too valuable, ecologically speaking, for any logging at all. In other forests, fairly extensive cutting will be appropriate.

---

the forest stewardship *principles*

To qualify for the FSC seal of approval, forest management must satisfy certain basic criteria. These include the following:

- encouraging the optimal use of forest products and services to ensure economic viability (in other words, so it produces a sustainable harvest)

- maintaining the critical ecological functions of the forest and minimising adverse impacts on biodiversity, water resources, soils, non timber resources (such as plants and wildlife)

- respecting rights of indigenous people to their own land

- supporting the social and economic well being of forest workers and local communities

- conducting regular monitoring of the forest condition, of timber yields, and of the chain of custody and management operations

- not replacing natural forests by tree plantations

---

changes in the air *– and on the ground*

Of course, a switch to certified well managed forestry isn't going to happen over night. In many areas of the world, forests are still being damaged or destroyed by unsustainable logging.

But change is in the air – and on the ground, too. A growing number of forestry companies across the world are transforming their practices. Some already operate to very high environmental standards. Others have a long way to go. And one of the most powerful tools to encourage them to change is customer opinion.

THE PRICE OF CHANGE

We've become accustomed to the idea that we have to pay extra for products which don't destroy the environment. In a few cases that's true. But it need not be so for wood and paper.

At Sainsbury's, we are confident that our commitment to buying products from well managed forests will not in itself result in significant price increases for our shoppers.

This is one area where environmental protection should not come at a price.

FORESTS WITH A FUTURE

This is just the start. Making sure that all the timber in our products comes from well managed forests will be a long process. But it's one to which we are thoroughly committed.

The overall aim is a simple one:

To ensure that our customers can continue to enjoy a wide range of high quality wood and paper based products – and that the world's forests still have a future.

16.04.96

J Sainsbury plc Stamford House Stamford Street London SE1 9LL
Printed on paper with a 100% recycled fibre content.

Sustainability *on the shelves*

Some of the wood stocked by Sainsbury's come from forests that meet all the criteria of sound environmental management. The two case studies on this page are examples of what can be achieved.

A Greener Barbecue

This spring sees the introduction of a range of barbecue charcoal from South Africa, representing a real success story in sustainable forestry.

The charcoal is the by-product of an imaginative, and highly successful scheme, to halt the spread of coastal sand dunes. Casuarina trees are planted in areas prone to erosion by the wind and the sea. The trees help to nourish the soil, and prevent it being blown or washed away. The ever present threat of soil erosion means that a charcoal producing operation is the

only form of production suited to the site. The charcoal production brings economic benefits to the local community.

Sustainable Chopping ...

Certified natural beech forests are our source for a range of chopping boards. These well managed forests are allowed to regenerate naturally after felling. They provide an important habitat for wildlife, and are being managed for both commercial and environmental purposes.

Menominee Tribal Enterprises: Sustainable Forestry to Improve Forest Health and Create Jobs

CASE STUDY

PREPARED BY:
CATHERINE M. MATER

A Case Study from "The Business of Sustainable Forestry"
A Project of The Sustainable Forestry Working Group

The Sustainable Forestry Working Group

Individuals from the following institutions participated in the preparation of this report.

Environmental Advantage, Inc.

Forest Stewardship Council

The John D. and Catherine T. MacArthur Foundation

Management Institute for Environment and Business

Mater Engineering, Ltd.

Oregon State University
Colleges of Business and Forestry

Pennsylvania State University
School of Forest Resources

University of California at Berkeley
College of Natural Resources

University of Michigan
Corporate Environmental Management Program

Weyerhaeuser Company

The World Bank
Environment Department

World Resources Institute

CCC 1-55963-624-6/98/page 9-1 through page 9-19
© 1998 John D. and Catherine T. MacArthur Foundation
All Rights Reserved

Menominee Tribal
Enterprises: Sustainable
Forestry to Improve
Forest Health
and Create Jobs

CASE STUDY

PREPARED BY:
CATHERINE M. MATER

A Case Study from "The Business of Sustainable Forestry"
A Project of The Sustainable Forestry Working Group

Contents

Menominee Tribal Enterprises: Sustainable Forestry to Improve Forest Health and Create Jobs

Introduction

The Menominee Tribe has lived in northeast Wisconsin and on Michigan's Upper Peninsula for generations, where ancestral tribal lands once encompassed more than 10 million acres. Following several treaties and land cessions, the Menominee people established a Reservation in 1854 totaling 235,000 acres of predominantly timber land. Since then, the backbone to the economy of the Menominee Nation has been its forests and the industry surrounding the sustainable management of that resource.

The Menominee Tribal Enterprises (MTE) has been an engine of the Menominee economy over the last 140 years and, within the last 25 years, has pioneered the implementation of sustainable forest management (SFM) throughout the Menominee Forest.

Today, the Menominees remain the only Native American tribe to have their forestlands independently certified as being sustainably managed. They are also the only forestlands operation in the United States and Canada that holds dual environmental certification from both the Forest Stewardship Council-approved SmartWood and Scientific Certification Systems (SCS).

The concepts of sustainability in forest ecosystems and surrounding the communities that the Menominee have practiced for so many years include three components of a sustainable forest system:

1. The forest must be sustainable for future generations.

2. The forest must be cared for properly to provide for the many varying needs of people over time.

3. All the pieces of the forest must be maintained for diversity.

Looking closely at what MTE has accomplished in SFM and product development during the last twenty-five years provides unique insight into the economic opportunities and constraints that face other forest products operations considering SFM practices. With a twenty-five-year track record, MTE is one of the few examples in the world where realized forest management performance over time can be compared with intended results to determine whether SFM actually does what it is purported to do:

- Increase the *quality* and *volume* of wood grown in a forest system over time.

- Provide more *consistent* and *stable* annual harvested timber volumes while maintaining or improving forest ecosystems.

- Maintain or improve *a forest ecosystem health* that recognizes the value of multiple uses of a forest.

- Sustain *communities* that surround the forest through job generation and the creation of educational opportunities.

- Increase the *value per unit* of wood products produced from SFM forest resources through documented performance in the marketplace.

MTE's forest management choices may not apply to all forest products concerns. MTE's management and decision-making structure does not appear to be well suited to the management of larger private forestry operations in North America and Europe. It could, however, be applicable to forest businesses owned and/or operated by other tribal or native entities throughout North and South America, and smaller privately-owned forest products concerns worldwide. Equally important, MTE's process of managing tribal forests and the techniques it uses may be well suited for managers of public forestland throughout the world, especially those required to balance the multiple use of forests and deal with the issues of community and public stakeholder trust in the management of the forests.

The Business of SFM

MTE incorporates both forestry and forest products operations in its total business package. However, to understand the business definition and strategic intent of sustainable forestry and sustainable forest products development at MTE, it makes sense to evaluate the forestry operations separately from the sawmill facility. The criteria that provide the foundation for business operations for each are detailed below.

FROM THE FOREST

In 1884 about 1.2 billion board feet of standing timber was documented on the Menominee Forest. Since then the amount of billion board feet of standing timber has increased 40%. Moreover, since 1854, even though over 2.25 billion board feet of timber has been harvested off the Menominee Forest, the number of high-quality and large-diameter timber has significantly increased.

The Menominee forest contains a higher diversity of tree species than surrounding forests, which makes using SFM practices more challenging than in other less-diverse areas. The Menominee lands contain approximately eleven of a total of fifteen plant associations found in the entire state of Wisconsin. The major soil types range from dry, nutrient-poor sites that grow poorer-quality tree species (scrub oak, jack pine) to moist nutrient-rich sites that support the growth of very high-value species such as sugar maple and basswood. More than 9,000 distinct timber stands make up the Menominee Forest.

The forest management objective of the Menominee Tribe is to maximize the quantity and quality of sawtimber grown under sustained yield management principles *while maintaining the diversity of native species.* The Tribal leaders have recognized the need for economic survival but only at a speed or intensity that allows for the forest to regenerate itself. Unlike more traditional forest management

---

MTE Forest Facts

Menominee's Land Includes:

• 235,000 acres of which 94% is covered by productive forestland

• Dominant tree types in the forest are varied and include: northern hardwoods such as maple, red oak, and basswood; hemlock; three types of pine; aspen; and lower quality oak

Source: MTE

Figure 1

---

regimes, this policy promotes a timber harvesting practice that removes timber according to the vigor of the trees rather than their saleable size only. Vigor is defined as the measure of the growth potential of an individual tree. It describes the health and ability of trees to respond to management. Today, Menominee forestlands retain a larger variety and volume of larger-diameter trees on a per acre basis than surrounding forest systems, including federal forest systems next to the tribal lands.

In accordance with their directive to maintain the diversity of native species, the Menominees employ a new Forest Habitat Classification System that provides them with a method to accurately assess the productivity of forest sites and identify the best forest cover for each site, regardless of the tree species or tree quality currently growing on the site.

Traditionally, forest management decisions are based on the current appearance or condition of a forest stand without considering past events such as clearcutting practices. The silvicultural system used by the Menominees recognizes that disturbances on a site and less appropriate harvest practices of the past have dissipated native species and allowed other lower-value scrub species to overtake prime forestland. The system also recognizes that while many species of trees will grow on multiple sites,

Examples of the Most Desirable Forest Cover Types on Menominee Forests

Featured Cover Type	Objective Species	Associate Species
White Pine Mid-Tolerant Hardwoods Red Oak	White Pine Red Oak White Ash Basswood	Red Maple White Birch Quaking Aspen Pin Oak White Oak
Sugar Maple	Sugar Maple	Yellow Birch Hickory White Ash Red Oak Basswood Hard & Soft Elm

The matching of tree species or cover types to a particular habitat type is based on: a) sawtimber growth potential in quality and quantity; b) biological/ecological suitability to the site; and c) competitiveness with other tree species commonly associated with it.

Source: MTE

Figure 2

woods and pine but were altered by past management practices that included clearcutting for railroad logging and uncontrolled slash fires caused by poor logging practices. Restoring the 60,000-plus acres currently growing below maximum potential is one of the tribe's primary forestry objectives. The benefits of achieving this objective are listed here.

- The low-value material growing on the site is usually converted into pulpwood that is processed off the reservation. Few on-reservation jobs are supported by this low-quality material.

- The tribal forestlands are converted back to native conditions with larger-diameter, larger-volume, more valuable material constituting the forest landscape.

- Higher-value pines and hardwoods are all processed in the Neopit, Wisconsin, sawmill that is owned and operated by tribal members.

- The higher value species can be better directed toward not only higher profit-per-unit secondary wood product markets, but also markets that prefer certified wood over noncertified wood for product development and may pay a premium for that certified wood source.

Over 6,000 acres are selectively marked for harvest each year. Reentry for cutting in the hardwood stands occurs every twelve to fifteen years. The MTE "order of removal" tree marking rules ensure that a wide diversity of tree species in all size classes and ages are maintained throughout the forest.

SFM Forestry Objectives Produce Impressive Results

With twenty-five years of SFM practices in the forest, the Menominees performance in achieving their intended results ranks high. According to an independent evaluation conducted for FSC-approved certification (SCS and SmartWood) on the Menominee forest in 1994, independent evaluators stated:

they achieve their best form and quality on only one or two habitat types. The Forest Habitat Classification System can predict the forest cover type best suited to a site by identifying the existing forest and linking these tree species with a specific habitat. Some species, however, grow better on more than one or two habitat types. White pine is one of those that can thrive on five habitat types.

Using this new system, MTE has identified over 66,000 acres of forestlands that are currently growing below maximum quality and quantity potential. Examples include acreage of aspen, white birch, red maple, and scrub oak growing on sites with habitat types more suitable for pine and quality hardwoods. These acreage once supported high-quality hard-

Aesthetically, the Menominee Forest has no equal among managed forests in the Lake States region, although its total productivity measured in the value of the products removed greatly surpasses the adjacent Nicolet National Forest which has more than twice the acreage of commercial forestland.

Because of the unique SFM practices employed by the Tribe over the last twenty-five years, complete forest inventories conducted in 1963, 1970, 1979, and 1989 indicate that the Menominee Forest contained more volume of timber for each successive inventory period, even though over 600 million board feet of timber had been harvested off the forest during that period of time. Equally significant, evidence is clear that the quality of the timber stands and the value/volume of wood growing in the Menominee Forest improved dramatically in several critical areas.

Although the annual allowable cut (AAC) in the Menominee Forest actually decreased from 29 million board feet (mmbf) in 1983 to 27 mmbf in 1995, as noted in Figure 3 the annual harvest remained fairly constant during a thirty-year period between 1962-1995. In addition, SFM techniques used by the Tribe during that time actually increased the *quantity* and *quality* of trees left standing in the forest.

Between 1963 and 1988, forestry data documents the conversion of targeted acreage to higher-quality native species matching the appropriate habitat type conditions. The MTE continuous forest inventory (CFI) information shows that the acres covered by lower-quality, lower-value species, such as aspen, decreased, while the acreage covered by the native higher-value species, such as northern hardwoods, increased (see Figure 4).

Average Annual Harvest from the Forest (mmbf)

	Sawlogs	Pulp Conversion	Total
1962-69	20.2	4.9	25.1
1970-79	16.1	7.2	23.3
1980-89	13.4	8.4	21.8
1990-95	15.7	6.7	22.4

Source: MTE

Figure 3

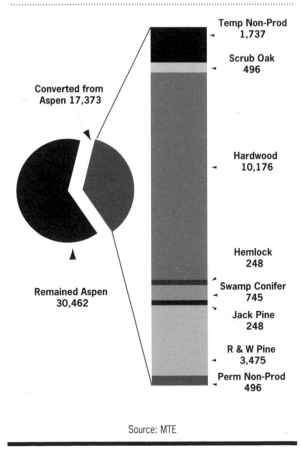

Cover Type Conversion Trends (1963-1988) on the Menominee Forest (in acres)

Temp Non-Prod 1,737
Scrub Oak 496
Converted from Aspen 17,373
Hardwood 10,176
Hemlock 248
Swamp Conifer 745
Jack Pine 248
Remained Aspen 30,462
R & W Pine 3,475
Perm Non-Prod 496

Source: MTE

Figure 4

During the same period, 1963 and 1988, the quality of almost all species represented in the forest system improved. In those same years, the total volume of sawlog (vs. pulpwood) timber resources growing in the Menominee Forest increased by over 200 million board feet, as well.

Through SFM practices, MTE's performance in increasing overall diameter sizes of standing timber within their forest system is exemplary. The quality and grade of sawlogs is often a direct correlation to the diameter size of the tree. Within limits, the larger the diameter of a tree at breast height, the better the grade of lumber material that will processed from that tree with correlating less defect. Measuring by the criteria of diameter at breast height the amount of acreage containing large northern hardwoods sawtimber increased 30% on MTE land between 1963 and 1988 (see Figure 5).

The increase in the diameter size of standing timber has translated into a greater volume of higher-grade wood that is processed into product at the Menominee sawmill. Grade 1 sawlog, for example, is often processed as higher-grade lumber for value-added products such as furniture. Lower-grade material from Grade 3 sawlogs might be processed into pallet stock. Between 1963 and 1988, MTE has increased their Grade 1 sawlog volume from 25% of their total growing stock to over 30% and has decreased their Grade 3 sawlog volume by an almost equal amount.

MTE has been so successful in increasing higher-grade material in its forest that it is currently considering developing three new classifications exceeding standard Grade 1 status. Currently, Grade 1 sawlog applies to at least a 14" diameter log with less than 13% defect. MTE is looking at new Grade 1 classifications that separate up to a 24" diameter.

One of the benefits of SFM practices is the ability for the Menominee Forest to provide more consistent flows of wood volume to the MTE sawmill, although *flow value* can vary from year to year. The sawmill knows in advance what volume it will be processing for the year, and what the grade or quality of material is likely to be. It is then contingent upon the marketing acumen of the sawmill operations to identify the best product distribution channels for both higher-value and lower-value material that can produce the best bottom-line results. This is particularly important with SFM harvests. These harvests avoid high-grading of the forest, or harvesting only the high grade material. Instead, they harvest a balance of lower-grade and higher-grade material. The annual sawlog harvest volume shown in Figure 6 highlights MTE's performance in this area. It shows that between 1970 and 1990 the mill had a more even sawlog harvest volume flow as a result of SFM practices.

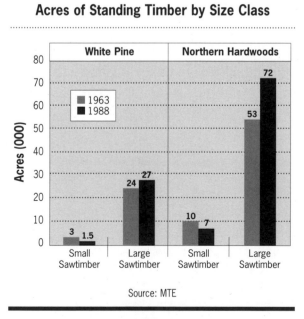

Acres of Standing Timber by Size Class

Source: MTE

Figure 5

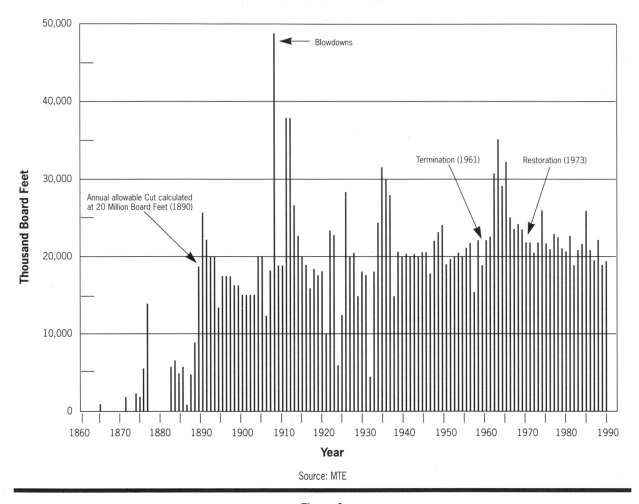

**Menominee Indian Reservation
Annual Sawlog Harvest Volume, 1860-1993**

Source: MTE

Figure 6

Unlike most other Native American Indian Tribes, and because of its proven track record in sustainably managing its forestlands over the years, the Menominee Nation, as part of its restoration process, was successful in negotiating an agreement with the federal government in 1975 that granted the Tribe the right to manage its forests. Typically, the Bureau of Indian Affairs (BIA) is assigned the duties and responsibilities of managing tribal forests. The tribe is provided a $1.3 million annual subsidy from the BIA to manage its lands. However, that agreement did come with a provision that the Tribe cannot sell or trade their forestland without congressional approval. The land cannot, therefore, be used as a financial asset, which obviously limits MTE's business financing options.

SFM Logging Practices Prove a Cut Above

Performance requirements for loggers contracting to harvest from the Menominee Forest must follow stipulated performance requirements not traditionally seen in logging contracts. The tribe incorporates both financial incentives and penalties for performing or not performing to required standards which ensure sustainable management practices in the field.

Those incentives and penalties include a bonus of $2.50/mbf of logs that are delivered to the MTE for the logger who successfully cuts 100% of the contracted harvest area in accordance with contract specifications. This is a critical incentive as the harvest from the forest includes all grades of material. It is essential to harvest lower-quality material to increase the growing space for higher-quality material in a sustainably-managed forest system. Traditionally, loggers will desire only to harvest and haul larger-diameter, higher-grade material because they get higher prices for it at the mill, or to harvest and haul straight pulpwood because they do it quickly with little regard for damage to the log. Under conventional contracting methods, it is less financially advantageous to the logger to incorporate both.

If a contractor fails to harvest the entire area, barring natural forces or conditions beyond human control, the contractor will forfeit all or a percentage of the performance guarantee based upon the percentage of the remaining contract that is unfulfilled.

Other financial penalties that are designed to ensure sustainable practices in accordance with contract requirements address the skidding and marking of logs, and excessive damage to logs during harvesting as indicated in Figure 7.

Also unique to MTE forestry practices, the logging contracts provide a $5/mbf incentive for the logger to use new thinning techniques in the field that require the logger to extract approximately thirty more "stems" (waste trees between 1"-5" diameter). This allows better growth potential for seedlings.

Examples of MTE Logging Performance Penalties

Action	Penalty Imposed
Unskidded or unmanufactured logs; logs manufactured into pulp	Double the logging rate of Agreement
Cut or girdled unmarked sawlog size tree	$250.00 per tree
Excessive damage to sawlog size tree	$125.00 per tree

Source: MTE

Figure 7

These incentive/penalty stipulations in the logging contracts have helped produce an exemplary record for reduced damage to the forest due to poor logging practices. A report to the Menominee Tribal legislature in 1984 conducted by the Wisconsin Department of Natural Resources found that the Menominee Forests experience 1.9 trees/acre of logging damage compared to 13 trees/acre on national forest systems in the region. This same report concluded a net board foot growth rate per acre per year in the Menominee Forest of 244 board feet compared to 235 board feet of national forest systems in the area.

Logger's "Lottery"—A Winning Solution

MTE also uses a unique contracting procedure for issuing logging contracts. The new "Loggers Lottery" as MTE forestry officials call it, is intended to increase performance on logging contracts and increase the responsibility on individual loggers to complete all their contracts during the year. The lottery process deviates dramatically from traditional logging contract bids, which normally operate on a closed-sealed bid basis, or are assigned by the forest manager.

In the MTE process, however, instead of assigning logging contracts, information on all the sites to be harvested based on certified sustainable management practices is released to all qualified loggers in advance. A date is set for an open bid process. The open bid does not include any price considerations, but concentrates only on the desire of loggers to receive their choice of sites for contract award. If more than one logger desires a particular site, ping-pong balls, one with each logger's name, are placed in a hat and the winning ball is drawn. Once site selection has occurred, negotiations on price begin with the winning contractor.

These negotiations usually end in an agreement. If no price agreement is met, the initial logger is rejected and a new bid process begins. Loggers that successfully gain a contract have a set period of time after the bid process to return selected contracts if they no longer want to handle the site. When this happens, MTE initiates new bids on the returned contract.

MTE gives bidding preference to Menominee Indian contractors. MTE currently has twenty-nine contract loggers who are prequalified to bid on MTE harvest contracts, nineteen of whom are tribal members. Even with the more difficult performance standards attached to logging practices in the Menominee forest, each year there is a growing list of tribal and nontribal contractors who want to contract with the MTE.

The purpose of this new approach is twofold:

• The open no-price bidding process allows contractors to discuss among themselves prior to bid time who should take what contract based on the capability of the contractor to meet performance requirements and avoid penalty payments. More important, the system is intended to ensure that full harvest amounts are completed each year, which is most critical to overall forest health and the viability of the sawmill operations.

• Because contractors select their own sites rather than being assigned sites, they have greater responsibility to perform according to contract stipulations rather than blaming nonperformance on being assigned a "bad site."

The process has worked reasonably well, although there have been problems with some contractors returning contracts at the last minute in hopes of being able to renegotiate a higher price to harvest more difficult and/or lower-timber-quality sites. In recent years, some acreage has not been harvested as a result. In 1995-96, for example, over 1 million board feet of timber was not cut as scheduled. This created a serious shortfall for the mill and negatively affected the health of the forest. In 1997, MTE was changing contract-return times and limitations to correct the problem.

To the Mill

The tribe began operating a small sawmill in the late 1850s. Today, the tribe owns and operates one of the largest sawmill complexes in the region, which employs approximately 160 people in Neopit, Wisconsin. The operation also employs a large team of forest management professionals at its Forestry Center located in Keshena, Wisconsin.

The mill manufactures primary commodity lumber products in both hardwoods and softwoods grown from the Menominee Forest. Standard lumber sizes produced by the mill range in traditional length and width lumber pieces. The mill also sells graded sawlogs and high-grade sawlogs for the manufacture of quality veneers. The approximate percentage splits for resource distribution and product production at MTE are as follows:

• 75%—Logs for lumber production

• 16%—Logs for direct log sales to lumber customers

• 7%—Veneer log sales

• 2%—Log inventory carry-over

The sawmill does not pay stumpage costs to the forestry operations for the logs harvested from their forests. This is a typical operating cost for many private forest products operations that do not own their own forestlands. However, all forestry costs are paid by the mill through product sales.

The sawmill receives no federal subsidies for its operations. Its success is critically dependent on the steady flow of timber from forest to market. The mill only produces from Menominee Forest resources and relies on no outside timber for lumber production and veneer and sawlog sales.

Although new dry kilns were installed in the mill operations in 1996, which provide much needed and valuable increased drying capacity and quality, MTE sells a significant portion of its lumber as green (nondried) to its customer base. Lumber drying is costed at MTE as a service to the client, rather than a high-value dried lumber product offering.

The mill follows product standards and traditional grading rules for both hardwoods and softwoods. Development of custom grades that could convert traditional lower-grade, defect material into higher-value character wood custom grades remains an unexplored opportunity for the mill.

Constraints to Production Using SFM-Harvested Material

While SFM practices produce positive, stabilizing results for a forest system, short-term impacts to a sawmill relying on SFM-harvested material prove to be significantly more challenging. Converting forestlands back to higher-grade native species and avoiding high-grading logging practices in the forest are but two SFM mandates that can wreak short-term havoc on a sawmill. The variations in species mixes and in total volumes per species per year that those practices can produce can have dramatic impacts on the business bottom line. Between fiscal years 1995 and 1996, for instance, the MTE sawmill experienced a 17% reduction in high-value hard maple harvest from its forest and a 47% increase in much lower-value aspen harvest, which resulted from an SFM forestland practice of converting acreage to mixed native species. Since the market price for hard maple is about $1,100/mbf more than for aspen, the short-term constraints experienced by the sawmill are dramatic (see Figures 8 and 9).

Differences in the mix of species as indicated in Figures 8 and 9 are especially important considering that white pine prices are historically higher than those for hemlock lumber, and hard maple and basswood are preferred hardwoods for product manufacturing—a fact that is reflected in the consistently higher prices those species command over

MTE Specie Mix as a Percentage of Production

	1995-1996	1996-1997 (Projected)
Total Production:	10,798,482 (bf)	9,266,940 (bf)
White Pine	26.6%	7.4%
Hemlock	4.0%	14.6%
Hard Maple	14.1%	36.9%
Basswood	6.7%	16.3%
Aspen	6.3%	4.2%

Source: Mater Engineering based on MTE data

Figure 8

MTE Specie Volume Harvest Variations

		% Difference from Previous Year				
		1991-92	1992-93	1993-94	1994-95	1995-96
Hard Maple	Base line yr.	−41%	+5%	+51%	−17%	
Aspen	Base line yr.	+49%	+7%	+<1%	+47%	

Source: Mater Engineering based on MTE data

Figure 9

other hardwoods such as aspen. In addition, hardwoods, in general, fetch higher prices per unit then softwoods. For MTE, this means that softwood lumber sales for fiscal year 1996–97 should decrease in overall dollar volume, but hardwood lumber sales (depending on the grade being offered) should increase rather significantly.

These species variations from year to year due to SFM practices have substantial impacts on the financial viability of a milling operation and underscore the need to maximize efforts to market and add value to the wood available for harvest in any given year. More clearly put, to afford SFM practices in the forest, it is essential for an operation to practice good business in production, marketing, and sales.

Selling Certified Wood—Not As Easy As It Seems

Although market research conducted by Mater Engineering in the United States during 1996–97 has documented a demand for certified wood that exceeds supply, it appears that MTE finds the selling of certified logs and wood products not as easy as it would seem.

Even though all MTE products come from a certified source, only 4% of hard maple veneer quality sawlogs sold by MTE are actually sold as certified. However, those certified veneer logs command a 10% free-and-clear (above cost plus standard markup) premium. The premium is clearly noted on the customer invoice with the following notice.

> 10% charge for product delivered from certified well-managed forestry operation. SCS does hereby certify that an independent chain of custody has been conducted at MTE and that this facility has been shown to meet certification requirements.

Hard maple and basswood lumber have also been sold as certified but with little documentation showing any premiums attached to those sales. This may be because the accounting procedure for lumber sales does not break out certified sales from noncertified sales, as does that for veneer log sales. However, an MTE sales manager stated that certified lumber sales make up about 5% of MTE's total annual lumber sales. Some of that material fetches premiums of approximately $50/mbf. However, it appears that only the hardwoods can command the premium, which equates to about 4% to 5% for certified hardwood.

THE CHALLENGE OF FILLING ORDERS

Both the sales managers of veneer and lumber at MTE report an increase in the number of inquiries asking for certified logs and lumber coming from all over the world. For different reasons, however, they face challenges filling those orders.

Veneer Logs

The problem stems *not from a lack of orders but from an abundance of orders and volume that MTE does not have the wood volume to fill.* Unlike high-grading practices that are often employed in standard forestry operations to meet demand, MTE's sustainable practices of pulling set volumes of high-to-low grade material from its forest on an annual basis prevents substantial sales of certified wood that could get 10% premiums. During the research for this case study, for example, MTE received a sales inquiry from a substantial U.S. veneer operation that wanted to purchase 4 million board feet of certified veneer logs annually. Based on SFM practices, the total volume for all veneer logs sold by MTE during 1995–96 was only 540,720 board feet. The projected volume for all veneer logs to be sold in 1996–97 is 814,550 board feet. In essence, the demand for certified veneer logs by one inquiry alone exceeds MTE's capability to supply it annually by almost 400%.

Because of the importance of offering certified veneer log sales to MTE's bottom line (the value per unit of veneer sawlog sales compared to lumber sales is 2:1), veneer sawlog sales are usually negotiated a year in advance, with first buying preference given to certified veneer sawlog buyers.

Lumber

Excluding the exceptions mentioned above, many of the inquiries that MTE receives are for small orders that are not cost effective to supply. Although, as with MTE veneer sawlog sales operations, between six and ten inquiries for certified lumber are received each month, inquires for certified lumber often range below 1,000 board feet per order. Without evaluating more creative product transportation options such as freight load matches and reverse shipping capability at different locations throughout the United States, manufacturers typically prefer full truckload orders to keep per-unit prices competitive for consumers. Standard truckloads volumes per species are 11,000 board feet for hardwoods, and 20,000 board feet for softwoods.

Another problem appears to be MTE's ability to supply consistent volumes of certified wood per grade requirements *on a year-to-year* basis for manufacturers that want to use certification as a marketing strategy for their products. Since MTE is only one of a handful of sources for certified timber in the entire northeastern section of the United States, and with sawlog grades for sustainable harvesting varying from year to year, lack of other certified wood resources in the area that a manufacturer could rely on to ensure consistent quality and quantity of certified resource must be a strong consideration. One advantage, however, of certified forestry is that those harvest evaluations are done several years in advance. This means that volumes and grades of material scheduled for harvesting are also known well in advance. This does allow a level of resource information that MTE can rely on over time for sales and marketing.

NEW OPPORTUNITIES REQUIRE DOING BUSINESS DIFFERENTLY

Due to the visibility MTE has received for its sustainable forestry practices in the Menominee Forest, the Knoll Group, a well-known furniture manufacturer, approached the sawmill in 1991 to consider becoming independently certified. The Knoll Group wanted to purchase and use Menominee Forest hard maple in a commercial furniture line that would be marketed as furniture made from "sustainably-managed maple."

The Knoll Group paid FSC-approved SCS, out of San Francisco, to undertake an independent assessment of the MTE forest management practices. Based on that in-depth assessment, both MTE's forestry operations and the sawmill were certified. MTE became the first forestry and forest products operation to be awarded certification status in the United States. In 1994, the operation underwent another assessment by SCS and a new assessment by SmartWood, another FSC-approved certifier located in Vermont. Based on those assessments, MTE became the only forestry and forest products operation in the United States to receive dual certification. It remains so today. It is also the only Indian tribe in the United States and Canada to hold certification for its forestlands and forest products operations.

While opportunities for selling certified wood do open, as evidenced with the Knoll Group, MTE has discovered that paying for sustainably-managed forests and taking advantage of a certified product offering difference in the marketplace puts added demands on the sawmill and requires management to do business differently. For MTE, this may mean shifting a percentage of its product offering to a new customer base, which is a risky proposition for any operation. MTE has relied most heavily on customers located in its region, and has only recently engaged customers in other parts of North America, Asia, and Europe. Since the demand for certified wood product is greatest in European markets, MTE needs a stronger marketing effort for that part of the world.

Market Analysis Leads to Changes

The mill, which recognizes the need to improve sales and product offerings, mounted a significant mill modernization program in 1995 that has included a complete evaluation of its manufacturing productivity and a new focus on implementing value-added manufacturing in the milling operations in Neopit. The mill also conducted a marketing analysis to improve its performance with existing customers and increase its customer base. A portion of the markets analysis conducted in 1995 was targeted to expanding opportunities for certified MTE wood. The purpose of the survey was to identify ways to:

- Increase the value of certified wood resources to MTE consumers. Up to this point, MTE had not campaigned actively to sell certification of their wood as a unique product offering to customers.

- Identify if an opportunity existed to create custom grades of lumber from wood that is classified as defect material by standard grading rules. This issue was important for evaluating MTE's older-growth hard maple that was scheduled for harvest under certified harvest plans. The old-growth hard maple has a unique brown color configuration ribboned in its sapwood. This "brown stain" is graded as defect material under traditional grading rules based on appearance only (hard maple is often used in sports flooring which requires clear, vertical-grain lumber). MTE wanted to find out if there were opportunities to create a custom grade of lumber from this material.

- Identify if opportunities existed to incorporate added-value in the mill's product manufacturing that could meet customer needs and increase value per unit of timber produced to the mill.

Approximately 20% of the total MTE customer base was directly interviewed for this survey, which produced the following results:

- All of the surveyed MTE customers indicated a desire to continue or increase business relationship with MTE.

- 55% of the surveyed customers indicated an immediate interest in evaluating potential for MTE to provide value-added product to them rather than just commodity lumber. Customers had the most interest in MTE's ability to add either fingerjointing and/or edge-gluing operations to their existing production line. Customers also indicated a strong interest for MTE to increase their kiln-drying capacity for lumber products sold.

- 32% of surveyed customers indicated an immediate interest in evaluating brown stain maple for their high-value products such as flooring, door parts, and panel products.

- 55% indicated an immediate interest in working with MTE to develop visibility for their product manufactured from MTE certified wood, although the majority of customers were unaware they were receiving certified wood from MTE. Customers rated high-value products such as face veneers, flooring, furniture, and specialty products such as cutting boards and tool handles as those having the best product potential for certified wood.

- 23% of the surveyed MTE customers stated an immediate interest in engaging in all three activities with MTE: a) receiving value-added product; b) marketing certified wood products; and c) working with brown stain maple.

Recognizing an Advantage

Although a significant number of MTE's existing lumber and veneer customers were either unaware of or did not use the certification status as a market advantage for their product development, there were some exceptions.

Conner AGA

Located in Amasa, Michigan, this long-time sports flooring manufacturer has been a consistent buyer of MTE's high-grade hard maple. The company had been tracking MTE's certification process and, during 1996, had its own operations assessed by the SmartWood program to receive chain-of-custody certification, which would allow the company to process certified wood from MTE and sell wood flooring as certified. When Conner received SmartWood certification, the company developed a brochure to tell clients and customers about the "new" product offering.

During the research for this case, Conner AGA informed MTE that it had just secured a contract from Walt Disney World, Inc., for hard maple wood flooring for Disney's new facilities in Florida. Conner indicated that the certification status of its product was a key asset in winning the contract, which will use over 165,000 board feet of MTE SmartWood-certified hard maple.

Gibson Guitar

Using MTE's certified maple, Gibson Guitar has released its first line of wood guitars made from SmartWood-certified wood. Two years in the planning, the first Gibson SmartWood Guitar (a Les Paul standard model), made its debut in a widely-publicized October 1996 concert held in New York City to benefit the Rainforest Alliance-SmartWood Program.

Certified SFM—Meeting the Bottom Line

With the exception of 1996, MTE has operated at a profit since 1991 (see Figure 10) even though it has relied primarily on substantially lower-value green lumber sales; has appeared to price some products at below-published price structures; has not yet taken advantage of custom grade development opportunities; and is still in the process of evolving additional value-added production on site. MTE has done so while pioneering SFM practices that have set standards for forest management that are now being reviewed and implemented by other forest products operations throughout the world.

BEYOND THE BOTTOM LINE

While not meeting traditional industry standards of reaching 20-25% net profit as a percentage of total sales for comparable sawmill operations, MTE's mill

MTE Financials

	1991-92	1992-93	1993-94	1994-95	1995-96
Total Sales ($)	$9,388,258	$10,840,269	$11,528,901	$12,610,480	$11,214,027
% increase/(decrease) in total sales from previous year		+15%	+6%	+9%	(–12%)
Net Profit/(Loss)	$718,942	$776,849	$1,679,780	$1,274,083	($402,507)
Net profit as % of total sales	8%	7%	14.5%	10%	0
Volume bf produced	10,909,368	12,081,244	10,065,446	10, 460,992	10,798,482
Sales price/bf of production	$.86/bf	$.90/bf	$1.15/bf	$1.21/bf	$1.04/bf

Source: MTE

Figure 10

has, except for the 1996 fiscal year, operated at a profit while employing more than double the number of personnel traditionally used in an operation of this annual volume production.

The MTE sawmill, which produces between 10-12 million board feet of lumber annually, employs about 160 people. Typical hardwood milling operations in the private sector for that size of annual production would employ between 80-90 full-time people. Since labor costs typically constitute over 40% of total mill operating costs, this one deviation in operations makes a substantial difference in the financial viability of a milling operation. However, unlike typical private concerns where increasing the bottom line is the prime objective, MTE objectives are intentionally different. MTE strives to achieve, within certified SFM practices, as many consistent, full-time, family-wage jobs for tribal members as possible while maintaining a positive bottom line from year to year.

President Larry Waukau, in the annual President's Report for 1996, stated those goals clearly:

> Our long-term approach to integrated sustainable forest management and wood products manufacturing (a 150-year horizon) precludes MTE from capitalizing upon market peaks and dips. Instead, our approach sets a long-term course which at times requires us to forfeit short-term gain for long-term sustainability.

In addition, the forestry operations incur additional logging costs by providing financial incentives to its contract loggers for harvesting Menominee Forest wood sustainably. These include premiums paid for increased stem thinning in the forest, and premiums paid for harvesting 100% of the contracted volume.

Finally, unlike traditional forest products operations that use intensive high-grade forestry operations, the variation in grade of logs that MTE must harvest from year to year based on sustainable management practices poses additional challenges to achieving more typical profit levels. Over the last

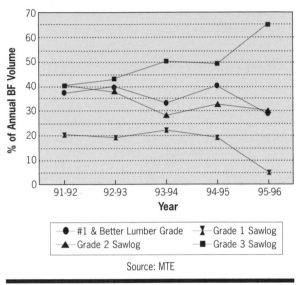

Correlation of MTE Hard Maple Sawlog Grade to Lumber Grade When Processed at the Mill

Source: MTE

Figure 11

five years, the sustainable annual allowable cuts from the Menominee Forest have produced dramatic shifts for the mill operations.

Using hard maple volume data from the Menominee Forest as an example, Figure 11 shows that Grade 1 (good quality) logs harvested from the Menominee Forest dropped dramatically between fiscal years 1994-95 and 1995-96. Conversely, Grade 3 (poorer quality) logs increased significantly as a percentage of the overall log volume brought to the mill. The volume of Grade 2 logs remained constant. The impact on the mill is clear: The volume of Grade 1 and better-grade lumber manufactured by MTE, which are used in high-value products, decreased markedly.

While other factors, such as efficiencies in the mill and the accuracy of log grading certainly affect the overall mill production, the short-term variation in harvestable grades necessary to achieve long-term sustainable management practices is a key constraint for traditional forest and mill operations. It clearly illustrates the need to match sustainable forestry practices with other profit-oriented solutions.

The financial loss evidenced by MTE in 1996 was primarily due to weather conditions and unfulfilled logging contracts that left over 1 million board feet of timber standing in the forest that should have been harvested under the certified allowable cut for this last year. However, both forestry and mill operations continued to operate at full employee levels as if 100% of the certified harvest volume had been brought to the mill.

GOOD BUSINESS PRACTICES A KEY TO SFM SUCCESS

It is clear that to financially sustain a certified operation, there can be no substitute for matching SFM practices with other profit-oriented business solutions such as increasing mill efficiencies, adding value to commodity products, developing custom grade material, and marketing products made from certified wood better.

As they are in most contemporary sawmill operations, efforts to improve production and increase profit margins at MTE are ongoing. MTE has a number of potential opportunities that it could consider to increase sales and improve production.

Evaluating Product Pricing Practices

Product prices on a per-unit basis can vary due to many factors. Long-term relationships with customers, distance of the mill from customer location, and fluctuations in current market values are but a few significant factors that can affect pricing decisions. Even so, keeping track of averaged pub-

Comparison of MTE Selected Lumber Prices to Standard Published Lumber Prices over Time (prices are for non-dried 1" thick lumber)

	July 1994-June 1995			July 1995-June1996		
	Published Avg. ($/mbf)	MTE Avg. ($/mbf)	% Difference	Published Avg. ($/mbf)	MTE Avg. ($/mbf)	% Difference
Hard Maple:						
FAS	1132	1063.5	(–6.5%)	1141	1027.2	(–11%)
#1C	761	715.3	(–6.4%)	756	683.6	(–10.6%)
Red Oak:						
FAS	1354	1276.6	(–6%)	1252	1304.76	+ 4%
#1C	884	881.6	(–.3%)	818	943.78	+15%
Basswood:						
FAS	769	801.6	+4.1%	749	818	+ 9%
#1C	383	412.1	+7.6%	365	409	+12%

Source: Mater Engineering based on published lumber prices

Figure 12

lished market prices can provide some interesting insights into possible missed opportunities. For this case study, published hardwood prices in Hardwood Review during fiscal years 1994-95 and 1995-96 for high-grade lumber were analyzed and compared to MTE pricing data for the same period of time. While MTE appears to be capturing above published prices for certain species such as basswood and red oak, they appear to have also consistently priced below published prices for higher grade hard maple (see Figure 12).

This action is noteworthy considering that hard maple has represented a significant portion of MTE's annual hardwood lumber sales. For 1996-97, hard maple is projected to make up almost 40% of MTE's total lumber sales.

Converting Green Lumber Sales to Kiln-Dried Lumber Sales

Adding value to raw resource is a critical area most mills are exploring today along with MTE. Currently, only about 25% of the lumber MTE sells is sold as dried in its own kilns, the rest is sold as green. However, MTE charges for that lumber drying as a service to the customer at an average rate of between $50-100/mbf of lumber dried. This value is then added on to a green lumber price. The missed opportunity to increase gross revenue to the mill is best seen by comparing published green hardwood lumber prices against published prices paid for kiln-dried lumber as a product offering (vs. a service offering). For the hardwood species such as hard maple, red oak, and basswood that MTE sells, the average percentage increase between green and dried lumber indicates a missed gross profit opportunity of approximately $300/mbf; a 40% average increase in gross revenue per thousand board foot of hardwood lumber sales (see Figure 13). MTE's projected 1996-97 value of $100/mbf added for lumber drying as a service appears to represent only a fraction of the actual income that might be generated by offering dried lumber as a product.

Developing Custom Grades for Targeted Species

Although as yet unexplored by MTE, as it is with many other traditional wood product producers, developing custom lumber grades can present significant opportunities for a sustainably-managed forest products operation that must rely on varying grades of material from year to year. MTE's 1995 customer survey indicated interest in this area, especially for hard maple with brown ribboned character in the wood. Under normal conditions, the configured wood would be either considered waste or converted into economy grade material often used in pallet production. Pallet stock usually sells between $125/mbf to $200/mbf. Converting the wood into a custom grade to be used in the flooring markets, for example, could increase prices to over $500/mbf.

Non-Dried (Green) vs. Dried Lumber Comparison ($/mbf; 1" thick lumber)

	Green (FAS; #1)	Kiln-Dried (Sel & Btr; #1)	% Difference
Hard Maple:			
Sel & Btr/FAS	$1,126	$1,454	+29%
#1C	744	1,079	+45%
Basswood:			
Sel & Btr/FAS	756	1,0571	+40%
#1C	380	607	+59%
Red Oak:			
Sel & Btr/FAS	1,340	1,693	+26%
#1C	869	1,161	+34%

Source: Mater Engineering

Figure 13

FUTURE BUSINESS STRATEGIES

Since its increased visibility as a certified sustainably-managed forest products operation, MTE has initiated additional marketing and sales activities that are intended to help maintain its SFM practices in the forests and increase certified wood products sales. Those activities are noted below.

1) Improved Production Goals

MTE has identified the following mandate to complete in fiscal year 1996-97.

- Improve lumber inventory control systems to increase production efficiency and provide better information for consistent product delivery to MTE customers.

- Complete energy audits and related development improvements in energy system designs and development. These improvements will reduce overall production costs and provide additional revenue to sustain a certified wood products operation.

• Continue the study and planning for development of a log sorting yard, whole log chipping, and boltwood processing facilities. These activities should greatly increase MTE's ability to make more with less and help to convert traditional wood waste to wood profits.

2) Increased Marketing and Public Relations Tools

MTE has developed a number of devices that highlight its certified wood product offerings and SFM practices in the Menominee Forest.

• *Brochure visibility.* MTE's facility brochure highlights its commitment to sustainable forest products both on its cover and in the text of the brochure which states:

> Another added feature for MTE is the recognition by Scientific Certification Systems, more commonly known as "Green Cross Certification," for incorporating sustainable management practices into the overall management of their forest and the contribution protecting habitat specie biodiversity and timber resource base of the forest in which they are harvesting.

• *Video visibility.* Within the last year, the demand for information about MTE's SFM practices has prompted the development of two videos that are now distributed to MTE customers, interested parties, and the public worldwide. *Listening to the Forest* is a fifteen-minute documentary on the management of the Menominee Forest and the way that management has improved the health of the forest and the economy of the Menominee Nation. A more technical video, *Sustainable Management of White Pine Stands,* addresses interests specific to forest managers and concerns about the sustainable management of white pine stands in the Great Lake States area.

• *Establishment of public demonstration sites on sustainable forestry and certified wood products from MTE.* Three sites within the Menominee Forest have

been designated as sustainable forestry demonstration sites for the public. The sites will include signage, designated guided tours, safety requirements, and equipment and materials used in sustainable forestry operations. At the main sawmill in Neopit, a certified wood products showcase is to be established, starting with the MTE building itself, illustrating the certified wood products used in the construction of the MTE administration building.

• *Establishment of the MTE World Wide Web site.* Initiated in 1996, MTE has its own Web site to expand Menominee's sustainable forestry and certified wood products visibility worldwide (http://www.menominee.com).

• *Bumper stickers for cars.* The tribe has bumper stickers to distribute to the public that advertise MTE as offering hardwood and softwood. They read *Native American Sustainable Resources.*

• *Annual marketing workshops for MTE customers.* During 1995, MTE initiated its first annual marketing workshop for MTE customers. The annual workshop provides current marketing information about the specific products manufactured by MTE's customers and incorporates a full section on growing markets and product distribution systems for moving certified wood products in international markets.

SFM PROMPTS SOCIAL PRIDE AND CONTINUING INTERNATIONAL EDUCATION

The changes that have occurred in the Menominee Nation through the implementation of SFM in the Menominee Forest have provided significant social and educational benefits to the People of the Menominee Nation. The visibility of Menominee's work in SFM has strengthened social pride within the Menominee People and increased opportunities for new programs that generate added income to the tribe. Some of these changes and benefits are described here.

Presidential Recognition

In 1995 MTE was one of a handful of U.S. organizations to receive the first-ever United States President's Award for Sustainable Forestry Development. The event, which was one of great pride to the Menominee Nation, honored MTE for its pioneering achievements in SFM practices.

Development of a Sustained Development Institute

Prompted by the success of SFM practices employed in the Menominee Forest, in 1993 MTE and the College of the Menominee Nation jointly established the Sustained Development Institute (SDI) for the Menominee Nation to promote the Menominee Forest and its management, and to educate the public and especially the Menominee children about the resource. The Institute was also chartered to further develop and describe the principles of sustainable development, apply them to different environmental situations, and then to educate the public about the principles and how they are applied. The institute's specific tasks include:

- Providing tools useful for those who want to study and work toward sustainable communities.

- Developing on-reservation demonstration areas to teach visitors about the major elements of Menominee sustainable development.

- Developing a curriculum, working in concert with the U.S. Department of Agriculture, that teaches the elements of Menominee sustainable development and relates those to issues, efforts, and movements in other parts of the world.

- Developing an Internet-based program in concert with the Land Tenure Center of the University of Wisconsin that can lead to a four-year degree in sustainable development.

- Developing short seminars on Menominee sustainable development for visitors, students, professional foresters, environmentalists, and others.

Initiation of Sustainable Forestry Demonstration Workshops

In October 1996, MTE gave its first sustainable forestry demonstration workshop for public and private forestland managers. The conference and field demonstrations, which concentrated on timber harvesting systems for SFM, had over 120 participants from 10 U.S. states and other countries. About 20% of the participants came from the provinces of Canada, with private forestland owners from as far away as South America also attending.

College Curriculum Development

College curriculums are now offered at College of the Menominee Nation detailing cost/benefit economic analyses of sustainable forestry issues. The designed coursework is not only offered to Menominee Tribal students, but is also extended to Department of Natural Resources state forestry staff and U.S. Forest Service personnel. The course is also offered through the Internet with "virtual students" registered from as far away as Sweden. Case studies used in the training and analyses are actual Menominee Forest management cases.

Development of Student and Youth Programs

The MTE has also initiated new programs aimed at educating tribal youth about job opportunities and the values inherent in SFM practices, which include:

- Student-to-Work Programs that target high school students interested in working within the MTE Forestry Division to give them experience and high school credits for hands-on experience in Menominee SFM practices.

- Youth learning programs that allow younger tribal students to spend time in the field with MTE foresters learning about SFM practices. The students are then required to transfer what they have learned to global students through the Menominee Web site (http://www.menominee.com/camp).

Lessons Learned

The experience of the Menominee Nation illustrates both the opportunities and constraints that can arise in the practice of SFM.

The evidence clearly indicates that SFM practices can work—in the forest, the community, and for the bottom line. Since 1854 The Menominee People have lived on their Reservation and intensively managed the land—yielding billions of board feet of timber for production. The wise use of the resource has created hundreds of sustained, full-time, family-wage jobs for tribal members for almost 100 years. But equally important, SFM has increased *quantity* and *quality* of remaining standing timber in the Menominee Forest.

The experience of MTE, however, also suggests that SFM practices come with a price. Short-term gains may need to be sacrificed for long-term sustainability goals. The short-term challenges of producing a profit underscore the importance of conducting good business practice *in tandem with* SFM practices. For MTE, this may mean:

• Reevaluating product pricing strategies.

• Increasing product diversity through increased product offerings (such as dry-kilned lumber).

• Developing custom grades that better market character wood.

• Increasing value-added product offerings that convert wood waste into wood profits.

• Implementing a log sort yard to take advantage of multiple log buyers' needs

Using SFM practices in the forest does make for good business in the mill. While critics may point to the business benefits of being an Indian Tribe (government subsidy; no payment of taxes, etc.) as key components to financial success, it is clear that the Menominees have made specific choices in the way they run their business. This includes *double* the typical level of employees needed for similar-sized forestry and wood processing operations. When one considers that labor wages for a typical wood processing operation the size of MTE account for about 30% of the total operating costs—with taxes accounting for about 8%—the impact of MTE's increased employment goals on its financial performance is clear.

It is unclear whether any premiums attached to the sale of Menominee's certified wood and wood products are, or will be, a growing or even consistent trend to the company or others marketing certified wood products. However, it is evident that in certain product offerings, the current annual demand for certified products exceeds the supply. For this reason, expanding the amount of certified wood available is vital to opening up and gaining access to markets for certified products.

Nonindustrial Private
Forest Landowners:
Building the Business
Case for Sustainable
Forestry

CASE STUDY

PREPARED BY:
MICHAEL P. WASHBURN
STEPHEN B. JONES
LARRY A. NIELSEN

A Case Study from "The Business of Sustainable Forestry"
A Project of The Sustainable Forestry Working Group

The Sustainable Forestry Working Group

Individuals from the following institutions participated in the preparation of this report.

Environmental Advantage, Inc.

Forest Stewardship Council

The John D. and Catherine T. MacArthur Foundation

Management Institute for Environment and Business

Mater Engineering, Ltd.

Oregon State University
Colleges of Business and Forestry

Pennsylvania State University
School of Forest Resources

University of California at Berkeley
College of Natural Resources

University of Michigan
Corporate Environmental Management Program

Weyerhaeuser Company

The World Bank
Environment Department

World Resources Institute

Nonindustrial Private Forest Landowners: Building the Business Case for Sustainable Forestry

PREPARED BY:

MICHAEL P. WASHBURN

STEPHEN B. JONES

LARRY A. NIELSEN

A Case Study from "The Business of Sustainable Forestry"
A Project of The Sustainable Forestry Working Group

Contents

(continued)

Contents

Nonindustrial Private Forest Landowners: Building the Business Case for Sustainable Forestry

Introduction

No discussion of sustainable forestry would be complete without considering the unique aspects of nonindustrial private forests (NIPFs). Owners of these forests control 58% of the commercial forests in the United States. East of the Mississippi this type of ownership accounts for more than two-thirds of the region's timberland, whereas west of the Great Plains the majority of forests are in public ownership. The 261 million acres in NIPFs protect watersheds, provide wildlife habitat, offer scenic beauty, and supply 49% of the timber harvested in the United States, according to the U.S. Forest Service. This supply is critical for many large wood products manufacturers. Weyerhaeuser Co., for instance, harvests 58% of its timber supply from NIPFs nationally, and 90% are in the South.

The ten million NIPF owners—a diverse group including individuals, partnerships, estates, trusts, clubs, tribes, corporations, and associations—confront a variety of challenges that can complicate the practice of sustainable forest management (SFM). Many are not well informed about the economic value of their resource or the importance of consulting professional foresters when making management decisions. Annual property taxes and capital gains taxes can be disincentives to sound, long-term forest management. Without proper estate planning, owners can be forced into making decisions that may prevent them from passing forest land from one generation to the next, and may lead to the conversion of the forest to other uses. Equally important, the objectives of the owners combined with their individual financial circumstances are determining factors whether forest land will be managed sustainably or not.

The cases of seven NIPF ownerships presented here range from a small family forest that is managed for amenity values to a large tract managed for timber and investment. They are located in the Northeast, Pacific Northwest, and Southeast, which represent very different timber-growing regions. Although all these owners use professional forestry advice, and all the properties have been in family ownership for decades, they are indicative of the wide range of NIPF owners' backgrounds, objectives, and financial circumstances. They also illustrate how a diverse group of private landowners has addressed issues of forest sustainability. A section on certification examines three innovative approaches now underway to certify NIPFs: a certified resource manager, a chain-of-custody certified manufacturer, and a single forest owner seeking certification.

A Portrait of NIPF Lands and Owners

The NIPF category includes properties not held by government or forest products manufacturing firms. As Figure 1 indicates, 90% of the NIPF owners hold less than 100 acres. These small parcels account for 30% of NIPF acreage. Just 3% of private owners hold about 29% of the private forest acreage in parcels greater than 1,000 acres. This includes forest products companies and some large NIPFs. (The data below treat both NIPF and forest industry lands as "private." The "corporations" category includes forest industry and companies that own land but do not manufacture or sell forest products—for example, land-holding and investment firms. Timber harvesting is the primary objective on all of these lands.)

The number of NIPF owners continues to grow, increasing by 27% from 1978 to 1994, according to a study done by Thomas Birch of the U.S. Forest Service, Northeast Forest Experiment Station in Warren, Pennsylvania (see Table 1). More than 40% of current NIPF owners acquired their property since 1978. However, during the same period there was a drop in the number of large tracts, over 1,000 acres, which indicates that private forest lands are

Distribution of Private Forest Ownerships, by Size and Class of Ownership in The U.S. in 1978 and 1994

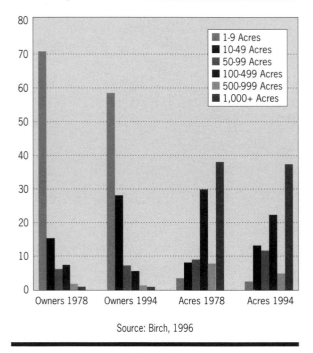

Source: Birch, 1996

Figure 1

Estimated Number of Private Ownership Units and Acres of Forest Land Owned by Type of Ownership, United States, 1978 and 1994

Owners	1978		1994	
Form of Ownership	Thousands	Percent	Thousands	Percent
Individual	6,793	87.6	9,319	94.1
Partnership	484	6.2	289	2.9
Corporation	237	3.0	157	1.6
Other	246	3.2	136	1.4
Total	**7,760**	**100.0**	**9,901**	**100.0**

Acres	1978		1994	
Form of Ownership	Millions	Percent	Millions	Percent
Individual	183.5	55.1	232.3	59.0
Partnership	35.8	10.7	29.7	7.5
Corporation	101.1	30.4	107.1	27.2
Other	12.7	3.8	24.3	6.3
Total	**333.1**	**100.0**	**393.4**	**100.0**

Source: Birch, 1996

Table 1

becoming increasingly fragmented. Tract size is an important criterion in the NIPF sustainability equation. Small forest tracts produce less timber, which can force heavier cutting in the short term to meet immediate financial needs.

In general, recent studies by Birch show that the "new" NIPF owners who have acquired land since 1978 are younger, better educated, and have a higher income than the average owner of 1978. Birch also found that the proportion of retired owners has increased: About 20% of owners are retired, which raises serious questions about the continuity of management philosophy and the need for estate planning. Smaller parcels cannot be

managed as efficiently as large ones. And the disposition of lands to several heirs or outright sale to pay estate taxes are major contributors to the fragmentation of forest lands, and the parceling of larger tracts into smaller ones.

NIPF owners hold their land for a variety of reasons. About 40% cite recreation or hunting as the primary reason for owning forest land, according to Birch (1996) and Jones et al. (1995). Ownership may be incidental to other uses. For example, forest land may be part of the farm. In suburban areas, forests are often conveyed with homes as part of subdivisions. For many, however, their ownership is by design. Nine percent of NIPF owners

NIPF Owners Reasons for Owning Forest Land

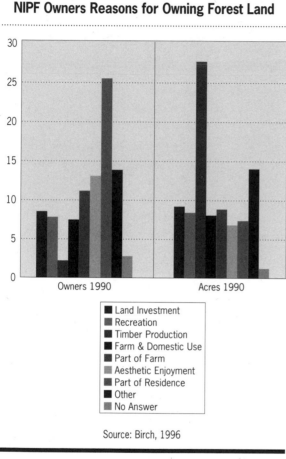

Source: Birch, 1996

Figure 2

NIPF owners have diverse reasons for owning land, and desire a variety of values, both amenity and commodity. Most owners are not well informed about the economic value of their resource or the importance of consulting professionals when making management decisions. Knowledgeable landowners, and those who use natural resource professionals when making decisions, tend to make decisions more consistent with principles of sustainable forestry.

A 1993 study by Andrew F. Egan, Assistant Professor of Forestry at West Virginia University, and Stephan B. Jones, director of the Alabama Cooperative Extension System, found that landowners' forestry decisions vary directly with their knowledge of forests and forestry: Informed landowners are more likely to make decisions that result in sustainable practices. This finding suggests that education should be a component of efforts to promote NIPF sustainability.

Although a number of studies have characterized the "typical" landowner, each *real* landowner brings a unique set of demographics, motivations, understanding, and objectives to managing an individual property. The seven individual properties in this case each encompass a unique set of circumstances. Collectively, they express a wide range of ownership characteristics that illuminate the broad NIPF case and justify general conclusions about building the business case for sustainable forestry. The properties represent a continuum from the low-intensity approach of a retired brother and sister who manage their 171-acre tract for scenic values, wildlife habitat, and timber production (within the context of maintaining continuous forest cover), to the income-driven activities on a multigeneration, family-owned property of 1,728 acres. The properties and their distinguishing features are listed in Table 2. The key forestry terms used in the cases are explained in the glossary at the end.

(10% of the NIPF acreage) purchased their land as an investment (see Figure 2). The reason for ownership plays a critical role in landowners' forest management decisions. However, landowner behavior is not always consistent with their attitudes. Despite citing other objectives as more important, about 50% of owners, representing about 75% of the acreage, have harvested timber at some time during their land-owning tenure, according to 1994 U.S. Forest Service estimates.

Key Characteristics of the NIPF Properties

Property/Total Acreage	Ownership Type	Owners' Objectives
Brent Oregon, 171	family, brother and sister	scenic values, wildlife habitat, timber
Freeman Pennsylvania, 639	family, husband and wife with three sons	forest stewardship, community education, timber
Trappist Abbey Oregon, 1,350	community, 37 brothers	self-sufficiency, timber, aesthetic quality
Cary Florida, 2,634	individual	maintain ownership through land and timber sales
VanNatta Oregon, 1,728	family, four generations	sustaining family farm business
Lyons Pennsylvania, 2,000	family, two brothers	timber, investment, recreation
Frederick Alabama, 12,768	corporate trustee	asset growth, investment, estate planning contingencies

Table 2

THE BRENT TRACT: A PERPETUAL FOREST OF DOUGLAS FIR

The Brent Tract, 171 acres of farm and forestland located in Oregon's Willamette Valley, is covered with 40- to 80-year-old-stands of high quality Douglas fir. The owners, a brother and sister, Matthew Brent and Virginia Picht, value the land not only for the timber income it provides, but also for its scenic beauty, and the ecological and recreational services it provides. Maintaining the health of the forest is the owners' primary objective. That is why they favor selective thinning and natural regeneration of the forest over the clearcutting, industrial-style techniques typically practiced on neighboring properties. Today the Brent tract is one of few examples of uneven-aged management of

Douglas fir in western Oregon. The management techniques used on this property offer a practical alternative to clearcutting for the Douglas fir region and make it a good site for further study, demonstration, and refinements of selective thinning.

Family Heritage

Virginia Picht and Matthew Brent consider their property a family heritage and feel a responsibility to be stewards of the land. Their parents acquired the property as a farm and moved there with the two children in 1920 to raise their family. The Brents farmed the property until 1958, when neighboring farmers leased the farming rights.

Today Picht, age 76, lives in Corvallis and is responsible for most of the management of the property. Brent, age 78, lives on the farm. Picht has a daughter who currently lives in Germany but hopes to live on the farm one day. They are striving primarily to maintain a perpetual forest and keep the property in the family. They take a long-term view of management, even though they realize that neither may live to see the final results of their actions.

In the Shadow of Mary's Peak

Picht and Brent own the 171 contiguous acres in a single joint undivided interest. The property is located five miles southwest of the small sawmill town of Philomath, and 15 miles from Corvallis. The property, which sits on the eastern edge of the central Oregon Coast Range, consists of gently rolling hills that reach an elevation of between 600 to 800 feet. It lies near the base of Mary's Peak, the highest point in the Oregon Coast Range. Rock Creek, a source of drinking water for Corvallis, flows through the property for over half a mile. The 10,000-acre Mary's Peak Watershed (Suislaw National Forest and City of Corvallis ownership) borders the Brent property. The reservoir for the city's water is located upstream.

The Brent Tract is located among public, small private, and industrial ownerships. In the foothills region bordering the Willamette Valley, the fertile valley bottoms are farmed, whereas most sidehills and ridges are forested. Grass seed is the primary crop on the valley bottoms, and Christmas tree farms of Douglas fir and noble fir dominate agriculture in the foothills. Much of the heart of the Coast Range is included in the Suislaw National Forest, which until 1993 was managed primarily for commercial timber production. Recently, much of the Suislaw, which contains numerous stands over 100 years old, was taken out of production as part of efforts to protect the northern spotted owl under the Endangered Species Act.

Intermingled with the national forest land are large tracts of industrial forest, mainly Douglas fir, that are managed as short rotation (40-60 years) plantations. At the valley margins most forestland is held by nonindustrial private owners. Since the NIPF land contained younger timber than the surrounding industrial lands, NIPF lands were largely unmanaged until the 1980s, when rising timber prices led to increased clearcutting on those properties.

About 136 acres of the Brent Tract are wooded; the remaining 35 are in fields with a residence. About 111 of the wooded acres are well suited for commercial timber production. The remaining 25 acres are lowland stream areas stocked with mixed hardwoods and conifers. The deep rich soils and abundant rainfall of the Coast Range, which reaches between 50 and 60 inches a year, provide good forest-growing conditions. The average site index for the property is 125 (50-year basis), which means the land has the potential for timber growth of nearly 1,000 board feet per acre annually. The

property lacks a current timber inventory, but timber volumes are estimated to be near 1.55 million board feet (mmbf), much of that export grade. Based on an average stumpage value of $700 per 1,000 board feet (mbf) the total standing timber value would be $1,085,000. The total annual growth is estimated to be 95 mbf, or about $66,500 based on 1996 prices.

By conventional even-aged management standards, many stands would be considered understocked. Regular thinning has kept the stands relatively open, but it has also encouraged very high-quality growth. Thinning has also allowed the firs and hardwoods to regenerate naturally and develop significant vertical stand structure.

The Brent Tract is in the *Tsuga heterophylla* (Western hemlock) zone of the western Oregon Coast Range. Forests are typically dominated by Douglas fir *(Pseudotsuga menziesii)*, with Grand fir *(Abies grandis)*, big leaf maple *(Acer macrophyllum)*, and Oregon white oak *(Quercus garryanna)* common associates. The property provides habitat for deer, coyote, beaver, hawks, owls, and trout. Three spotted owl pairs have been identified on the nearby Corvallis Watershed or Suislaw National Forest, although no owl-related restrictions affect the Brent Tract.

After settlement until the early 1900s, most of the foothills were small farms and ranches. As marginal agricultural land was abandoned beginning in the early 1900s, it reseeded to fir. Nearly all of the region has been harvested and managed for timber since the 1940s or before.

Prior to the 1960s, forest management of the Brent Tract was limited to isolated small sales for farm use or income production. Cutting occurred where trees were the largest. After a windstorm felled a large number of trees, the property was commercially thinned in 1964-65, then again in 1980, 1985, and 1987, each time under the supervision of consulting foresters.

These thinnings consistently used an individual tree selection technique called high thinning, with limited use of shelterwood and overstory-removal harvests to promote natural regeneration. In the even-aged management techniques used conventionally in the region, low thinning is the primary intermediate harvest method. Low thinning selects trees for cutting from among the subdominant size class to give growing space to the dominant trees of the stand. In contrast, with high thinning, a small number of the dominant trees are selected for harvest, based on how saleable they are, and how much space their removal would provide to neighboring subdominant trees.

The property has been managed sustainably since 1965. Only recently, though, have the owners understood effective uneven-aged management techniques. Previously, low log prices and lack of knowledge and experience with alternative management regimes kept the owners' focused on commercial thinning. Rising timber values were a strong incentive to pioneer high thinning practices.

Sustainable Management for Douglas Fir

Like most NIPF owners, Ms. Picht and Mr. Brent have diverse objectives. They want periodic income, but no family member relies on the property as a primary income source. Timber harvesting is acknowledged as a source of income and a means to promote the health of the forest, but both Picht and Brent want to maintain a perpetual forest and the aesthetic values of the property. They wish to avoid clearcutting whenever possible.

The owners have no formal written management plan or timber inventory. However, they have relied on advice from professional foresters for all management activities since 1964. The foresters have also advised the Brents on tax issues involving timber, and an accountant has provided additional tax advice.

As long as the forest stands are healthy and attractive, timber volumes and values increase, and har-vests occur when income is needed, the owners feel their objectives are being satisfied. Regular (annual or biannual) meetings and walks with the forester keep the landowners apprised of actions needed and progress toward fulfilling goals.

The years of SFM have created a healthy forest. Generous riparian set-asides provide wildlife habitat for a variety of species, add to diversity within the landscape, and protect the water quality. Stand management practices enhance the diversity of the tree stands and the landscape. Hardwoods and minor conifers are encouraged, although fir is clearly the dominant species. The owners tolerate brush and shrub growth. The trees are grown to large sizes (larger than 24" in diameter), and old ages (more than 80 years old). Selective thinning has increased vertical stand structure, so much so that in some places it resembles the uneven-aged conditions characteristic of a natural forest.

Costs and Rewards of Conservative Management

Property taxes represent the main fixed cost. Annual taxes on the woodland portion of the property have been between $600 and $700 through the 1990s.

Timber sale administration represents the sole variable cost of management. Forester fees were initially 10% of the gross stumpage proceeds. Since log values have risen, however, forester fees have dropped to the current 7%.

Much of the property is zoned exclusively for forest use, which prohibits development. Zoning would allow division and/or establishment of one additional homesite, valued at $80-100,000. Thus, keeping the property in its present use represents an opportunity cost of $80-100,000.

Timber Sales

	Harvest Volume	Gross Stumpage	Fees	Net to Owner
1987	229 mbf	$26,376	$2,638	$23,738
1985	259 mbf	$26,144	$2,614	$23,530

(no data for sales in 1980 or 1965)

Since 1985 the conservative management used at the Brent Tract has generated over $47,000 of net stumpage income. Although there have been no recent appraisals of timber or land value, the total current property value is estimated to be more than $1.5 million (standing timber $1,085,000; bare land $171,000; location value $250,000). With recent increases in stumpage prices, a conventional regime of heavy cutting and conversion of "underproductive" stands would not have performed nearly as well as the more "conservative" regime chosen.

The property is enrolled in Oregon's "timberland deferral" classification, which allows a lower annual property tax, with a severance tax of 3.8% on the net stumpage proceeds at harvest. There has been no income from nontimber sources.

Careful management of both the timber resource and family estate tax affairs could enable Picht's daughter to live on the land as she one day wishes to do. However, there is no estate plan in place yet. The owners do not fully understand the monetary value of their forest asset, nor the significant role estate planning plays in the ability to maintain and perpetuate that asset. The lack of an estate plan may threaten Picht's and Brent's abilities to achieve those long-term objectives.

Practical Alternative Management for Douglas Fir

Low log prices through the 1980s probably deterred harvesting on the Brent land. Because the owners' financial needs were low, the resulting conservative management has been beneficial. Thinning stands that might otherwise have been candidates for clearcutting has allowed both regular harvests (and income), and a significant increase in the volume of standing timber. The growth in timber coupled with rising prices for stumpage have greatly increased the value of the property. At the same time the property remains an attractive place to live and enjoy, while generous stream area set-asides promote wildlife diversity.

Much of the Brent Tract is gently sloping, which makes it well suited to ground-based logging. Most NIPF timberland in western Oregon lies on similar sites, but much of the industrial timberland in the Northwest does not. Significant modifications would be needed to manage mixed-age stands in those areas on an industrial scale and would probably require the use of more expensive cable logging.

One stand on the Brent Tract was among four sites chosen for a 1995 study of selective thinning and uneven-aged management for Douglas fir by consulting forester Mark Miller. The preliminary results show that 30 years of selective thinning and natural regeneration are leading to uneven-aged stand structures.

Currently the city watershed has no active forest management plan. Such stewardship forestry next to the watershed could provide an example of the successful use of alternative management techniques for Douglas fir, which might be applicable to other sites in the area.

THE FREEMAN FARM: STEWARDSHIP IN ACTION

George Freeman, a retired Quaker State Oil Company executive, and his wife Joan purchased the first 93 acres of the property with a partner who supported strip mining in 1964. The mining company went bankrupt and the coal mining stopped. By 1971 the Freemans were sole owners of the property and have since managed the 639 acres in Pennsylvania, which consist of highly productive stands of mixed hardwoods and a high proportion of red pine and Japanese larch, as a productive forest.

The Freeman Farm is a fine example of steward-ship forestry. The Freemans, however, do not rely on timber as a sole source of income. Because they have a variety of objectives, they are willing to accept a disparity between income and costs. Although timber income covers the costs of many of their projects, it does not cover all the costs of running the farm. Essentially, timber income allows the owners to enjoy their retirement, practice sustainable forestry, and provide environmental education, which they consider worthwhile.

Freeman Mission: Sustain and Share the Forest

The Freemans have a clear mission for the property. They intend to sustain the land as forest and pass it on to their three sons. The Freemans derive great enjoyment from caring for the forest, which they work in almost every day of the year. Both cite "sanity" as the most important output they get from managing the farm. According to George Freeman, "Some of the best relationships come from working together in the woods." Equally important, the Freemans want to share the forest with as many others as they can, so that others "may learn to preserve these resources for future generations."

The Freemans accomplish their second objective by hosting hundreds of visitors each year who attend workshops that cover topics ranging from timber taxation to forest stewardship. The farm also includes a 12-acre Stewardship Demonstration Area, which is managed with Pennsylvania State University and the Pennsylvania Bureau of Forestry. It shows how six different forest management techniques are applied on two-acre parcels. The farm is also part of the Pennsylvania Forest Stewardship Program sponsored by the U.S. Department of Agriculture to encourage SFM on NIPF lands, and the National Tree Farm system, a private program started in the 1940s to ensure future timber supplies. The Freemans have received a number of awards for their conservation efforts. Most recently, in 1995, they were awarded The Three Rivers Environmental Award for contributing to environ-

mental education in Pennsylvania. George also belongs to a variety of forestry and environmental groups, including the American Tree Farm System, American Forestry Association, and the Pennsylvania Forestry Association.

The couple is implementing an estate plan, which demonstrates their commitment to passing the forest lands on to their three sons. Ninety-three acres of forest land was given to the three sons in 1995 as part of a lifetime giving approach to avoid estate taxes. As a Stewardship Forest, the farm is under a management plan prepared by a professional forester. These factors increase the potential for sustaining the forest for the long term.

A Farm of Hardwood Forests

The 639 acres are located in Richland Township, Clarion County Pennsylvania, just south of Interstate 80, which makes the farm easily accessible—and hard to miss. The Freemans proudly display the "Stewardship Forest" and "American Tree Farm" signs on the interstate. The land surrounding the farm, which is located in Pennsylvania's ridge and valley province, is made up of farms and gently rolling forested hillsides. The farm itself lies at the northern end of the oak-hickory forest type range that stretches south into Virginia and the Carolinas.

The terrain ranges from gently rolling to flat. Ridge tops are narrow but there are no geological features that hinder forestry activities. Forest stand types include northern hardwoods (birch, beech, maple, and pine), mixed oaks (white, black, and chestnut oak, hickory, black cherry, and red maple), and some red pine, Japanese larch, and Norway spruce plantations. Grapevines had invaded many of the stands and were contributing to a decline in tree growth. Due largely to George's efforts of cutting and applying herbicides, the grapevine is under control on 430 acres. In some areas, though, the grapevine has been left as food for wildlife. Many eastern songbirds occupy the farm in summer, and rabbit, whitetail deer, red squirrel, turkey, and coyotes live there year-round. The diversity of wildlife is a primary source of the Freemans' enjoyment of the land and an added incentive for them to practice sustainable management.

Forest Management

Almost all of Pennsylvania's forests were cleared around the turn of the century, primarily for lumber, fuel, and agriculture. The Freeman Farm was cleared for agriculture. As lands became marginal and society urbanized, farmland was abandoned and reverted back to forest. Most of Pennsylvania's forests are currently about 70 years old.

In the case of the Freeman farm, 60 acres is younger because it was actively farmed until 1965. The Freeman forest has been actively managed since 1970. Sixty acres had been clearcut right before purchase. That is now a lush mixed hardwood stand containing mature oaks, tulip poplar, cherry, and an occasional cucumber tree.

The site quality is good over most of the property. Some 497 of the 639 acres are commercial timberlands. The Freemans manage the forest to encourage the growth of high-value commercial species and regenerate these species after harvest. They conducted their first thinning in 1973 to remove "mature and defective" stems. They held timber sales in 1981, 1986, 1992, 1993, and 1995. Although clearcutting has been used in the past, new selective cutting techniques such as crop tree harvests, which retain the better growing stock to mature and regenerate on the site, are now implemented to regenerate oaks and black cherry while maintaining the aesthetics of the property.

In addition to native stands, the farm has several plantations devoted to red pine and Japanese larch. These stands are pruned regularly to encourage the growth of more valuable clear wood. Competing vegetation, especially the grapevine growth, is controlled by hand sawing and the careful use of herbicides.

During harvest, care is taken to protect water resources and trees that provide habitat for wildlife. Roads and trails are designed to minimize erosion and sedimentation in streams. Culverts are strategically placed to protect existing water courses. Harvesting is done with chainsaws and skidders. Roads and landings are reseeded after harvest to minimize erosion and to benefit wildlife. Herbicides, which are applied directly to avoid unintended damage and to protect surface waters, are used to control ferns on 15 acres and to control grapevine.

The Freemans base their management decisions on professional advice. They use a private consulting forester and take advantage of the advice of extension and state service foresters. The Freeman's Forest Stewardship Plan was prepared with the cooperation of a forester with the Pennsylvania Bureau of Forestry. The Freemans have had professional legal help. George handles the finances, except for taxes, which are prepared professionally. The family's involvement with Pennsylvania State Forest Resources Extension programs and the Tree Farm and Stewardship Programs also gives them access to natural resource professionals.

The Freemans cite property taxes as a major impediment to sound, long-term forest management. Owners may occasionally receive a return on timber, but taxes are due every year.

The Freemans also face an ongoing issue with the local utility. High-tension wires cross the property on a right-of-way for which the Freemans were paid $700 in 1966. They have no rights to use the area, yet they must pay property taxes on the approximately seven-acre parcel that the utility uses. They have also had to negotiate with the utility over the use of herbicides, and the utility's need to access the parcel across the Freeman's other lands. The utility now consults with the family before it enters the parcel or conducts any activities there.

The Freemans propose two remedies for these issues. First, that the utility pay the taxes for that portion of the property it uses. Second, NIPF owners should be given special consideration in taxation. George recommends incentive programs that would allow reductions in property taxes in exchange for commitments to practice and achieve sustainable forestry.

The owners realized early that the property had the potential to pay for itself. Timber production is a primary objective. The Freemans have effectively managed timber while enhancing wildlife habitats and aesthetic values, protecting watersheds, and maintaining biodiversity.

The management strategy is aimed at keeping the farm intact and forested. Timber harvests are sporadic. They are scheduled based on a combination of factors: when commercial quality timber is ready to harvest, when improvements are needed to encourage the growth of commercial species, and when market conditions are favorable. The estate plan is designed to pass the farm on to the next generation, but this does not ensure that the land will stay in forest use.

The Freemans own the land because they enjoy caring for the forest and working to develop a high-quality timber stand. But the farm also provides recreation and an opportunity to educate the public about forests and forestry. The objectives of recreation and occasional and incidental income from timber reflect those that are most common among NIPF owners. The Freemans feel that their objectives are being met. The objective of producing high-quality timber is evidently achieved, as indicated by the timber sale figures below.

Financial Rewards of Stewardship Forestry

The single largest fixed cost is payment on an equipment loan. For the period 1993-1995, annual payments averaged $3,203.61. Combined taxes on the property are the second largest fixed expense. For the period 1990-1995 payments

averaged $2,703.07. Taxes on the property have increased by $132.05 in this five-year period, which translates into a cost of $9.24 per acre per year (based on 1995 figures).

The variable expenses on the farm reflect the diversity of activities. They include machinery maintenance, fuels, forester fees, vehicle and facilities maintenance, and insurance. Timber sales far exceed variable costs over the span of ownership, but they do not exceed fixed costs. This indicates that as a sustainable business venture, the farm would not survive in its current form. Timber management alone, however, would be self-sufficient.

Recently the Freemans refused an offer to locate a radio transmission tower on the property, which would have eliminated several acres of forest land. That and their refusal to strip mine the property for coal represent significant opportunity costs. Although the Freemans have received some revenue from selling corn grown on open fields and limited coal mining, most of the income produced from the property comes from timber.

Timber Sales

There have been five timber sales since 1980 totaling $111,857.84.

Sale	Total	Freeman	Commission Consulting Forester
1995	$ 2,240.00	$ 2,000.00	$ 240.00
1993	6,971.89	6,431.89	540.00
1992	26,113.98	22,980.30	3,133.68
1985	62,177.52	55,405.00	6,772.52
1980	14,355.45	12,483.00	1,872.45
Total	$111,857.84	$ 99,300.19	$12,558.65

This represents $11.67 per acre per year of gross timber revenue. Roughly $1.31 per acre per year has been paid to a consulting forester to administer the sales, which leaves $10.36 per acre per year to the Freemans. These estimates were calculated from the table above, by averaging the totals over 15 years, and do not account for expenses other than the consultant's fees.

The complexity of the venture makes calculation of changes in investment value difficult. The land is not currently threatened by localized development pressures. Clearly the value has increased if only in terms of the quality and value of standing timber. Nonmarket benefits include enjoyment of the land, hunting, wildlife observation, and the satisfaction of realizing a dream.

Forestry Incentive Program monies helped pay for some early timber stand improvements. The farm is not currently enrolled in Pennsylvania's "Clean and Green" program, which allows forest owners to have their lands taxed at a low current use rate. The Freemans have not taken advantage of this program because it would not significantly reduce their current assessment. The program also requires owners to "buy out" by paying back taxes at the higher rate if they should choose to convert their lands to other uses at a later time.

The Freemans operate the forest as one would operate a small business. This allows them to depreciate buildings and equipment. Not doing so would neglect tax benefits associated with expenses that are necessary for careful forest management.

External Costs and Benefits

The Freemans feel that they get little recognition for their efforts by their immediate neighbors. In fact, these neighbors have had some negative effects on the farm. One neighbor has occasionally hosted concerts. In the past these have led to trespass, burglary, and fire damage to the Freeman land, and have even led to legal action against the neighbor. In addition, the farm's proximity to the intestate sometimes brings unwanted visitors, especially during hunting season. To prevent the intrusions, the Freemans have had to buy gates for many of the access points, which they consider a necessary cost to protect their investment in the land.

Within the context of alternative land uses, the Freemans's efforts provide an ecological benefit. By foregoing options for other types of development and making a commitment to long-term forest management, the Freemans have improved wildlife habitat, and protected watershed and the aesthetics of the forest, which could otherwise have been lost.

Deciding Factor for Forest Stewardship

The Freemans are financially able to practice sustainable forestry. It is questionable, however, whether they could support the multiple objectives they have on timber income alone. They have invested a lot of their own time in the management of the forest, as well. Many NIPF owners are unable to do so. Changes in taxation could improve prospects for sustainability.

On the other hand, it is common for forest lands to be part of a diverse income stream. The Freemans have assured that their efforts can continue through careful estate planning. The prospects that the Freeman Farm can be sustained for the next 50 to 100 years are good.

The primary indirect benefit of the farm is the ongoing education it provides. The workshops and demonstration sites set a strong example for forest stewardship and help to spread the word to other landowners and the public about the benefits of sustainable forestry.

There is no doubt that if the Freemans did not actively manage the land it would look quite different. The grapevine control campaign George has waged has made the tree stands far more productive than they would otherwise have been. Using a professional forester assures that payments for timber are fair and that sound forestry principles are used in harvesting.

Sustainable forestry means different things to different people. The Freemans are committed to sustaining the forest. By viewing their land as a forest, as opposed to simply a financial asset, they have consistently made decisions that promote sound forestry. They have also chosen to invest their own capital in making the farm work. There is no current threat to the farm from urban sprawl or rapidly increasing land values, which threaten many other NIPF land holdings in Pennsylvania. If there were, the Freemans' commitment would be the deciding factor between sustainable forestry and something else.

THE TRAPPIST ABBEY FOREST: EFFECTIVE CONFLICT RESOLUTION

The Trappist Abbey of Our Lady of Guadalupe, a monastic community of men, owns 1,350 acres of forestland near Lafayette, Oregon. The Abbey purchased the property when it moved from New Mexico to the present site in 1953. The forest is an essential part of the monastic community. It provides the physical and spiritual setting for the community, acts as a buffer to neighbors, and allows for cloistered retreat. It is also the basis for a forestry enterprise, one of the four cottage industries that members run, which enable the Abbey to be self-supporting. To date, each of the other industries (book bindery, fruitcake bakery, and wine storage) has provided more income than forestry. In the future, however, the forestry program is expected to provide an increasing share of the Abbey's income, as members age and income from other industries declines.

In 1996, 37 members with an average age of 67 lived communally at the Abbey and worked collectively in the four cottage industries. A forestry crew of three work year-round to manage the land, including planting, thinning, timber stand improvement, and harvesting. A forestry advisory committee, made up of members within and without the Abbey, was created in 1994 to address forestry issues. The forestry decisions, however, are made democratically, and the entire community is involved in planning discussions.

The forestry experiences of the Abbey illustrate just how much time and effort it takes to build a productive, sustainable forest from the ground up. The Abbey Forest is also a good measure of the financial returns that are possible through continued forest improvement and management. The Abbey community has learned to resolve conflicts over forest use in its own ranks, in part through the use of sustainable forestry. The Abbey's approach to resolving the inevitable conflicts that arise over forest management may offer a useful example for forest use conflict resolution using SFM.

The Abbey Forest

The Abbot of Trappist Abbey, Inc. (ATA), a 501-C3 not-for-profit corporation owns all the land. The various enterprises of the Abbey are organized under the Trappist Monks of Guadalupe, Inc. (TMG), a 501-d corporation. Similar to a partnership, each monk, as a member of TMG, receives shares of income, and pays individual federal and state taxes. TMG receives timber harvest income and pays expenses, including property tax.

The Abbey owns a contiguous block of 1,350 acres on gentle slopes and low hills at the west side of the Willamette Valley, three miles north of the small agricultural town of Lafayette and 30 miles southwest of Portland. The forested area is divided by central fields, the residential and commercial area, numerous stream areas, and a county road.

The soils of the Abbey Forest vary from shallow, rocky, dry southwest-facing ridges and wet lowlands to productive uplands. Elevation ranges from 200 to 1,000 feet. These are seasonally dry sites, since they are in the rainshadow of the nearby Coast Range. Rainfall averages just 40 inches per year. The site quality on average is fair to moderately good, with a 50-year site index of 100 to 125, which means that on average trees grow 100 feet every 50 years.

Of the 885 forested acres, approximately 125 are either too wet or too rocky to grow commercial

Brother Clarence and
consulting forester
Scott Ferguson,
Trappist Abbey Forest

timber. Of the remaining 760 acres about 150 acres are covered with maturing fir stands, 230 acres with mixed oak-fir stands, 330 acres in 10-26-year-old plantations, and 50 acres in plantations less than 10 years old. Based on a 1996 cruise of the land, timber volumes exceed 7.8 million board feet, with annual growth of 500,000 board feet. At a 1996 stumpage rate of $550/mbf, present timber values are near $4.3 million, with annual growth of $275,000. Most agricultural land has been leased to neighboring farmers since the 1960s, when the community gave up farming.

The Abbey Forest is considered a mixed oak-conifer forest type of the interior Willamette Valley. Forests here are typically dominated by Douglas fir *(Pseudotsuga menziesii)*, grand fir *(Abies grandis)*, Oregon white oak *(Quercus garryanna)*, bigleaf maple *(Acer macrophyllum)*, in either pure or mixed stands. Most of the moderately sloping sites in the area were at one time farmed or grazed. As such, the forests are relatively young. Most trees are less than 100 years old, although a

few isolated oaks over 150 years old are still standing. The property provides good to excellent habitat for deer, game birds, songbirds, and a host of other species, but contains no known rare or endangered species.

The Abbey Forest is set in a primarily rural, agricultural area of the densely populated Willamette Valley. Small farms and woodlands dominate the immediate landscape. Agricultural lands are generally productive, raising a variety of crops, including vegetables, small grains, fruit and nut orchards, and wine grapes. About half of the surrounding foothills are forested, mostly in small nonindustrial private ownership.

Forestry on Abbey Lands

Most of the property was logged just before the Abbey purchased it in 1953. At that time most fir greater than 10" in diameter was cut. In the mid-1960s The Abbey started active forest management when a forestry crew and chief were appointed. Initially, they concentrated on replanting cutover stands and surplus agricultural lands. From 1969 to

1981 all suitable timberland was planted, totaling 320 acres. The first commercial harvests began in the late 1960s, but extensive annual cutting did not occur until the mid-1980s, when many plantations were ready for thinning.

The Abbey has used a variety of management techniques over the years in response to changing market opportunities, and the growing expertise of the forestry committee. An early harvest of oak veneer used some of the older high-quality trees left after the early 1950s harvests. Early work with a consultant in 1978 thinned maturing stands. Commercial harvesting in the 1980s thinned young plantations and converted less productive, poor quality residual stands to plantations. In conjunction with these harvests, defective fir and oak was cut for firewood. During the 1988 to 1995 period over 100 mbf were cut annually.

Several early plantations failed on droughty sites. At the time—the 1960s—there was a relative lack of knowledge of how to reforest harsh sites, and few private woodlands in the area were under any type of stewardship management. The Abbey forestry committee learned by trial and error, refining their methods as they gained more experience. In later plantings other species, such as Ponderosa and Radiata pine, Leland cypress, and poplar, which were better suited to the adverse sites, were used in addition to Douglas fir.

Early management needs were easy to recognize, and with the exception of droughty sites, relatively easy to implement. Only in the late 1980s and early 1990s, however, were all parts of the property brought under management. It has been difficult for the Abbey to achieve integrated, sustainable management of the entire ownership due to the lack of a thorough management plan, and until recently, a lack of consensus among the community on how to proceed.

As harvests intensified in the mid-1980s, small clearcuts (5-6 acres) became more common, and

resistance in the community to the visual and aesthetic impact of logging grew. In 1989 when a site near the edge of a common picnic area was clearcut, the brothers became openly divided over the forestry program. Some insisted that the cutting stop. Others wanted the program reconsidered. From 1989 until 1994, small clearcuts continued, but they were out of sight of the residence. In 1994 a new Abbot, Father Peter, took over. He created the forestry committee and the community decided to hire an outside forest manager to draw up a long-term plan for the Abbey's woodland, and oversee future harvests.

Abbey's SFM Strategy

Conscientious stewardship has always been a primary goal of the Abbey community in all of their endeavors. In the forestry program, maintaining the spiritual and aesthetic values of the forest are top priorities. Annual timber harvests are planned to help provide for community financial needs, as much as possible without conflicting with other forest uses. The forestry program also aims to increase understanding of forest management activities within the community, and through planning and record keeping, to pass the program on to future members.

Over the years the Abbey has increasingly sought help from outside experts. In 1995 the Abbey contracted with consulting forester Scott Ferguson (ITS Management) to help develop a forest policy, prepare a detailed long-term management plan, and assist in future timber sales. The Yamhill County extension forester serves on the Abbey forestry advisory committee. The Abbey has also used outside financial and legal advise in many of their community endeavors.

In 1996 the Abbey adopted specific forest policies for SFM and integrated them into a long-term management plan to minimize conflicts between the community's financial needs and their needs for the forest's other amenities. Specific policies address ways to maintain diversity in tree species and ages; landscape and preserve unique trees and areas, such as old oak or trees near streams; implement suitable harvest methods, such as thinning; limit the size and use of clearcuts; and maintain the productivity of the soil.

The Abbey is currently implementing these new management policies. Harvests in 1996 used a combination of individual tree selection, high thinning, and regeneration cuts using group selection, and scarification to mechanically expose mineral soil, which helps the forest regenerate naturally. The advisory committee will provide oversight to help ensure that management activities conform to specific policies, and that any conflicts over forestry activities are resolved.

The long-term prospects for the Abbey forestry program are favorable. The program has an impressive record to date, and both timber values and sustainable harvest levels will increase significantly in the future. The long-term success of the Abbey forestry program will ultimately be determined by its ability to provide a greater proportion of the community's financial needs as the brothers get older. Meeting those needs may eventually force changes in the forestry program. In the future, the community will face a major challenge in maintaining current levels of management, whether by new young initiates or by outside contractors.

Forestry Balance Sheet

Property taxes on the 885 forested acres totaled $1,800 in 1995, or $2.03 per acre. Taxes are paid by TMG which also pays a modest rent of $750 per year ($0.85 per acre) to ATA for use and operation of the forestlands.

The Abbey forestry crew does all logging, road building, timber stand improvement and tree planting. Trucking is contracted. Operating expenses for 1995 were $29,741, in these main categories:

Depreciation	$2,811
Administration	$4,696
Insurance	$ 365
Gas, oil	$ 603
Rent	$ 750
Repairs	$3,604
Service center allocation*	$8,799
Supplies	$2,488
Property, fire taxes	$1,954
Trucking logs	$3,000
Miscellaneous, other	$ 671

*reflects wages, equipment, etc., for items used by forestry but not dedicated to forestry.

Based on 1995 harvest of 30 acres (26 acres thinned, 4 acres clearcut), operating expenses were $1,142 per acre harvested. However, the above costs include timber stand improvement and maintenance expenses (plantation, equipment, forestry facilities) for the entire property. With low annual harvest levels, expenses per harvested acre are high. In 1995 logging and management costs were $292/mbf.

There have been no sales of nontimber products from forestlands. Income from rental of agricultural lands and crop shares provides an average annual income of $6,000.

The forestland provides important functions for the community that would not be served by liquidation, development, or alternative uses. As such, there are no opportunity costs associated with forest land ownership and management.

Timber Sales

Timber sales began in 1966. Annual sales began in 1987, with annual harvests averaging $46,743 (gross). In recent years, per acre harvests proceeds (mill delivered values) were $2,306 (1994) and $1,419 (1995—includes significant hardwood).

Year	Harvest volume	Gross income	Expenses	Net operating income
1987		$ 20,020	$ 19,329	$ 691
1988		$ 37,456	$ 21,783	$ 15,673
1989		$ 78,143	$ 32,215	$ 45,928
1990	144 mbf	$ 57,383	$ 30,829	$ 26,554
1991	122 mbf	$ 40,555	$ 32,197	$ 8,358
1992	127 mbf	$ 38,338	$ 37,053	$ 1,285
1993	88 mbf	$ 30,584	$ 26,881	$ 3,703
1993Q4*		$ 29,665	$ 6,785	$ 22,880
1994	114 mbf	$ 57,660	$ 26,769	$ 30,891
1995	102 mbf	$ 42,574	$ 29,741	$ 12,853
Total		$432,378	$263,582	$168,796

* "short year" reflects change of fiscal year from ending 9/30 to 12/30; records prior to 1987 are incomplete.

Bare land value has appreciated from $125 per acre in 1953, to at least $500.00 today (as zoned for forestry, actual market value probably higher). Timber value on the tract in 1953 was zero. Today stumpage values exceed $5,500 per acre (10 mbf/acre x $550/mbf).

The forest management program provides other benefits. It gives year-round employment to at least three community members. This work is both physically and spiritually satisfying to the forestry crew, a factor the community feels is nearly equal in importance to the financial considerations. The Abbey also lets community members, guests, and retreatants use an extensive network of forest trails and several retreat cabins at no charge.

There has been no government cost sharing used for forest management. All forestland is enrolled under the Western Oregon Small Tract Optional Tax (WOSTOT), a program for NIPF owners, which features a low annual property tax based on forest soil productivity for lands dedicated to timber production and no severance tax.

Benefits to the Environment

By maintaining the fields and forests in an attractive and productive state, the Abbey helps preserve the aesthetic values of the area. Since this part of Yamhill County has little nonfarm development, the aesthetic qualities probably add little monetary value to surrounding properties today. As residential development spreads throughout the Willamette Valley, however, the Abbey's productive forests, species diversity, and maturing stands will increasingly act as an ecological benefit to the area. Projected regional population growth promises to increasingly fragment forest areas and convert them to other uses. As forest management intensifies in the region, rotations will probably shorten and diversity diminish. That will make the Abbey Forest more unusual—and ecologically valuable—in time.

Lessons from the Abbey Forest

Stewardship has always been part of the Abbey values, so striving for it in the forest management program was automatic. The Abbey has overcome its early forestry difficulties, which stemmed from a lack of expertise in small woodland management techniques within the community and in the field of forestry, and its later conflicts over forest use. The Abbot is excited about sustainable forestry and the prospects of certifying sustainable forestry practices. He sees third-party certification as a way to validate the goals and achievement of the forestry program.

The disagreement over logging practices began here: the picnic area in the Trappist Abbey Forest

Having a professionally prepared stewardship management plan has been instrumental in reassuring the entire community (and its visitors) that the Abbey is taking a balanced, defensible approach to resource management, and allows the community to make more informed judgments about the future of the forestry program.

The members of Trappist Abbey live within a defined structure, share many basic beliefs and philosophies, and have made a long-term commitment to living and working together. This has helped the community adopt SFM. Although the Abbey's structure and the community's values are an advantage for sustainable forestry, 37 individuals still must always agree on a plan, on a year-to-year basis. Annual meetings on each year's harvest plan take place, with a democratic vote, either up or down. The forestry committee's oversight and a concerted effort to educate the community have made this process more effective in recent years.

The Abbey offers several lessons for other private forest owners. The long-term commitment required for forestry is especially true for sustainable forestry. Owners may have to bypass immediate benefits to ensure optimum conditions in the future. Even with a 30-year management history, the Abbey's new management plan recommends conservative harvest levels for the next 10 years. Restraint now will enable growing stock *and* sustainable harvest levels to *double* in the next decade.

Conflicts are inevitable in forestry settings where there are several individuals or groups with differing interests in a piece of forestland. The Abbey with its forestry committee has been able to resolve those conflicts. When crisis became evident, talks involved all members of the Abbey community, as well as individuals from outside the community, in frank discussions of intent and need. Policies based on sustainable forestry were an instrumental part of that process and the solution.

THE CARY PROPERTY: THE IMPACT OF ESTATE TAXES

The 2,364-acre Cary property consists of several tracts of pine forest near Pensacola, Florida, that was inherited by the wife of a career military officer in 1947. Until she and her husband retired she engaged the services of a consulting forestry firm to manage the property. In 1962 the owner, Mrs. Cary, and her husband, Col. H.T. Cary, then retired, moved to Pensacola, Florida, and took over management. For 50 years, using the advice and help of professional foresters, the Cary's managed the land with the goal of building a productive forested property that would provide income. In the process they participated in conservation activities that included membership in state and regional forestry organizations and contributions to legislative lobby groups representing landowner interests. When Mrs. Cary inherited the land the timber volume was light and the stocking poor. Over the years through sound management, the land produced regular income, and the timber volume and stocking increased to near optimum levels.

Mrs. Cary's death in 1992 forced the single heir, a daughter named Ann Veldy, to heavily cut the timber to pay estate taxes. Current management is aimed at building the volume of longleaf pine, which the owner understands will require time and patience before the timber reaches a good level of productivity. Ms. Veldy hopes to keep the property, then pass it on to her two sons. As the experience of the Cary property indicates, landowners can successfully use their timber like a savings account, but time is the crucial ingredient in building timber value.

Fifty Years of Consistent Management

For nearly 50 years little changed on the thirteen separate parcels making up this 2,364-acre ownership. Most of the parcels are within a three-mile radius located in two counties of northwest Florida. The topography is typically level to gently rolling lower coastal plain and ranges from river bottom hardwood land to deep sandy soil. Most areas are a sandy loam with good site index in the range of 75 to 90. Longleaf pine grows on the sandy uplands and slash, while loblolly pines are found on the heavier soils and lower slopes. Elevation differences in this lower coastal plain are minimal. The streams, creeks, and river bottom areas range from mixed hardwood and pine (including some spruce pine), to cypress, tupelo, gum, bay poplar, magnolia, and many other species in the deep swamps and river bottoms. Approximately 15% of the property is planted pine with the remainder natural stands.

Over the past 150 years the land has been used for timber growing, naval stores operations, grazing, land clearing for some oil, gas, and mineral exploration, and some recreational activities. The original old growth timber was most likely removed before 1925. Although grazing and uncontrolled fire inhibited forest regrowth for many years, starting around 1950 improved fire control, along with owner protection and improvement activities eventually led to a well-stocked and productive forest. In the late 1940s management began with a cruise to estimate timber volumes on each 40 acres. This provided information for management decisions and led to the first improvement cuts that were designed to remove slow-growing, defective, or poor quality trees and encourage regeneration, better quality, faster growth, and good stocking. Other early efforts included clearing up title, boundary line, and encroachment problems. The planting of small, marginal cropland fields and severely understocked areas, combined with the girdling of scrub hardwoods (killing a tree, but leaving it standing to reduce costs, provide structural diversity in the stand, and enhance wildlife habitat) where needed, was followed by annual fire line plowing to protect plantations and other vulnerable areas. Regular patrols and boundary line maintenance during this period provided added protection while the timber stands developed.

Cutting plans were developed and periodically revised based on the forest condition, silvicultural needs, and owner's changing cash flow requirements. Some years modest annual cuts were made. At other times, several years of logging significant volumes of timber were followed by periods of little or no cutting. Occasionally, hurricanes forced the salvage of damaged timber and required revisions in management plans, activities, and cutting schedules. The level of silvicultural activity and other management work also varied, depending on the owner's situation and wishes as well as the condition of the forest.

When the owner died in 1992, the average volume of timber was at a good level to continue periodic cuts for income. However, Ann Veldy decided, after considering various options for paying estate taxes, to use timber to pay them. After cutting most of the sawtimber to raise the estate taxes, the volume of timber on the property dropped to near the 1947 level. Ann Veldy is now faced with rebuilding the forest. In the meantime, she has built a house on the main tract and plans to sell outlying tracts of land to produce income.

Forest Management Strategy

Currently, the management objectives are aimed at producing periodic income, building the timber volume, and improving timber quality, with emphasis on managing longleaf pine in natural stands and working for natural regeneration. As a result of the heavy cutting to pay estate taxes little or no income from timber sales can be expected for at least 10 years. Eventually the young stands will need thinning and in other stands the residual seed tree or shelterwood overstory trees may be removed. Primary products will be longleaf pine sawtimber and poles with pulpwood as a by-product only. This cutting is expected, provided that current market conditions continue.

Forest management for the past 50 years has met the needs of the owners without damaging the long-term productivity of the land. Under the cur-

rent system, annual tree growth can be removed either each year, or occasionally. Prescribed burning and removal of selected trees are the most useful practices to manage stands of even and uneven aged trees. Once a target volume is reached, it can be maintained indefinitely unless the owners' objectives change or natural disaster disrupts the forest condition.

For the near term, the management needs are basic protection and prescribed burning. Periodic timber estimates and assessments of forest conditions, particularly stocking and growth on the recently cutover areas, will be used to judge progress toward long-term goals.

Forest Management Analysis

In this case, as in most cases where value is being accumulated, income must be deferred. Timberland owners often experience long periods of low cash yield and are acutely conscious of opportunity costs and options foregone. Their land and timber could at any time be cashed in for another investment or use.

Selective marking has been the predominant method of designating trees for sale on the Cary property; however, clearcuts have also been used on occasion. Over the years competitive sealed bid timber sales have been made based on prices per cord, per thousand board feet, per pole, per ton, and per tree. Because of 631(b) federal tax code requirements, use of the safest, surest, most lucrative method of lump sum sales has been limited. Stumpage prices and land values for the property have increased 15 to 50 times over the nearly 50-year management history. This amounts to a compound annual rate of price appreciation only, unadjusted for inflation, of about 5% to 8%. In addition, the timber also grew and significant income was produced. Management costs were expensed and some outlays were capitalized.

With the exception of ad valorem taxes, management costs for the property are all variable. The general categories in which expenses are budgeted include timber sale activities such as sale planning, marking, inviting bids, and monitoring cutting, patrol and protection, record keeping, boundary line maintenance, and prescribed burning. The owner has the options to elect the timing for most of these costs. Over the years, overall costs have increased from less than $1 per acre to $5 or more per acre. At present, with the owner living on the property and able to carry out management functions and the forest in a stage of regrowth, costs should be minimal.

Excluding mineral income, nontimber income has amounted to about 5% of the income derived from the land. Occasionally significant income has come from seismic permits or mineral leases, but there has been no oil, gas, or mineral production. The other income has come from a variety of sources over the years, including hunting agreements since 1980; areas leased for grazing; one field for farming; and an area near a river landing for fishing camps.

The owner derives nonmarket benefits from the land, including the pride of ownership in family land, aesthetic enjoyment, personal recreation, and the satisfaction of providing stewardship for a natural resource. She considers this some consolation for the heavy logging required to pay taxes on the property her mother cared for and improved over half a century. She considers herself fortunate to be able to live on the land, ride horses, and spend time in her woods.

In the late 1950s some government cost-share programs were used in connection with planting open fields and understocked areas and for hardwood (competing species) control. Again in the 1980s, some open lands were planted to trees under the Conservation Reserve Program (CRP). Seedlings from the state nursery were also used and the State Forestry Commission from time to time helped with plowing protective fire lines as well as detecting and suppressing wildfires.

Care has been taken to assure that, to the extent possible, income qualifies for capital gains treatment

and all costs were expensed. The property has been taxed at current use value, but because of the rural location this probably has not yet provided an advantage over market value.

External Costs and Benefits

There are several environmental benefits from the continued maintenance of the Cary tract as productive forest. Adjoining timberland properties have probably benefited from having a similar use on the contiguous land. In general, water and soil have been protected, which are ecological benefits. The good volume of sawtimber on the property provided the liquidity to pay estate taxes, which has allowed the heir to keep most of the land, although in a largely cutover condition. The ability to keep the land, even in this condition, could be considered a benefit for future generations.

Lessons from the Cary Property

As the experience on the Cary property indicates, land and timber investment, over the long term, can provide appreciation and protection from inflation, but it is a low-yield investment and requires active management. A good volume of standing timber can afford landowners liquidity. A low volume of timber produces poor cash flow and little liquidity. With patience and attention timberland can be a good investment and produce the extra dividend of owner enjoyment. Timber growing, however, requires an appreciation of the dynamic nature of a forest. The condition of timberland may change from cutover to well stocked or vice versa, yet economic productivity and environmental values can continue. Successful private forestland management, however, depends on meeting the owner's objective and recognizing the long-term nature of timber growing.

The case also highlights the dramatic changes that can result from failing to plan for successful intergenerational continuity of an estate. Had timber volume not been high, Ann Veldy might have been forced to sell most or all of her land. If so, the land might have converted out of forest use.

THE VANNATTA FAMILY TREE FARM: A LIVELIHOOD FOR THREE GENERATIONS

George VanNatta and his late wife, Irene, purchased the 1,728-acre VanNatta Family Tree Farm in the northern Oregon Coast Range, which was then a cattle ranch, in 1940. They wanted to raise their three sons in a safe, rural setting. In the mid-1960s timber harvesting replaced cattle raising as the primary farm occupation.

The VanNattas express their management objectives simply: They want to make a living from forestry, have a nice place to live, and keep the farm family-owned. The family hopes to perpetuate the Tree Farm and the family business indefinitely for future generations.

Timber management at the VanNatta Tree Farm has provided annual income since 1966, and nearly every year the harvest volumes and values have increased. Annual timber harvests and other forest management activities are carried out by VanNatta Bros., the family timber management business. The VanNattas have used conservation-based timber management strategies since they bought the farm. Although the family does not necessarily embrace "sustainable forestry," they are vocal advocates for a range of progressive and conventional management techniques. As a result, the VanNatta property is one of the few examples of uneven-aged management of Douglas fir on high-quality sites in western Oregon and demonstrates a successful alternative forest management system for the Douglas fir region.

Currently, four generations live on the property, with timber harvest revenues providing most of the income to support three households. George VanNatta, born in 1907, practiced law in the nearby county seat of St. Helens. His oldest son, K.C., who holds a degree in biology, lives on the farm and manages the family tree farm. He is the prime operator of VanNatta Bros., the family timber management company, which also does contract logging for other landowners. K.C. is active in a variety of forestry organizations and currently serves as director of both the Columbia County Small Woodland Association and the Columbia Soil and Water Conservation District. A second son Fred lives in Salem where he owns and manages a public relations firm. A third son, Robert, maintains the VanNatta law practice, lives on the farm, and is K.C.'s business partner in VanNatta Bros. K.C.'s son, Jeffrey, 36, also lives on the farm with his wife and son. He works for the family business and may one day take it over.

Productive Timberland in the Oregon Coast Range

The VanNatta Tree Farm is wholly owned by five family members in divided interests. Of the 1,728 total acres, fields, roads, residences, and a power line encompass approximately 68 acres. The remaining 1,660 acres are wooded. The bulk of the ownership consists of one contiguous block of 1,578 acres outside of the central Columbia County community of Apiary, 40 miles northwest of Portland. A second 150-acre parcel is located 3/4 mile east. The property is located in a remote part of the northern Oregon Coast Range, on gently rolling hills at an elevation of 1,125 to 1,300 feet. Ninety percent of the property can be logged using ground-based machines.

Of the 1,660 wooded acres, approximately 1,650 are suited for commercial timber production. In Columbia County annual rainfall of about 55 inches and deep fertile soils combine to create some of the most productive timber-growing ground in the nation. The Site index on a 50-year base averages 130 (high Site II), which means that the land has an annual growth potential of over 1,000 board feet per acre.

A 1984 timber inventory estimated net timber volume of 27.5 mmbf, with annual growth of 1.3 mmbf. Since 1984, 7.8 mmbf have been harvested, an average of 649 mbf/yr, which is 50% of growth. As of September 1996 the standing timber volume was nearly 36.2 mmbf. At a 1996 stumpage rate of $500/mbf, that timber was valued at about $18.1 million, with an annual growth value of $650,000-850,000.

The VanNatta property is in the western hemlock zone of the western Oregon Coast Range. Forests here are typically dominated by Douglas fir (*Pseudotsuga menziesii*), with western hemlock (*Tsuga heterophylla*) and western red cedar (*Thuja plicata*) also commonly growing on the same site. Douglas-fir plantations managed intensively on 30-50 year rotations dominate much of the country landscape. The property provides good to fair habitat for deer, elk, songbirds, and a host of other species. No rare or endangered species are known to exist on the property or nearby.

The VanNatta Tree Farm is centered in what is sometimes termed temperate coniferous rain forest. This broad forest region extends most of the length of Oregon and Washington, and is one of the most important timber-producing regions of the Northwest. Though 40 miles inland from the coast and east of the Coast Range divide, the wet, mild, maritime climate encourages lush growth of numerous tree, plant, and animal species.

Forest Management on the VanNatta Farm

Old growth Douglas fir forest with stump diameters of 3' to 6' originally covered the VanNatta property and much of the surrounding county. This forest was logged between 1920 and 1928. Wildfires were common after this logging, and burned much of the area in 1932. After the fire, the VanNatta land went back to Columbia County for nonpayment of taxes, and was grazed extensively by sheep until purchase in 1940. During that time Douglas fir was beginning to seed in from a remnant stand on a distant ridge. George VanNatta kept a heard of cattle, but gradually the forest began to dominate the landscape again, and in 1965 the VanNattas began logging, shifting the focus of the farm from cattle to forestry.

Early management focused on replanting understocked areas and commercially thinning overstocked areas. In the 1970s a year-round network of roads was built for access to the timberlands. Initial

thinning involved primarily the selective cutting of the largest trees. The trees that were initially logged were some of the only ones that were saleable at the time. They often were the "roughest" trees, which had large limbs, high taper, and wide rings. This type of thinning released the second tallest size class in the stand from competition, and resulted in their rapid growth into high-quality timber. Encouraged, the VanNattas continued to practice "high thinning," which involves removing the tallest, largest trees. The VanNattas still maintain a small herd of cattle that they graze throughout the forest during the summer to reduce dry grass fire hazard, control brush, and aid in natural regeneration.

The VanNatta family considers tax laws and increasing regulations as the greatest impediments to sustainable family tree farm management. Commonly, when an owner dies, most saleable timber must be sold to pay estate taxes, which often makes sale of the land inevitable to pay taxes. The family also fears increasing restrictions on logging to protect endangered species.

Inversely, however, they also view *effective* tax policies as potentially the greatest encouragement for sustainable practices. Reduction of estate taxes on family-owned businesses would allow family tree farmers to make management decisions based on maintaining a healthy forest, rather than on tax management needs. They also support a reduction in the capital gains tax, and allowances for writing off reforestation and forest maintenance as expenses to encourage additional investments in sustainable practices.

Increasing log values over the past decade have aided sustainable practices by offering dependable markets for forest products. Markets for a wide variety of forest products are located nearby the VanNatta farm, including log export facilities at Longview, Washington, 25 miles north.

The VanNatta family believes that family expertise in logging, biology, and legal fields leaves little need to regularly seek outside professional assistance.

Two family members have completed Master Woodland Manager training (Oregon State University Extension Service program). The VanNattas have allowed their property to be used as a study site for research on root rot control and uneven-aged management. The timber management company has been trained and is certified under the Professional Logger program of the Associated Oregon Loggers (1995). In fact, family members frequently give advice about forestry, logging, and legal issues to others. The VanNattas have used very little government technical assistance.

VanNatta SFM Strategy

The family's primary objective is that management activities and the cash flow resulting from them support the family and allow it to live on the tree farm long term. As a secondary objective they hope to provide an example of good sustainable family tree farm management.

The VanNatta family currently uses a variety of progressive and conventional management techniques. Most harvests in healthy young stands are conducted at 6-12 year intervals and use selective high thinning methods. At harvest, competing vine maple is mechanically cut and/or uprooted, which helps fir and hemlock to regenerate naturally. In stands thinned repeatedly and/or heavily, natural regeneration and isolated enrichment planting is leading to uneven-aged stands.

Annual harvest levels have fluctuated based on family financial needs, market prices, and outside employment. Each year harvest priorities are influenced by which market may be paying premium prices, such as those for domestic logs, export logs, pulp, or hemlock/alder. The current and long-term financial needs of each family owner also influence the annual harvest on each site, as well as long-term management of those lands.

To combat the laminated root rot (Phellinus wiereii), a native disease common on rich Coast Range sites, clearcuts are used to salvage infected and at-risk stands. These clearcuts are replanted with resistant species (cedar, white pine, alder). Excavation of infected roots has also been studied as a potential control measure.

The family has no formal management plan or rigid harvest schedule, although complete background information and records are kept. Harvest levels are maintained at 50-75% of annual growth, with a constant goal of increasing the vigor of tree stands and their growth rates.

The family's estate planning is informal but thorough—two family members are lawyers. They are aware of the value of their assets and have structured land ownership and plan harvest operations to best meet the family's long- and short-term financial needs. Recent logging, for example, has focused on George's (father) land, because he has higher financial needs, and needs to keep his asset value down for estate tax purposes. Meanwhile, K.C. has delayed harvesting on his small noncontiguous lot, allowing volumes and values to accumulate. He assumes this parcel will be sold at some point to help pay estate taxes. The family has not invested in insurance for estate tax purposes because they calculate that the growth of their trees provides better returns.

Forest Management Analysis

The property is paid for. Fixed costs are primarily property taxes, which average $4.35 per acre for undeveloped forest land, or approximately $7,200 per year. In 1940 the property tax rate was under $0.10 per acre.

The bulk of property taxes on timberland are paid as a severance tax of 3.8% on net stumpage proceeds at harvest. The family calculates management costs to include harvesting labor and equipment expenses, as well as all associated road building and reforestation costs. In 1994, payroll, logging operations, and depreciation costs were $191,137, which averages $235/mbf. Records of harvest acreage have not been kept, however, so accurate per acre cost estimates are unavailable. The Oregon Department of Revenue uses $190/mbf as a standard for regional cutting/yarding/hauling costs, which would be at least 10% higher for uneven-aged management.

There are no significant opportunity costs involved with the current management situation. The property is zoned for exclusively forestry use, restricting development to one house per 160 acres. The property's remote location precludes development potential. Timber volume and value growth rates exceed comparable alternative investments, so there is little incentive to liquidate and reinvest. The annual volume growth is 3.5-4%, log grade appreciation 1%, and real stumpage value appreciation 1-2%.

Annual sales of livestock and mushrooms make modest contributions to the farm income, averaging $2,500 and $250 respectively.

Timber Sales

Timber sales are perpetual on the tree farm, with a full time crew of four. In 1995 timber sales grossed $578,000, from 781 mbf of export and domestic sawlogs, and 1316 tons of pulp (equivalent to 120 mbf). In 1994, timber sales grossed $481,213, from 742 mbf of export and domestic sawlogs, and 136 tons of pulp (12 mbf equivalent). Sales have generally increased every year for 30 years. Historically 25% of the harvest volume has been export quality. In recent years this percentage has risen, as the VanNatta timber has increased in size and quality, and export specifications have loosened.

There has been a timber sale every year since 1966. Since then, sales have totaled 12,561 mbf. During the period the sales price for Douglas fir has increased from $50/mbf in 1966 to $650/mbf in 1995. Sales in the period 1990-1995 averaged 730 mbf/year. Bare land value has appreciated from $1/acre in 1940 to over $300/acre in 1996. Timber value on the tract in 1940 was zero. Today stumpage values likely exceed $10,000 per acre (20mbf/acre x $500/mbf).

Cost data for the operating company VanNatta Bros. for 1994 (the most recent available) are as follows:

Payroll	$ 70,110	wages, taxes
Logging operations cost	$ 97,777	supplies, repairs, fuel, etc.
Depreciation	$ 23,250	
Stumpage paid to landowners	$171,220	all to VanNatta family
Payments to partners (VanNatta Bros.)	$ 72,000	most reinvested in equipment
Ordinary income (pretax profit)	$ 36,650	distributed to partners

The VanNatta family places significant value on the tree farm as a good, rural place to live. They use harvest residues for fuel wood to heat the four households. They do not use government cost-sharing support, nor are there any tax incentives for sustainable management. There is a current use (timberland deferral) assessment for property taxes, but due to zoning restrictions and the remote location of the tract, growing trees is the only likely use of the property. The severance tax is viewed as a beneficial form of taxation, because it delays the bulk of tax payment until the time of harvest, when cash is available. It also helps to reduce loss in the case of natural disaster. (A severance tax is a one-time tax paid at time of harvest.)

The family takes full advantage of all timber and business-related tax provisions, with business activities timed and structured to minimize tax consequences. The management company, VanNatta Bros., receives gross timber revenues, assumes all relevant costs (including equipment, wages, road building, reforestation, etc.), and pays net stumpage to the five property owners.

Through continued management the property should be able to provide a similar living for successive generations, if estate tax burdens and regulations don't become confiscatory.

Lessons from the VanNatta Farm

The VanNatta experience demonstrates that although it is difficult to finance the necessary land, knowledge, and equipment to start and maintain a sustainable forestry enterprise from normal family income, it can be done. Conservative harvesting strategies have led to significant increases in timber volumes over the ownership period. Rising timber values have resulted in significant growth in the value of the family's woodland assets.

K.C. VanNatta attributes the family's success, in part, to its attention to natural processes, a practice he advises other landowners to follow: "Study mother nature and duplicate what you see, and be very careful about major species changes." K.C. VanNatta also considers attention to government regulations and policies essential for successful management. "Political discussions about management practices and animal populations can have very definite effects on your tree farm," he cautioned.

The history and physical condition of the VanNatta land contributed to the success of the family's management. The mixed-age nature of the existing stand, and the VanNatta's ability to successfully use natural regeneration is based, to some extent, on the fire and grazing history of the property. Douglas fir regenerates well on mineral soil (as after fire) and when brush is controlled, through grazing, for instance. It is unlikely that the efforts would have been so successful without the careful control of both overstory stocking and competing vegetation. Much of the VanNatta land is gently sloping, which makes it well suited to ground-based logging. Although much NIPF land in the region is similar, most of the industrial land in the Northwest is not. On those lands, significant modifications would most likely be needed to manage

mixed-age stands with cable logging. For these reasons wholesale application of these methods in the region might not be so successful.

Nevertheless, the management techniques used by the VanNatta family offer alternatives for conventional management in the Douglas fir region, which is generally synonymous with clearcuts. Experience has shown that natural regeneration and selective thinning can work well on high-quality sites, if thinning intensity and competing vegetation are controlled.

THE LYONS FAMILY TRACT: INCOME THROUGH SOUND FORESTRY

As far as Paul Lyons is concerned, "you'll never find a better investment than a good piece of timber." The Lyons family owns about 2,000 acres of timberland in northwest Pennsylvania and southwest New York State. Mr. Lyons acquired the tracts mainly in the 1940s and 1950s as an investment and managed them, periodically removing selected trees. To ensure a smooth ownership transition and reduce the estate tax burden he has transferred ownership to his four sons.

Although they have no firsthand knowledge of timber management, the sons share their father's dedication to properly managing the woodlots. Their current goal is to manage the property aggressively for maximum timber production using the best forestry principles, selling when the market is strong, and cutting with an eye to future timber harvests. The brothers intend to pass the land to their heirs and are working to instill in them a strong management ethic. The sons recently turned down an unsolicited offer of $500,000 for their 200-acre tract in New York. Instead, they want to acquire other tracts of timber.

The Lyons's management provides an excellent example of landowners who are generating income while practicing sound forestry, so much so that the family was featured in the *Forbes Magazine* 1996 Money Guide issue of June 17, 1996.

Authors Michael Washburn and Jamie Ervin on the Lyons Property

Paul Lyons graduated from high school and attended two years of college before being pressed into service during World War II. He was, and still is, an astute businessman. The eldest son, William Lyons, who holds a B.A. and an M.A. in education, taught in the public school system, and served as an assistant principal. Since retiring from public education he has pursued real estate full time as a salesman and owner/manager of income properties. Lynn Lyons, who holds a B.A. and an M.S. in education, spent his professional career as an educator and administrator in a large Florida school district.

Cress Lyons attended college for two years, and later purchased Central Distributors, a chain saw and small engine retail and repair business, from Paul Lyons. Sam Lyons, who holds a B.A. degree, currently owns and operates a marina in Florida.

Hardwood Forest in the Allegheny Foothills

The 2,000 acres are owned in eleven tracts that are located in western Warren County, eastern Erie County, Pennsylvania, and southwest Chautauqua County, New York. The tracts are located on the eastern foothills of the Allegheny Mountains. They range from flat to gently rolling with few steep slopes. Blue Eye Creek dissects the Blue Eye property and is listed as an "exceptional quality" stream by the Pennsylvania Fish and Boat Commission. Two other properties border on French Creek and the south branch of French Creek. The parent soil is generally of glacial origin. No geological features exist that limit silvicultural activity.

The properties contain mostly northern hardwoods, mainly sugar and red maple, tulip poplar, black cherry, American beech, and ash. The Blue Eye tract has a significant red oak component. Eastern hemlock and white pine are a minor component. Approximately 20-25 acres are in larch plantation on the Blue Eye tract. The forests support white-tailed deer, black bear, turkey, beaver, songbirds, and birds of prey. The creeks contain native trout; French Creek even supports a population of fresh water mussels.

Management for Regular Income

Site quality is very good over all of the properties. A high percentage of the sawtimber is of veneer quality, and nearly all of the acreage grows commercial-grade timber. All of the tracts appear to have been commercial woodlots for a long time. There is no evidence of grazing, mining, or other nontimber-related activity. Like most of the region, these lands were heavily cut over at the turn of the century. The Blue Eye tract did have approximately 50 acres disturbed by a fire in the early 1950s, but

Logging with horses minimizes damage to the forest on the Lyons Property

it was not hot enough to cause extensive mortality. Likewise, the Greenfield tract was extensively damaged by wind storms in the early 1970s and a thorough salvage operation recovered most of the economic value of the fallen timber. Salvage is only viable when the value of the timber exceeds recovery costs.

Both generations agree that the current tax laws are a deterrent to forest ownership. Pennsylvania's Clean and Green tax program has given some property tax relief on the Blue Eye tract. However, the owners view estate and capital gains taxes as obstacles to good long-term management.

Although timber production is the primary owner objective, the family (now three generations) enjoys their properties for other values as well. Several of them hunt deer in the fall and will travel from Florida to spend a week hunting with friends. They also enjoy just being in the woods, and spend time controlling unwanted grapevines and other competing vegetation. They take satis-

faction that their activities are not only compatible with other uses but in many cases help to enhance those activities. Skid trails, for example, make excellent hiking paths and increase access for hunting. Satisfaction also comes from knowing that they are helping to assure future generations of clean water and pure air as well as providing wood products from the trees they produce.

The current goal is to provide regular income from the harvest of high value hardwoods. Those harvest areas are selected through a stand analysis to determine management options and when markets are strong to ensure top dollar for the timber harvested. The Lyons will harvest no areas (except salvage) unless it is silviculturally prudent and market conditions are favorable.

The Lyons currently use SILVAH, a computerized stand inventory and analysis tool, that can show how various levels of harvesting will affect a stand. The program accounts for species composition, wildlife habitat quality, and the ability of the stand to regenerate to a new forest. Prior to cutting, a SILVAH stand analysis is taken, and the subsequent recommendation is used as a guideline in marking trees to be cut. The Lyons intend to cut each tract/stand when appropriate both from an economic and silvicultural standpoint. While being aggressive, they would rather err on the side of caution, cutting less and waiting for markets to improve.

Each time timber is harvested, care is taken to minimize disturbance to the site and ensure the appropriate reclamation. Locally rare species of trees have been left in certain areas. Wet areas and stream seeps are protected. Typically, rubber-tired skidders and chainsaws are used for harvesting. On some wet sites, horses have been used to minimize the impacts. Temporary plank roads have been built with rough cut wood to protect soil. Potential and certain den trees and snags are identified and retained for wildlife.

The Lyons are careful to ensure that each stand is in a healthy and vigorous condition following treatment. As some of the stands near maturity, the owners plan to use proper regeneration techniques including herbicides to control undesirable competing vegetation. The use of herbicide is expensive but, in some cases, the failure to use it results in a loss of tree cover due to an invasion of ferns and grasses.

The Lyons base their management decisions on professional advice. They rely on a consulting forester for silvicultural and financial advice. They do not use any on-site assistance from the government with the exception of the timber stand improvement (TSI) project on the Greenfield tract. Two of the brothers have attended the SILVAH training sponsored by Pennsylvania State Extension and the U.S. Forest Service Northeast Forest Experiment Station in Warren, Pennsylvania. The Lyons also use the services of an attorney and a professional accountant.

The Lyons have prepared a formal estate plan to pass the land on to the third generation, which is the goal of the four sons. A trust has been set up to hold the land for Paul Lyons's six grandchildren.

Returns from Forest Management

The only fixed cost of ownership is property taxes. In the period 1990-1995, property taxes averaged $5 per acre per year. ($10,000/year total) Variable costs include site inspection visits, consultant fees, and miscellaneous expenses for management activities. Three of the four brothers travel to the property twice a year at a cost of approximately $1,200 per year total. During these trips, they conduct activities such as boundary line confirmation and painting and grapevine removal. These expenses are estimated to reach $500 per year. Consultant fees for the same period have totaled $79,812.00 total for four timber sales. This is equivalent to $7.98 per acre per year for this period.

By not leasing for natural gas production on selected sites, the owners have incurred a significant opportunity cost in the interest of sustainable forestry. Similarly, rejecting the offer of $500,000 demonstrates the family's interest in practicing forestry.

Timber Sales

There have been 10 silvicultural treatments since 1983, 9 of which were commercial.

Sale year	Revenue	MBF
1983	$125,000	122
1984	$ 12,000	35
1986	$ 54,900	239
1992	$ 90,559	252
1994	$ 55,500	55
1994	$220,015	340
1996	$ 61,450	103
1996	unavailable	
1996	unavailable	

The appreciation in the value of the Lyons's property has been dramatic. The stateline tract was purchased in 1933 for $500 for the 237 acres. In 1996 they turned down an unsolicited offer of $500,000! The Blue Eye Tract was purchased at tax sale for $5/acre. The current offer far exceeds that amount. The Pennsylvania Game Commission (the adjoining owner on the east) will pay as much as $400/acre for cut-over land. It is not uncommon for individuals and clubs to pay substantially more than that just to own a property for hunting in this part of Pennsylvania.

The Lyons completed a TSI cutting on their Greenfield tract in fall 1996. This was the first time they had ever used any cost-share funds. It was done through the federal Forestry Incentive Program (FIP). They would not participate in the Forest Stewardship Program due to the 10-year commitment it requires to the plan's objectives. They consider such written plans too inflexible to be useful.

External Costs and Benefits

H.B. Lyons was in the timber and mill business. His son, Paul, followed in his footsteps. Paul's four sons have a deep and abiding affection for the forests they now tend. They are introducing their heirs to the forests they will one day own, and have provided a trust to help ensure that the family legacy lives on. This gives the Lyons tract a multigenerational tie.

Lessons from the Lyons Tract

The owners cite current federal tax policies as impediments to sustainable management. They also consider the inability to expense carrying costs of land and the current treatment of timber in income and capital gains tax structures as significant issues.

Despite those impediments, the Lyons family has demonstrated that a profit can be made from timber, under certain circumstances, without destroying the resource. Many owners, however, might have accepted the offer for $500,000 without regard for the future of the forest resource or generations that will depend on it.

The Lyons have an intergenerational link that should be the envy of most NIPF owners. They also have the satisfaction of seeing the positive results of their labor. This occurs in the form of a strong financial return along with recreational enjoyment.

For this landowner a strong financial return is the prime motivator. However, the Lyons case demonstrates that a strong financial return can be compatible with sustainable forestry. Patience, careful planning, and execution improve prospects for sustainability.

THE FREDERICK PROPERTY: AN INVESTMENT CORPORATION

The 12,768-acre Frederick property was purchased in 1967 with the proceeds from the sale of a family-owned sawmill. The owner bought it to provide long-term financial security for himself and his heirs. He had spent his career from the 1920s to 1960s managing the woodlands and forestry operations for a large family-owned lumber company, then conducted a forestry consulting business for private landowners. He had faith in the inherent value of land and timber as an investment and knew from experience that good management could return extra dividends.

That is why he purchased back 15,000 acres when the family company was sold to a major forest products firm. He managed the land through his consulting forestry business until his death in 1973. The property then became part of a trust, which is managed by a bank as the corporate trustee. The bank is charged with managing the land to provide income for his widow and two daughters as primary beneficiaries, and for grandchildren and other heirs as residual beneficiaries. This trust will last until the death of the owner's wife and daughters, which could be 60 years or more. Although not an absolute restriction, the trust directs the trustee to follow sound forestry practices, so that the timber volume will be increased and retain land and timber as a long-term investment. The terms of the trust also relieve the trustee of the duty to diversify.

Predominantly Pine Property

The owner originally purchased 14,597 acres located in four south Alabama counties. The property consisted of nearly 20 separate parcels ranging from about 40 acres to about 7,000 acres. The topography of the parcels is generally gently rolling upper coastal plain interlaced with branches, creeks, streams, which contain some swamp and river bottom areas. Soils range from heavy clay to loamy to light sand, with many gradations among the three. The majority of the forest is classified loblolly-shortleaf pine. Longleaf is a major species on some parts, and there are areas of slash pine and some spruce pine. Pines predominate in the upland areas, which usually have site indexes from 75 to 90. Hardwood-pine mixtures are found along the branches and creeks and on a few upland areas. Swamps and river bottoms support hardwood and some cypress trees. Approximately 14% of the currently owned 12,768 acres has been planted to pine.

Forest Management Situation

Timber harvesting, land clearing, farming, woods grazing, uncontrolled hunting and some turpentining were all part of the early land use history for this property. The land was logged initially in the 1800s and early 1900s, some for farming by original settlers and some for lumber export or use. Cutover areas regrew naturally despite uncontrolled fire and grazing by cattle and hogs. Forests regenerated on the farmland after it was abandoned.

From about 1900 the lumber company owners mined the land for timber. In the mid-1930s they began sustained yield operations, selectively marking trees and cutting less than growth. Later, forestry practices, such as planting open fields and girdling hardwoods overtopping pines, were followed and by 1967 the lumber company management had built volume to near optimum.

The new owner wished to leave this, his primary asset, to provide for his family. Familiar with the potentially destructive effect of estate taxes he began a planned program of removing most of the merchantable timber. He hoped this would allow the property to pass through his estate at more favorable cutover land values. Only a small part of the property had been cut when the owner died at age 69.

The corporate executor applied and received a special use 10-year payout of estate taxes as provided by IRS tax regulations. For the next 10 years cutting was designed primarily to provide funds to pay estate taxes. Some small land acquisitions were made before the owner's death, but total acreage has since been reduced to 12,768, following the sale of some tracts. This was an attempt to provide the trust with some diversity, block ownership, and in recent years, build liquidity for estate taxes due upon death of the owner's widow. The widow has also withdrawn some lands as gifts to daughters. All timber sale proceeds by the terms of the owner's will are allocated to income for the family. Upon sale of timberland, 80% of the land value is distributed as income and the balance retained as principal for the trust.

Strategy to Build Timber Volume

The trustee's overall objective is to meet the beneficiaries' needs for current income, while bearing in mind the rights of and obligations to future beneficiaries, all in keeping with the trustee's legal and fiduciary responsibilities.

Currently, the objective is to build the stocks of pine sawtimber to a target level of about 4,000 feet (Doyle log scale) per acre of pine sawtimber through a combination of area and volume control. This will involve cutting less than the annual growth while rebuilding stocking. The goal is to produce the higher value products of pine poles and pine and hardwood sawtimber, with lower value pulpwood as a by-product only. Even aged and multiaged stands are managed with prescribed fire and selective cutting.

There is no set age for harvesting; size is used as the measure. Most trees, however, are harvested by age 60. Both natural regeneration and planting are used to establish new stands, with emphasis on the former. Prescribed fire may be used several times during the life of pine stands to maintain favorable conditions for growth and reseeding. In those situations where hardwood competition or other conditions are judged to be too severe for good natural regeneration, site preparation and planting may be prescribed. Site preparation may include mechanical treatments such as shearing, raking, disking, or bedding to prepare the soil for planting, or aerial application of herbicides followed by burning. Depending on conditions, either hand or machine planting is done.

Timber sales have been conducted every year since 1967, but the size of sales has varied considerably. Whether thinnings, improvement, regeneration, or harvest cuts, sales are all designed to meet income needs and maintain the land in as well stocked a condition as possible. Current removals, forest condition, and estimated volume and growth are continuously tracked to measure progress toward the long-term goal of cutting annual growth once the target volume is reached.

The property has been managed by a consulting forestry firm for 30 years. Through its Natural Resource Department the bank trustee provides knowledgeable oversight for management. The services of accounting and legal advisors have been consistently sought.

Forest Management Analysis

Some lands have been sold to diversify and provide liquidity. However, the trust instrument does express the trust creator's strong feeling for timberland as the preferred investment. In spite of the sale of over 2,000 acres (14% of the original purchase), there has been a 700% gain in total asset value over the 30 years. Some of this is attributable to a 22% increase in timber volume, but most is due to land and timber appreciation. Net income has averaged less than 3% of current asset value, not including income from land sales.

Annual timber sales produce over 98% of the income, and since 1967 have been primarily to one buyer. When acquired in 1967, the property was subject to a forest products company's right of "first offer" on the sale of any land or timber. This restriction expires in less than a year and thereafter timber sale income is expected to be enhanced through use of competitive sealed bid sales

Historically, sales preparation (marking and tallying trees, drafting contracts, negotiating price) and inspecting, monitoring, and tracking income generally costs between 4% and 8% of sale proceeds. Over the 30 years, pine sawtimber prices have increased from about $50/mbf Doyle to over $400, pine pulpwood increased from $6.50 to $34 per cord, and hardwood pulpwood from $2.50 to $20 per cord. The average value of land in the area of the property also changed from about $50 per acre to $400 per acre.

Regeneration costs, which generally must be capitalized, account for the largest financial outlay, but may not be incurred every year. Fire protection, road and boundary line maintenance, lease administration, and office records are the other categories of expense.

Nontimber income has totaled less than 2% of all income. Since the 1970s hunting permits, leases, or agreements have produced regular income and seismic permits or mineral leases have produced income occasionally. A small amount of income comes from miscellaneous land use permits, such as gardens, driveways, or dog pens.

The bank trustee is entitled to a management fee, in this case calculated as a percent of asset value and a percentage of timber income, similar to that of an investment manager. This could be considered a fixed cost, but might not be classified as a forest management cost since it would be applied to any asset in the trust. Although money has been borrowed to meet cash flow needs, interest is not considered a fixed cost for this property. Ad valorem taxes are the only true fixed cost.

The State Forestry Commission has provided wildfire detection. Until the 1986 tax law changes, capital gain treatment of income and the ability to deduct most management costs as expenses were major tax incentives. The property has been assessed at current use value for ad valorem tax purposes, but due to its rural location, this has not provided any advantage over market value.

External Costs and Benefits

Forestry is the best possible use of the land for soil and water protection. Ecological benefits also accrue for the many plant and animal species that depend on a forest habitat. Because of the owner's faith in the value of land and timber as a secure investment and use of a long-term trust as a vehicle to hold and protect the asset, this particular property has provided an intergenerational benefit.

Lessons of the Frederick Trust

As this case demonstrates, land and timber can be a solid investment, and forest management and estate planning can foster resource productivity, preserve environmental values, and meet owner objectives across multiple generations. It also highlights the potential of even absentee owners to practice good forest management.

Certification of NIPFs

The certification of forest products has increasingly become a tool with the potential to provide forest owners with an array of benefits in return for sound management practices on their lands. However, NIPFs, which make up 85% of the forest parcels in the Northeast, of which at least one-third are in parcels of 500 acres or less, face a number of barriers to certifying their lands.

The cost of certification can range enormously, but traditionally has been disproportionately skewed to favor large land holdings. As the size of acreage increases, certification begins to become more and more cost-effective, and may, in parcels over several hundred thousand acres, cost only a few cents per acre. In contrast, the owner of a very small parcel may pay up to $5,000 for a single assessment, according to the Forest Stewardship Council (FSC).

The requirements for certification can be equally imposing, especially for owners of very small parcels—50 acres and less. In 1996 the FSC, the international accrediting body of forest management certifiers, required, for example, the following documentation from a forest operation: 1) environmental impact assessment; 2) guidelines for erosion control, reduced stand damage, water resource protection; 3) a comprehensive management plan; and 4) research on yields, growth rates, regeneration, environmental and social impacts, and cost and productivity of the operation. Such requirements, albeit "appropriate to the scale and intensity of the forest management," according to FSC documents, can be burdensome for many landowners, particularly when only a fraction of NIPF owners have management plans.

Even if an owner of a small parcel can afford the cost and can meet the requirements of certification, there are still many barriers that hinder effective marketing of certified forest products. Irregular and uneven flow of species and volume and lack of access to "certified chain-of-custody" processors, manufacturers, and distributors are likely to prevent the landowner from realizing any significant financial benefits from certification.

Many certifiers are aware of the difficulties that private landowners face in certification, and are developing new certification models to address these barriers. Traditional models focus on a single forest operation. New models for small landowners focus instead on creating an umbrella structure, which involves several forest operations and landowners. This structure relies on a single forest manager to be responsible for the management of a number of small forest holdings. This forest manager may be associated with a forest owners' cooperative, a land trust, or a chain-of-custody certified buyer or processor, according to 1996 FSC guidelines for such a structure.

Under such an umbrella structure, it is possible that each forest owner may have a separate management plan, or the entire umbrella organization can have a single, shared management plan. In either case, certification assessments are conducted by a random sampling method, rather than on every individual parcel.

One of the key distinctions between the traditional and the new certification models is the legal right to manage the forest resource. Traditionally, the certifier has had a contract directly with the forest owner. In the new model, the certifier has a contract with the resource manager, who, in turn, may have numerous contracts with their landowner clients. However, in both cases, certification still relies upon a commitment by landowners and managers to long-term management, not just a single harvest.

This new approach to certification raises a number of unresolved issues. What are the criteria for membership into the umbrella association? What is done if any one landowner fails to comply? How are landowners admitted into formal contracts, and how do they withdraw? How are changes of land ownership, governance structure, sampling frequency and intensity, chain-of-custody certification, and annual monitoring to be handled?

Although such issues would have to be addressed, an umbrella certification model could offer numerous advantages to the owner of a small forest parcel who is seeking certification. Many of the written requirements posed by certifiers and the FSC, such as management plans and environmental impact assessments, could be spread out among all participants in the program. Costs of the certification assessment, program fees, and audits could be similarly shared, thereby drastically reducing the cost to any one

landowner. In addition, a cooperative approach to marketing is likely to ensure an even flow of volume and species of certified forest products, and thus have a better chance of getting the product into a certified chain of custody, which potentially would allow it to command a market premium.

INNOVATIONS IN CERTIFICATION OF NIPFs

A number of experiments are underway that are taking innovative approaches to solve the certification hurdles that small landowners face. Three of these new programs are discussed here. Chip Chapman, owner of the Northeast Ecologically Sustainable Timber Company (NEST), is developing a system based on a single certified forest resource manager, who will provide certification services for a number of small, private forest landowners. Eric Bloomquist, president of Colonial Craft, a certified chain-of-custody manufacturer, is hoping to expand a certified line by serving as a contact point for small forest owners in the region, as well as for large distributors of certified forest products. James Drescher, owner of Windhorse Farm, would like to form a landowners' cooperative to gain access to the certified marketplace. Although all three of these examples are based on different certification models, they all present similar challenges and opportunities in certification for the small landowner.

Linchpin between Small Landowners and Market Premiums

In 1994, New Hampshire forester Chip Chapman, frustrated by years of working with landowners who were practicing sound forest management but were not receiving financial rewards for their efforts, started NEST. The majority of his clients, whose holdings average less than 200 acres, have been frustrated with the lack of incentives needed to practice long-term forest management on a small scale. "The problem," according to Chapman, "is that the market doesn't care if you liquidate or manicure your forest. You get the same price."

Author Jamie Ervin in
New Hampshire with Chip
Chapman, owner of the
Northeast Ecologogically
Sustainable Timber Company
(NEST), an organization that
helps landowners handle the
cost of practicing sound
forest management

Chapman initially planned to develop a regional sawmill that would be able to provide high value-added processing options, and a certified chain-of-custody linkage for certified forest operations throughout New England. However, starting up a sawmill is a high-capital, high-risk operation. Furthermore, many of his clients found certification to be an expensive and burdensome process. So Chapman created NEST and approached Rainforest Alliance's SmartWood program, a certifying body, with a proposal to establish a resource manager-based certification program.

In 1996 NEST became certified by SmartWood. The participating landowners, almost two dozen, are all committed to long-term ownership of their lands, have demonstrated sound forest management practices in the past, have developed at least some type of forest management plan and resource inventory, and are committed to work with a pool of responsible loggers. In some cases these landowners are a local neighborhood association; in other cases they are scattered throughout New

Hampshire, Vermont, and Massachusetts. Chapman hopes to expand the program and increase the land base by increasing participation from his existing clients, as well as by cooperating with additional foresters and their clients who agree to follow agreed-upon management practices.

With this structure, Chapman also hopes to be able to pay a premium of 10-20% directly to the landowners, as well as pay a competitive wage to the logger. He is banking on the efficiencies and shortcuts provided by certification in the "labyrinthine marketplace" of forest products. This includes enormous markups between felled log and the sorted, graded, and milled end product. He also hopes that his certification venture will pay for the cost of participation in SmartWood's resource manager program, and provide a modest income.

By mid-1996, investments, including fees, labor, and direct costs, have totaled close to $35,000, far outstripping any tangible benefits. Chapman, however, remained optimistic. "When you put your soul into something like this," he remarked, "there ought to be some recognition." In 1996 Chapman was asked to serve on a Northern Forest Lands Council committee in New Hampshire to develop sustainable forestry guidelines. According to Chapman, many of his clients have joined the program simply because "it was the right thing to do." A number of loggers are also approaching him, hoping to have access to well-stocked, mature stands in exchange for a meticulous logging job.

Perhaps the biggest challenge, and the greatest potential opportunity, comes in finding the loopholes in the "labyrinthine marketplace" by linking up with certified chain-of-custody dealers, such as neighboring giant Seven Islands Land Company. By providing Seven Islands, which manages nearly a million acres of certified forest lands in Maine, with an increased supply and diversity of certified products, Chapman could become the linchpin between small landowners and realized market premiums. In mid-1996 Chapman was negotiating with Seven Islands to sell certified wood from the landowners' parcels, but "so far nobody's cut me a check," he admitted.

A Manufacturer Willing to Share Certification Costs

Manufacturers that want to sell certified wood products typically face difficulties finding enough certified supply. Colonial Craft, a manufacturer of a variety of milled products, including picture frames, moldings, and window and door grills, is based in St. Paul, Minnesota. For two years, Colonial Craft President, Eric Bloomquist, has had a standing offer to share with landowners the direct costs of certification, in return for access to the supply of certified wood—an increasingly scarce commodity. So far, no landowners have taken him up on the offer.

Until recently, problems in establishing a certified line of products had been twofold: 1) finding forest owners interested in participating in certification, and 2) finding traders interested in purchasing certified products. In 1996, a barbeque grill manu-

Mark Miller, author and Certified Resource Manager, tagging certified logs in Maine

facturer approached Colonial Craft with an offer to purchase wooden grill handles. Its customer, a major British retail chain, is a member of the 1995-Plus Buyers Group—a voluntary consortium of British retailers and manufacturers committed to trading in FSC-certified products over the next several years. The buyer tracked down Colonial Craft through the Good Wood List—a list compiled by the Good Wood Alliance, now part of the Certified Forest Products Council, of all certified producers, manufacturers, distributors, and retailers. "The only reason they initially contacted us," said Bloomquist, "is because we're certified."

The customer is creating a major business opportunity for Colonial Craft, with six-figure sales and possible expanded products. To date, the grill handles are made from low-grade materials, such as 9-inch to 15-inch lengths of aspen and basswood. Bloomquist has not yet calculated the extent to which the order represents a premium over an equivalent product; the handles are a new product line and it is difficult to compare to noncertified products. However, Bloomquist noted that the buyer approached Colonial Craft expecting to pay a premium for certified wood.

Bloomquist hopes to expand Colonial Craft's line of certified products by increasing the number of landowners, the size of the land base, and by switching to landowners who are willing to have their forests certified. The biggest challenge, according to Bloomquist, is "finding an adequate and reliable supply of certified forest products."

Cooperative Certification

Jim and Margaret Drescher recently purchased the 140-acre Windhorse Farm in Nova Scotia with the intent of managing the farm for long-term objectives. The Dreschers are also managing a neighbor's property, an additional 150 acres. The farm has a board sawmill, dry kiln, planer, and custom molding shop on the property, and the Dreschers sell approximately 150,000 board feet of lumber annually. To date, most of the products have been sold locally. Although not certified, Windhorse Farm has established a regional reputation for high-quality, responsibly managed forest products.

Over the past year, Drescher has considered certification, both as a tool for improving forest management and as an external verification of sound forest management. Until recently, there had been no indications by the marketplace that such verification was needed. However, a Belgian buyer approached Drescher early in 1997 with an offer to buy a product that represented slightly more than the total annual yield from Windhorse Farm, provided the products carry an FSC-accredited certification label.

Consequently, Drescher is pursuing certification. More importantly, he is exploring with neighboring landowners the possibility of applying for joint certification, to reduce costs and to increase the supply of certified products, thereby leveraging their position in the certified marketplace. Five nearby woodlot owners are interested in certification if the cost to each owner is less than $1,000. The Dreschers currently pay a 10% premium for logs from several well-managed woodlots, and receive a 10% premium on "ecoforestry wood." They are hoping to double their lumber production over the next three years, which would take them close to their kiln and shop capacity.

FUTURE OF CERTIFICATION SCHEMES FOR SMALL LANDOWNERS

The innovations represented by Chapman, Bloomquist, and Drescher raise a number of issues, including which potential models for small landowner certification make sense, what future barriers to certification are likely to be, and the economic feasibility of certification on NIPF lands.

Several programs for small landowners already exist that might serve as an appropriate umbrella certification scheme. These include 1) the Tree Farm Program, in which landowners are required to have a management plan and comply with some harvesting restrictions; 2) logger education programs, in which loggers participate in a comprehensive training program in environmentally-sensitive logging; 3) cooperative management programs, such as the Connecticut/Vermont Project, in which landowners share management resources for shared management objectives; they might, for example, have a forester look at their lands jointly; 4) homeowners associations, in which adjacent landowners may cooperate to reduce costs, protect a forest resource, or have access to better markets; and 5) state and local land trusts holding forest conservation easements.

Creating an umbrella structure for certification may challenge certifiers, as well. Finding resource managers with the full array of skills needed to fulfill the certification requirements and to meet all of the administrative details may prove difficult. Costs are still likely to remain high, even with the economies of scale afforded by multiple landowners. Certifiers will be challenged to find ways to economize on the assessment and monitoring costs, while still providing a rigorous and reliable service.

It is far too early to predict whether certification of small forest lands will be a boom or bust: More likely it will fall somewhere in between. Provided there is an adequate organizational infrastructure, the benefits of certification are likely to be comparable for any size forest ownership. These may include receiving a premium for certified wood, better access to markets, enhanced competitiveness, improved quality of product over time, and improved overall efficiency. Similarly, as the economies of scale become more equitable for small landowners, the direct cost of certification is likely to become relatively equivalent for most size land holdings.

Even if the certification of small forest parcels does become competitive with larger ones, the question remains, can certified forests be competitive with noncertified forests?

Conclusions

The seven properties profiled in this case provide a brief abstract of the complexity surrounding the motivations of NIPF landowners. The overriding lesson from these properties is that the interests of NIPF owners cannot be captured in the explicit arithmetic of business or in the implicit heuristic of "love of the land." The objective function for NIPF management incorporates the apples of hard financial logic, both short and long-term, and the oranges of aesthetics, legacy, and land health. It is, in essence, the cleavage in attitudes towards the land that Aldo Leopold made famous in *A Sand County Almanac:*

In each field one group (A) regards the land as soil, and its function as commodity-production; another group (B) regards the land as a biota, and its function as something broader. How much broader is admittedly in a state of doubt and confusion. In my own field, forestry, group a is quite content to grow trees like cabbages, with cellulose as the basic forest commodity. It feels no inhibition against violence; its ideology is agronomic. Group B, on the other hand, sees forestry as fundamentally different from agronomy because it employs natural species, and manages a natural environment rather than creating an artificial one. (p. 221)

NIPF owners are knitting that cleavage, which is the essence of sustainability. The doubt and confusion remain, however, and may be an essential characteristic of NIPF management—comparing money and meaning does not reduce confusion. Without question, the motivations, skills, and capacities of NIPFs vary, among people and properties, and through time for any person and property.

NIPF ACCOUNT BOOK HAS VARIED ENTRIES

It would make no sense then to assess the likelihood for sustainable management of NIPF lands solely in terms of monetary returns or sociopsychological benefits. The NIPF account book needs to balance some financial costs against financial gains, but it also needs to acknowledge that owners would invest some wealth and personal energy in their land even if their was no financial gain: Many NIPF owners own and manage land for the same reasons that people carefully steward their homes and gardens, invest in community projects, or attend church. The inverse is just as true—most NIPF owners will trade these nonfinancial benefits for a financial return when the need or opportunity becomes compelling.

Furthermore, the NIPF account book must credit some external benefits that accrue to the larger society. NIPF stewardship is often closely tied to community and ecological stewardship, with owners creating recreational opportunities, educational sites, stable soils, and aesthetic viewsheds for the surrounding community. Sustainability of forested lands is good for owners and for communities.

In addition to the basic complexity in the motivations of NIPF owners, the properties described below illustrate a series of other characteristics that affect the ability of landowners to practice sustainable forestry

Cost of Buying Land Is Not Included in the Sustainability Calculus

None of the properties profiled here have mortgages. All have been owned by an individual, family, or group for at least 25 years, and purchase price does not enter into the financial equation of annual costs versus income. In general, these landowners can be less concerned about the flow of income from their properties than someone who is making annual mortgage payments. Many NIPF owners, therefore, are the equivalent of family farmers whose land has been handed down from parents or grandparents. They are "land poor," but they fully understand the value of their property as a capital asset for themselves and their heirs.

Annual Property Taxes Are a Major Concern

Although most NIPF owners aren't making mortgage payments, they are all paying property taxes. The fundamental issue is the burden imposed by an annual tax when income is generated only occasionally. For most NIPF owners, annual property taxes are the largest fixed cost and the fundamental impediment to sustainable management in the short-term. When cash flow is restricted for a landowner, standing timber may be the most liquid asset, requiring cutting that is suboptimal in the long term. Current use tax plans tailored to the income potential of sustainably managed forests are in place in some states and locales, but forest-friendly taxes are still the exception rather than the rule.

Changes in Capital Gains Taxes Worry NIPF Owners

The appreciation of timber over time represents a major capital gain for a forest landowner. In recent years, tax rates on capital gains have increased, reducing the overall returns for landowners. Similar to the issue of annual property taxes, the development of forest-friendly capital tax programs would go a long way to promoting sustainable forestry on nonindustrial parcels.

Estate Planning Is a Necessity for Sustainability

A primary goal of the landowners profiled here is the maintenance of their forestlands across generations. Sustainability in general also depends on continuity in ownership and management strategy. On several properties, management plans have been interrupted by the need to generate income to pay inheritance taxes. The problem has been handled adequately in these properties, but failures in sustainability are often ascribed to failures in estate planning. Properties are often divided into smaller parcels, sold, and/or harvested without regard for long-term effects when they pass between generations. Careful estate planning can prevent such problems, but the legal and land management activities may require decades to implement. Most landowners need the services of competent legal assistance to assure smooth intergenerational transfers.

The Intensity of Management Is Highly Variable

The properties profiled here include lands that are tended daily by a resident owner to those that are visited and manipulated only every few years. The intensity of management relates to a variety of characteristics, but is largely a function of landowner objectives. For the resident owners of relatively small tracts, land stewardship is likely more hobby than business. The owners may prune trees, control exotic plants, build trails, repair fences, and do dozens of other chores, some of which are almost entirely uneconomic. For the absentee owners of relatively large tracts, for whom financial profit is the major motivation, land management is most likely confined to assuring regeneration, performing standard silvicultural treatments, and harvesting. It is important to note that forest sustainability does not depend on high-intensity management, but landowner satisfaction might.

Sustainability Will Look Different on Small Properties Than on Large Properties

The conditions on any area of land will change through time. If the area is large, the "average" condition over the entire area is likely to be similar through time, but for a small area, the "average" condition will be much more variable. For NIPF lands, averaging a few dozen to a few hundred acres, many events can force changes to the land that will give the appearance of unsustainability. A storm, fire, or disease may necessitate unplanned timber harvest, as might the need to generate income. However, the long-term sustainability of the area may be preserved if the land is maintained in the same ownership and if the long-term motivations of the owners remain unchanged. The definition of sustainability must allow for the dynamic condition of small properties.

Professional Advice Is a Necessity for Sustainability

Most of the properties described here use formal professional advice to set their strategies and to oversee implementation. However, recent studies indicate that less than 20% of NIPF harvests involve a professional forester. The source of advice varies, including consulting foresters, forest stewardship advisors, and management committees. In the few cases where formal advice has not been sought, the property owners hold substantial technical expertise themselves. Professional advice becomes especially important in planning and executing for financial returns. a properly managed forest returns much more profit, in both the short- and long-terms, than a forest growing randomly and harvested by people concerned more with what is cut than what is left.

Landowners Who Are Better Informed Make Decisions That Favor Sustainability

Implicit in these cases is the evidence that points to landowner awareness and knowledge as essential components for SFM. Informed landowners make better decisions. Knowledgeable landowners recognize alternatives that sacrifice future potential for short-term gain. Professionally assisted landowners can make optimum decisions, for themselves and for society.

Certification of NIPF Lands Requires New Models

The final portion of this case explores the dilemma of certifying small tracts of NIPF lands. Because per acre certification costs rise greatly as parcel size decreases, certifying each NIPF parcel individually is impractical. New models are needed to allow efficient and affordable certification. The case explores three options, including certification of forest managers (who warrant that lands they manage are sustainable), chain-of-custody certification by manufacturers (who warrant that the wood they use is from sustainable forestland), and group certification of multiple tracts (which are managed cooperatively).

Glossary of Forestry Terms

Board Foot: Timber sales are often transacted in board feet with "mbf" (one thousand board feet) and "mmbf" (one million board feet). A *board foot* is one foot by one foot by one inch.

Conversion: The outright loss of forest land to other uses, which often creates permanent fragmentation. Conversion may also mean changing one forest type to another; for example, a stand containing several species may be converted to a single species plantation.

Even-aged Stand: a stand of a single age created through treatment such as a clearcut, or a natural event, such as a hurricane, that creates conditions that eliminate trees of varying ages and replaces them with trees that all begin to grow at the same time.

Forest Fragmentation: The reduction of overall forest cover and the isolation of forest patches.

Forest Management: All the activities done to tree stands over a period of time while timber matures to meet specific objectives. These objectives might include creating a stand of a given species, encouraging the growth of high-value species, creating a forest similar to a natural one, or to enhance aesthetic quality.

Forest Stewardship Program: a federal program that offers NIPF owners technical support and cost sharing for conducting forestry practices and, in some states, for preparing forest management plans. The plans are used to identify forest management objectives and strategies for meeting them. It is intended to give landowners tools to adequately consider the forest management options available within the context of professional standards. Professional foresters prepare plans with oversight by state service foresters.

Riparian Set-Asides: Reserves of trees that are not cut along water corridors. Leaving them prevents sedimentation of streams and maintains stream temperatures by providing shade, which is important for protecting aquatic life.

Rotation: Period of time between commercial harvests, which varies depending on growth conditions and is different for various regions, species, and timber management objectives.

Shelterwood Cut: A silvicultural treatment in which some mature overstory (the higher, older trees) trees are left on-site to provide shelter for a regenerating understory. This regenerating younger stand is essentially even-aged. Once regeneration is successful, this overstory shelterwood may be removed, or simply left. An uneven-aged stand exists while the shelterwood is still standing.

Silviculture: The science and art of cultivating tree crops to yield a harvestable resource or other forest values and benefits. It includes any mechanical and chemical treatments that may be involved in the process.

Site: The area (environment) in which stands or even a particular tree grows. Site resources, then, are factors such as light, heat, water, available space, and other nutrients that influence the growth of trees.

Site Index: a measure of the capacity for a given site to grow trees. It usually is a reflection of height growth that can be expected on a given site at some base age. Most commonly, a 50-year base age is used. So, for example, a site index of 100 at base age 50 means that dominant (tallest) and codominant trees of that species will average 100 feet in height after 50 years of growth.

Stand: a community of trees that grow together at a particular place and that foresters can effectively manage as a unit. Transitions between stands in a natural forest are often gradual, and can be the result of changes in site. For example, a slope may have varying water availability, thus creating a line between two forest types. Silviculture, however, can create "hard lines" between stands.

Stand Structure: The variation in size, age, spacing, and height of trees in a stand.

Stocking: The density of trees on-site as compared to the maximum possible for that site. Optimal stocking refers to a density at which trees are achieving the best possible annual growth. An understocked stand is one that has not yet fulfilled its potential.

The Forest Incentive Program (FIP): Offers federal money to cost-share activities that foster healthy forests. FIP was created along with the Forest Stewardship Program as part of the 1990 Farm Bill.

The Tree Farm Program: a program created by the forest product industry in the 1940s to encourage investment in forest land to help provide a continuing source of wood and fiber. The program continues today and is the oldest of the national programs designed to encourage NIPF forestry.

Thinning: The removal of selected trees to achieve the management objectives for a particular stand of trees.

Timber Stand Improvement (TSI): The collective term for thinnings and other management techniques that improve the condition of a stand of trees. A commercial thinning is one in which trees are actually sold. A pre-commercial thinning is a treatment in which no sale occurs, but rather a stand is being treated in anticipation of a later sale.

Uneven-aged Stand: a stand of trees of various ages. It is produced from smaller scale disturbances ranging from a single tree felling to groups of trees that are selected for harvest.

References

Birch, Thomas W. 1996. Private forest landowners of the United States, 1994. *Proceedings of the Symposium on Non-Industrial Private Forests,* Washington, D.C., Jan. 18-20, 1996, pp. 10-18.

Egan, A.F., and S.B. Jones. 1993. "Do landowner practices reflect beliefs? Implications of an extension-research partnership." *Journal of Forestry,* Vol. 91, No. 10, pp. 39-45.

Ervin, J. 1996. "Forest Certification and Conservation Easements: a Case Study in Richmond Vermont." Discussion Paper. Forest Stewardship Council, U.S. Initiative. R.D. 1, Box 182, Waterbury, VT 05676.

Forest Stewardship Council. 1996a. "Group Certification of Small Landowners: A Discussion Paper." Forest Stewardship Council. Oaxaca, Mexico.

Forest Stewardship Council. 1996b. "Principles and Criteria of Forest Management." Forest Stewardship Council. Oaxaca, Mexico.

Jones, Stephen B., A.E. Luloff, and J.C. Finley. 1995. "Another look at NIPFs, Facing our 'myths.'" *Journal of Forestry,* Vol. 93, No. 9, pp. 41-44.

Leopold, Aldo. 1949. *A Sand County Almanac.* Oxford University Press, p. 221.

Nyland, R.D. 1992. Exploitation and greed in eastern hardwood forests. *Journal of Forestry,* Vol. 90, No. 1, pp. 33-37.

Simula, M. 1996. "Economics of Certification." In: *Certification of Forest Products: Issues and Perspectives.* V. Viana et al., eds. Washington: Island Press.

U.S. Forest Service, Forest Resources of the United States, General Technical Report, RM 234, Fort Collins, CO, 1992.

Weyerhaeuser Company. Data provided by the Strategic Planning Director, Tacoma, Washington.

Case Writers

Ervin, Jamison. Forest Stewardship Council, U.S., Richmond, VT: certification models.

Larson, Keville. Larson and McGowin, Inc., Montgomery, AL: Cary and Frederick cases.

Miller, Mark: Two Trees Forestry, Coopers Mills, ME: Trappist Abbey, Oregon, and the VanNatta cases.

Washburn, Michael. Penn State University School of Forest Resources, University Park, PA: Freeman case.

Webb, Mark. Mark Webb Consulting, Union City, PA: Lyons case.

Parsons Pine Products:
Trash to Cash

CASE STUDY

PREPARED BY:
CATHERINE M. MATER

A Case Study from "The Business of Sustainable Forestry"
A Project of The Sustainable Forestry Working Group

The Sustainable Forestry Working Group

Individuals from the following institutions participated in the preparation of this report.

Environmental Advantage, Inc.

Forest Stewardship Council

The John D. and Catherine T. MacArthur Foundation

Management Institute for Environment and Business

Mater Engineering, Ltd.

Oregon State University
Colleges of Business and Forestry

Pennsylvania State University
School of Forest Resources

University of California at Berkeley
College of Natural Resources

University of Michigan
Corporate Environmental Management Program

Weyerhaeuser Company

The World Bank
Environment Department

World Resources Institute

Parsons Pine Products:
Trash to Cash

CASE STUDY

PREPARED BY:
CATHERINE M. MATER

A Case Study from "The Business of Sustainable Forestry"
A Project of The Sustainable Forestry Working Group

Contents

Parsons Pine Products: Trash to Cash

Introduction

Since the U.S. Congress passed the Endangered Species Act in 1973, and subsequently listed the spotted owl as an endangered species in 1990, the debate over the appropriate management of public and private forests has continued at a fevered pitch in the Pacific Northwest. The listing of the spotted owl has led to the loss of tens of thousands of jobs in the logging and forest products industry, which has leveled a heavy toll on many rural communities in Oregon, Washington, and California that have relied for decades on a robust forest products industry to sustain their economies. In 1992 in Oregon, for example, the wood products industry was nine times greater as a share of the total Oregon economy than the industry was as a share of the total U.S. economy. While heated debate in the press and at the grassroots levels continues surrounding these issues, many remain unaware of a fundamental shift toward value-added manufacturing that has occurred in the region's forest products industry.

WASTE CONVERSION AND VALUE-ADDED PRODUCTION

Since the late 1980s, employment in the secondary wood products industry in Oregon has increased from 27% to 40% of the total forest products workforce in 1995, according to the Oregon Employment Division. As indicated in Figure 1, total employment in Oregon for logging operations, sawmills, and veneer and plywood operations dropped between 1990-95, losing over 13,000 jobs. In contrast, the value-added and secondary wood products industry—furniture, millwork, cabinetry, and the like—actually generated 11% *more jobs* during that same period and outnumbered total employment opportunities by a 2:1 margin for sawmills, veneer, and plywood operations, and a 3:1 margin for logging operations. By 1995, the percentage growth rate for value-added wood production in Oregon outpaced the percentage growth rate of all other industry sectors in the state, including the burgeoning high-tech and electronics industry.

**Oregon Forest Products Employment
1990-1995**

(# of Full-time Jobs)

	Logging	Sawmills	Veneer/ Plywood	Secondary Manufacturing
1990	11,300	21,500	14,500	19,800
1995	8,300	14,600	10,800	22,000
% Change 1990-1995	(–36%)	(–47%)	(–34%)	+11%

Source: Oregon Employment Division

Figure 1

Although an apparent surprise to economists tracking the economic impacts of harvest restrictions in the Pacific Northwest, the growth of the secondary wood products industry has proven to be a stabilizing influence to the overall Oregon economy. It has done so by focusing on making more product out of existing, or in many cases less, resource. In effect, the mandated harvest restrictions provided a unique "two-by-four" incentive to the industry to figure out how to maximize production with available resources. The results were surprising.

Research by the Oregon Wood Products Competitiveness Corporation has documented that for every one million board feet of wood being processed into commodity lumber, on the average only three full-time, family-wage jobs are created. Full-time, family-wage jobs are year round positions that provide industry-competitive wage rates with benefits. If that same one million board feet in lumber were processed into component parts such as furniture blanks or table turnings, an additional twenty full-time, family-wage jobs could be created. And if that same one million board feet of wood represented in component parts were then processed into quality furniture for consumer use, another eighty full-time, family-wage jobs could be created.

Even so, industry adaptation to more value-added wood product manufacturing has been slow. Citing, in part, the difficulties in changing an industry culture and mind-set, Oregon's Wood Products Competitiveness Corporation determined in 1995 that *less than* 20% of the log volume harvested just in the central Oregon region alone found its way to secondary manufacturers in the Northwest. Eighty percent of the total lumber volume (approximately 1.8 billion board feet of timber) was processed into value-added product outside the western region. This equated to *between 4,000 and 25,000 missed job opportunities* for the region because commodity lumber was redirected elsewhere.

IMPORTANCE OF ADDING VALUE

Increasing value-added wood product manufacturing in forest communities throughout the world may be as critical for achieving sustainable forestry as implementing new forest management practices. Making more with less, maximizing on the resources sustainably harvested, and converting wood waste into wood profits and full-time, family-wage jobs are all fundamental components of value-added wood processing. They provide the framework for achieving sustainable forestry and sustainable community development.

Parsons Pine Products, located in Ashland, Oregon, a small community of 14,000 people based in the heart of spotted owl territory, has been a pioneer and a leading advocate of value-added wood processing for the last fifty years. Once considered, by many in the industry, a maverick operation that often challenged traditional production assumptions and standard lumber grading rules, today Parsons Pine Products has emerged as a unique example of sustainable forest practices that turn "trash" boards into cash rewards. Its experiences in sustainable forest management (SFM) can be instructive for an industry in transition.

Company Background

In 1946, James Parsons decided to do something with the growing amount of wood scrap and waste that sawmills in the Pacific Northwest were producing from their lumber processing operations. The scrap trim ends and shorts (produced from cutting lumber to prescribed industry lengths) were typically ground up or burned for fuel. Subscribing to the old adage of making a better mousetrap, Parsons determined that there was an unmet demand for wood blanks used in the production of mousetraps and that the scrap wood he was seeing chipped up or burned at these sawmills could be converted into that product. As a result, Parsons Pine became the largest mousetrap blank manufacturer in the world—a distinction then often overlooked by the traditional wood products industry in the Northwest accustomed to producing commodity lumber and processing large volumes of "real" wood. That view changed, however, when the traditional industry was experiencing severe cutbacks and employee layoffs resulting from restrictions on timber harvests in the Pacific Northwest during the early 1990s. Parsons Pine continued to *grow*, first expanding, then maintaining, a labor force of over 100 employees while adding processing lines to its production facilities.

In 1996 Parsons Pine celebrated its fiftieth continuous year of producing value-added wood products sold in a variety of niche markets throughout the world. The company still uses wood scraps, defect material, and wood waste and still practices the same management philosophy that created the company in 1946:

- Make more with less.

- Maximize on the resources available for production.

- Turn trash into cash.

MAKING A BETTER MOUSETRAP

Illustrating the importance of converting wood waste to useful products and demonstrating the value of those small mousetrap blanks was the first order of business for Parsons in training production employees. His training system started with a wood pyramid created from 250 mousetrap blanks. He would then knock the first 100 blanks off the top of the pyramid and state "That's $10,000 in sales lost if thrown away or wasted in our processing line." Today, Parsons Pine president Jerry Sivin continues that same direct hands-on approach to employee training, underscoring that every piece of wood scrap processed in the operation has an impact on the company's bottom line.

Creative thinking in product applications was and continues to be a desired and necessary element in the Parsons Pine Products operations. Again, the mousetrap blank leads the way. Once Parsons Pine had a substantial share of the mousetrap blank market, it looked for new market applications for the same wood blank. What it found surprised many. In Japan, that same wood blank was used to serve individual servings of sushi to customers at sushi bars found throughout the country. It also proved to service the needs of the Shinto religion whose followers use small wooden boards (mousetrap blank size) on which they write daily prayers. These prayer boards are displayed at Shinto shrines throughout the country.

Today, running a single shift operation in an 85,000 square foot facility, Parsons Pine processes approximately 10 million board feet of wood annually. Seventy percent of that volume is in softwoods—spruce, pine, fir—with the remaining 30% in hardwoods of maple, red and white oak, and alder. The company produces a wide variety of component parts and finished consumer products including knife blocks, wine racks, roll-top counter organizers, CD box components, shoe organizers, specialty drawers, door louvers and slats, toy blocks and Lincoln Logs, window jambs, and fingerjointed and edge-glued furniture components parts.

During the last fifty years, over 80% of all products manufactured and sold by Parsons Pine have been produced from wood materials considered to be *recycled*; materials purchased as defect, scrap, and waste material from other wood processing operations throughout North America and other regions worldwide.

Production capabilities within the operation have allowed for flexible manufacturing options and new product development, which are critical to the success of Parsons Pine. These include the production of profile moulding lines, radio frequency edge glue and end trim lines, long and short board cut lines, a fingerjoint line, short block equalizers, product assembly lines, and stretch and shrink wrap lines.

CHALLENGE OF USING WOOD WASTE

Producing products for niche markets using character wood (also known as wood with defects) and short pieces of wood is no easy task. Not every product manufactured into wood can be made from character wood, since the character often denotes some form of wood deterioration. Differing coloration of the wood grain, for example, can mean that a fungus has permeated the wood which reduces the structural integrity of the wood cells. Knots and other marks found in wood can also decrease the wood's structural applications. Also, consumers simply prefer some wood products to have a noncharacter wood look. Sports floors, for instance, have always required clean, clear, straight-grain hardwood lumber—usually hard maple. Producers such as Parsons Pine that work with character wood have unusual requirements to track marketing and consumer purchase trends to find applications where character wood can be used.

Typically the wood products industry and the equipment used to manufacture wood products have concentrated on processing longer pieces, so working with shorter pieces of wood poses processing challenges. It can often result in higher production costs due to the increased manual handling of the wood required during manufacturing. Processing technologies that can automate the handling of these pieces do exist, but until recently many of those advanced technologies have been designed for large, high-volume production with price tags to match. Adapting advanced technologies that have been designed for large-scale production can be difficult for companies that have small-scale production needs.

Flexible manufacturing capabilities, astute marketing acumen, and making what you know you can sell instead of attempting to sell what you make are some of the challenges that confront Parsons Pine. Competition in the international marketplace and even access to supplies of waste wood also prove to be significant challenges.

Seven Strategies for Success

Parsons Pine's abilities to take advantage of the market highs and successfully maneuver through market lows and resource constraints revolve around seven key strategies.

- Finding creative options to satisfy its production needs for recycled wood material.

- Developing marketing acumen matched to product design and development.

- Purchasing the appropriate equipment.

- Establishing a new customer base.

- Employing new product distribution channels.

- Rethinking production processing.

- Recognizing the value of the company's own wood waste stream.

SATISFYING PRODUCTION NEEDS FOR RECYCLED WOOD MATERIAL

The purchasing of lower-grade wood and wood scraps, trim ends, and shorts may seem a simple task on the surface, but traditional industry practices often prove to be challenges to Parsons Pine in this arena. Industry mind-set, traditional wood products facility designs, and alternative choices for the use of wood waste are challenges that have all demanded creative solutions from Parsons Pine.

Fighting Industry Mind-set

Getting traditional sawmills to recognize that their waste wood may have substantial added value for solid wood products requires new approaches that get around traditional industry practices. Both softwood and hardwood lumber grading rules that govern what happens to a log from the moment it is harvested work against the product manufacturer that sees profit in short pieces or the value of defect or character in the wood.

When Parsons Pine president Jerry Sivin saw the lumber crisis coming on in the Northwest and had to expand his search for recycled wood supply in his product development, he went north to Canada. There he found, not to his surprise, the mills were "throwing away everything that wouldn't make a window part—stuff under 24" in length." Sivin asked the mills what they wanted for a truckload of their wood waste. The Canadians said they'd sell for approximately US$9,000 per truckload. Sivin offered them US$22,000 per truckload with a guarantee to purchase thirty truckloads per month. Even at that price, the value of the wood being purchased was substantially lower than traditional lumber prices. For white pine, the purchase price for the waste and trim ends ranged from $175 per thousand board feet (mbf) to $275/mbf compared to the normal lumber purchase price of seven times that value. For hardwoods, the price differences were closer to tenfold.

Getting recycled supply from U.S. producers is no less difficult. The traditional industry mind-set on wood waste prompted Parsons Pine to initiate a unique lumber purchase policy in the early 1990s. Coined the 80-20 Rule, the policy requests that the sawmill consider selling 80% of desired purchase volume of lower grade material to Parsons before a 20% purchase of high-grade material could be negotiated with the company. In many cases, states Sivin, it was the only way to get the sawmills to recognize the seriousness of Parson Pine's request for lower-grade material.

Facility Design Constraints

Even if the logic of converting wood waste to value-added product is clear, implementing a process to accomplish that goal is often difficult. For example, traditional sawmills typically have conveyor systems in their processing line that convey the wood scraps and trim ends directly to a fuel bin or a chipping machine called a "hog." Convincing the traditional sawmill to retrieve those pieces from their waste conveyor system may not only be inconvenient for the mill, it can also create a difficult added material handling challenge. Conveyor systems are often installed below standard production equipment in dangerous and difficult-to-reach areas. Chipping hogs can be installed in small areas where added material handling for short piece extraction can also be difficult and dangerous. Modifications to accommodate waste recovery can be costly.

Nor can all wood waste be used in value-added product development cost-effectively. For example, wood trim ends and shorts that are less than 6" in length can be substantially less valuable to recover due to current limitations in processing equipment that size scrap pieces of wood to uniform blanks. Those wood blanks are then used in multiple product applications. The physical process of separating a wood waste stream based on piece length can require the installation of additional machinery, will often be labor-intensive, and can create an added bottleneck in the vertical processing line that could affect daily production rates in other processing stations upstream.

Identifying solutions to these technical production constraints can be impaired by the industry mind-set. One way Parsons Pine has solved the problem is to subcontract to waste recovery contractors who set up on-site to evaluate and implement the technique for wood recovery without creating additional process concerns to the sawmill. Having an in-depth understanding of milling processes and industrial product flow also proves to be a valuable asset for Sivin in discussions with potential wood suppliers.

Alternative Uses for Wood Waste

Adapting standard milling production to allow for scrap and trim end extraction for off-site use can also run counter to company operations that may choose to convert their wood waste supply to other uses. Many mills, for example, use cogeneration processing on site. Cogeneration processing burns the wood waste created at the mill to produce steam that is then converted to electricity, which is used to power the wood products facility.

The conversion of wood waste to needed electricity can produce substantial financial benefits to a mill. Electricity costs can run as high as 25% of total operating costs, although 5-8% is more the norm for a primary wood products operation in North America. In addition, any excess electricity created by the cogeneration plant can be sold back to regional power companies that in turn use it for other electricity consumers in their area.

Wood waste that is converted to chips or fiber for other uses such as composite board manufacturing can be a wise use of the wood, but may not be the highest and best use of the wood on a per unit basis. For example, softwood waste processed into chips for board production captures about $125/mbf compared to between $480-$1,000/mbf if it is converted into fingerjoint block. For hardwood, the difference in value is even more pronounced: Hardwood chips can be sold at $50/mbf compared to up to ten times that value for hardwood fingerjoint block.

DEVELOPING MARKETING ACUMEN MATCHED TO PRODUCT DESIGN AND DEVELOPMENT

If one were viewing a horizontal process flowline with "forests" on the beginning left end and "retailers" on the right end, it might be said that the traditional wood products industry moves from left to right; from resource to consumer. Typical wood products producers know what species they have traditionally used to manufacture product they have continued to produce over time. Without gauging the changes in consumer preference for products, the typical producer makes the product, then figures out how to sell it.

Parsons Pine essentially reverses the process. They first figure out what consumers want and what they are looking for, then produce the product to match consumer demand. *They make what they know can sell—rather than trying to sell what they make.* This marketing approach applied to Parsons' product development is not only critical for capturing niche product market share, but offers opportunity to use wood resources more efficiently and waste less. Parsons product design and development process has focused on matching market demand with a wood supply predicated on the use of short pieces and configured wood.

Organizing a Product Line

Several of the key product offerings from Parsons Pine help consumers organize space better. Products such as their roll-top counter organizer and their anywhere drawer address immediate organizational needs in consumers' kitchens. The shoe organizer helps consumers tidy up their closet space, and the disk and CD storage shelves respond to home office organization needs.

Casegoods & Shelving: Organizational Need (1993)

Area	Rank
Closet	1
Kitchen/Cabinet	2
Garage/Utility	3
HomeOffice	4
Bedroom	5
Bath/Vanity	6
Family/Rec. Room	7
Laundry	8

1 = most needed, 8 = least needed

Source: The Peachtree Group

Figure 2

The emphasis on organizational products is not off the mark when considering the market demand for these products. Based on recent consumer research conducted in the United States by the Peachtree Group in Atlanta, and as noted in Figure 2, consumers rank their closets, kitchens, and home offices as areas with the highest need for organizational products.

Although classified as a niche market area, demand for these types of specialty products is not small. In 1993, U.S. consumers spent almost $1.5 billion for the purchase of these organizational products (see Figure 3). And certain segments continue to grow. Consider the needs, for example, of the 41 million Americans who work either full-time or part-time at a home office.

Parsons wood louver slats product offerings also appear well grounded in consumer preference. Market research evaluating door manufacturing opportunities for wood product producers show doors made with wood louvers hold strong market demand with excellent projected growth.

Casegoods & Shelving: 1993 Market Size

Area	Size (Mfgr.'s Level)	Percent
Closet	$510 Mil.	30.7%
Bedroom	260 Mil.	15.6%
Family/Rec. Room	220 Mil.	13.4%
Garage/Utility	180 Mil.	10.8%
Bath/Vanity	175 Mil.	10.5%
HomeOffice	60 Mil.	3.6%
Laundry	55 Mil.	3.3%

Almost $1.5 billion dollars of U.S. consumer dollars went to the purchase of casegoods and shelving organizers during 1993

Source: The Peachtree Group

Figure 3

According to U.S. Department of Commerce data, the value of louver door shipments in the United States was $39 million in 1992. By 1995, that value increased to $46.5 million, and is expected to reach almost $50 million by the year 2000.

The growth of wine consumption in the United States matched with the fast-paced growth of the vineyard and wine production industry in Oregon makes the development of a ready-to-assemble wine and stem storage rack a natural product for Parsons Pine. In Oregon alone, the number of cases of Oregon wine sold to worldwide consumers increased by almost 130% during the last five years—from 322,000 cases sold in 1990 to over 734,000 cases sold in 1995, according to the Oregon Wine Council.

All products made by Parsons Pine are well made, nicely finished, and easy to assemble. And all can be made with wood pieces less than 24" in length, which is typical wood scrap size for traditional sawmill operations.

Staying on top of consumer product needs can be done through intensive and expensive professional market research, but Parsons Pine finds that its best product ideas often come from its own employees and customers.

SMART EQUIPMENT PURCHASING

Like many value-added wood product producers, Parsons Pine matches production needs with technology support. Producers wishing to be competitive cannot afford to remain low-tech in their production, but many find that high-tech is not necessarily cost-effective. The key is to be competitive in product offering while being flexible in production. The purchase of a new mid-tech versus high-tech router during 1996, as an example, allowed the company to convert a percentage of the louver wood slats, which were facing heavy competition from offshore manufacturers, to CD wood bookmarks. The new product line increased Parsons annual sales by approximately 6% and paid back the capital investment on the equipment within the first two months.

ESTABLISHING A NEW CUSTOMER BASE

Perhaps Parsons Pine's most notable high-risk factor is its recent move to establish a new base of customers. In traditional wood products operations, about 80% of the customer base is comprised of long-standing customers. With Parsons Pine, however, due to new product offerings and new product distribution channels, over 70% of the company's 1996 customers were new from just one year earlier.

The risks for changing a customer base can be as high as the rewards. Long-term relationships can be a stabilizing factor for a company in an industry that can experience dramatic price and market fluctuations about every seven years. Lumber producers in particular are often sustained by their long-standing customers during slumps in the market. The ability to access international markets, such as those in Asia, often requires a long-term approach to customer relations. At the same time, however, many market opportunities are lost within the industry because of an unwillingness to look in new directions.

EMPLOYING NEW PRODUCT DISTRIBUTION CHANNELS

Typically, manufacturers use brokers and whole-salers for product distribution. Parsons Pine, how-ever, deviated from this norm in 1995 when it sought new channels for distributing its products. The company's decision to target the fast-growing mail-order catalog business for several of its product lines, which help organize space better makes sense, especially when considering the following:

• An estimated 10,000 product catalogs are dis-tributed throughout the United States alone, serving more than 100 million Americans who shop by catalog.

• Since 1983, the U.S. mail-order business has increased on the average of 10% in sales annually, so that in 1997 it exceeded $220 billion. The mail-order catalog business has outperformed the retail business in both sales growth and net income as a percentage of sales. Between 1991 and 1992, for example, mail-order catalog sales increased by 32.6% compared to a 6.8% growth in retail sales. Net income for mail-order catalogs was at 4.4% of total sales compared with retail at 3.1% of sales.

• Of the total sales for mail-order catalogs, about 51% are for consumer products and 25% are for business products, with specialty vendor products making up the remainder. Several of Parsons Pine products adapt to both business and con-sumer needs.

• Since 1991, the top selling product areas in mail-order catalogs that have realized above-average sales include functional home furnishings and home office and school supplies.

• Internationally, the fastest-growing markets for mail-order product sales are in Germany and Japan, two countries where Parsons Pine current-ly sells product through other distribution chan-nels. In 1993, for instance, mail-order sales in Japan exceeded sales of product in both depart-ment stores and supermarkets. Between 1982 and 1992, catalog sales in that country nearly tripled from 640 billion yen to 1.84 trillion yen.

RETHINKING PRODUCTION PROCESSING

Traditionally, wood product producers, especially those that start with higher-grade material, cut defects out of the raw lumber first before further processing the material into the intended product. The rationale for doing this is to reduce the cost of value-added processing on waste material. But because Parsons Pine typically works with lower grade material and the products it produces are made with small pieces, the company has reversed that standard in manufacturing. President Jerry Sivin realized he could actually gain more processed product with less waste by training per-sonnel to slightly vary cutting pattern on process-ing defect and pull the waste material at the end of the process line. Rethinking the production process has rewarded Parson Pine with an estimated 40% increase in overall volume of production and a pro-jected 12% decrease in falldown or generated waste. It's simply another way of making those 250 mousetrap blanks go farther.

RECOGNIZING THE VALUE OF ITS OWN WOOD WASTE STREAM

Parsons Pine adheres to its own advice when working with lower-grade resources. A percentage of the raw material purchased by the company falls below the wood quality able to be manufactured into Parsons Pine product. So, rather than throw-ing the by-product away, the company locates other value-added wood product producers that can use the lower-grade material in their product manufacturing. These lower-grade resource sales generate approximately $100,000 a year for Parsons Pine and provide valuable supply to other wood product producers.

Financial Performance: Historically Strong but Changing

Parsons Pine has been manufacturing wood products for consumers for over fifty years. The company, which had $7 million in sales in 1996, now produces more than a dozen component parts and finished wood products for consumer use. The financial success of Parsons Pine has historically been founded upon its ability to purchase wood waste at substantially lower prices than those for traditional lumber, then convert that wood waste into high value-added products. Historically, the company has been successful on both counts. While still successful in converting wood waste into valuable wood products, growth in offshore competition in certain product lines and recent U.S.-Canadian trade agreements, coupled with continuing harvest restrictions in the Pacific Northwest, have had a notable impact to Parsons Pine's bottom line.

GAINING MAXIMUM VALUE FROM THE WOOD RESOURCE

Much of Parsons Pine's financial success depends on its ability to dramatically increase the value the company generates for each piece of wood waste they purchase and convert to consumer product. In 1995, wood waste originally purchased from Canada for $750/mbf was converted to $25,000/mbf when processed into a value-added consumer product such as a knife block. Even with higher wood purchase costs than in the past brought about, in part, by the 1996 U.S.-Canadian trade flow agreements, Parsons Pine has still been able to achieve notable conversion values. Figure 4 illustrates the value added to every thousand board feet of wood processed at Parsons Pine in 1995 from the price of initial wood purchased by the company to the eventual product retail value.

Offshore Competition Mandates Product Change

During the later part of 1995, however, offshore competition in its wood slat and millwork markets

Converted Value of Wood Per mbf

Item	Species	Lbr. Purchase $/mbf	Retail $/mbf
Knife Block	Oak	$2,200	$25,000
Wine Rack	Pine, White Spruce	900	8,000
Roll-Top Counter Organizer	Pine	900	10,500
Door Louvers	Pine - Western, White Spruce	900	1,700
Toy Blocks	Pine	600	3,500
Logs for Play Cabins	Pine	600	3,500
Computer Organizer	Pine	600	1,500
Shoe Organizer Anywhere Drawer CD Rack	Alder, Maple, Pine	900 to 2,200	5,500

Source: Mater Engineering

Figure 4

forced the company to modify product designs, establish new customer bases, and identify new product distribution channels.

LEGISLATIVE ACTIONS AND TRADE RESTRICTIONS POSE MAJOR CHALLENGES

While Parsons Pine is at the forefront in advocating value-added manufacturing, it is also first in line to underscore the fact that you cannot make something from nothing.

Two factors have recently made it far more difficult for the company to conduct business as usual. First, over 70% of the company's wood supply has historically come from the harvest off National Forests throughout the western half of Oregon. In the 1990s logging bans, court injunctions, and ongoing environmental protests surrounding harvesting from those forests have seriously impacted Parsons Pine's options to get recycled wood due to the continuing closures of sawmills. Many other sawmills outside the immediate region and across the United States produce a large volume of wood waste, of course. Although transportation costs can be a critical factor in evaluating whether it makes sense to purchase wood outside the region, the greatest limitation to making such purchases is the lack of information available to the manufacturer to assess such options. For companies like Parsons Pine, little information exists to help them identify other wood product producers that generate wood waste in the species, character, sizes, and volumes which might be compatible for value-added product development.

Second, during 1996, the United States and Canada entered into a major trade agreement that poses significant challenges to Parsons Pine. U.S. lumber producers have long claimed that Canada was selling its lumber in the U.S. at unfairly low prices. Canada's share of the U.S. softwood lumber market has grown to 36%, and the dollar value of the imports rose from $2.5 billion in 1991 to $5.6 billion in 1995, according to the office of U.S. trade representative Mickey Kantor, at a cost of approximately 30,000 jobs to U.S. sawmills.

In an effort to reduce government-subsidized Canadian softwood lumber imports to the United States, the Canada-U.S. Softwood Lumber Agreement went into effect on April 2, 1996. The agreement largely affects the Canadian provinces of Quebec and British Columbia—the province where Parsons Pine had a waste wood contract negotiated. The trade agreement stipulates that in exchange for a guarantee against U.S. trade actions for five years and an unprecedented U.S. government commitment to dismiss any new petitions for trade action, softwood lumber exports from British Columbia, Quebec, Ontario, and Alberta that exceed 14.7 billion board feet a year will be subject to a $50/mbf border fee for the first 650 million board feet (mmbf) and a $100/mbf border fee for greater quantities. The 14.7 billion board feet is almost 10% lower than the 16.2 billion board feet imported to the U.S. during 1995.

Sobering Results

For Parsons Pine, the loss of the Canadian contract for waste wood was vital. The ability to purchase lower-grade lumber and wood waste at reduced values is important since processing wood waste and smaller scrap and trim ends may actually cost more due to increased material handling and production costs. Gaining the highest value from the resource processed is a critical factor to Parsons bottom-line success. For Parsons Pine, that means stretching as much as possible that differential between the material cost as a percentage of the total sales price of the product. With decreased access to wood waste from Northwest wood processing operations and the loss of the Canadian supply, Parsons Pine has been forced to purchase standard lumber loads at competitive lumber prices. The results are sobering for the company. As noted in Figure 5, the costs for raw material as a percentage of total product price increased over fivefold for many of Parsons Pine products. Two years ago approximately 70% of the company's total raw resource came from recycled wood and wood waste, in 1997 just 30% came from that source.

Parsons Pine now purchases both recycled and traditional lumber from Mexico and offshore countries including New Zealand, Chile, Brazil, and South Africa—countries that may actually employ more environmentally-harmful harvesting and forest management practices than what would currently be allowed within the forest regions of the Pacific Northwest.

POTENTIAL MISSED OPPORTUNITY

During 1995, the Institute for Sustainable Forestry (ISF) located in Redway, California—a few hours from Parsons Pine Oregon-based operation—undertook research to determine the level of awareness and interest of retailers in purchasing certified wood and wood products from targeted wood product manufacturers and wood product buyers in Oregon, Washington, and California. The results of that research suggest opportunities for Parsons Pine.

Parsons Pine Raw Material Cost as a Percentage of Sales Price

Product	(a) Board Feet Per Unit	(b) Raw Material $/mbf Paid	(c) Total Raw Material Cost [a x (b ÷ 1,000)]	(d) Retail Sales Price (Product)	(e)* Manufacturer Sales Price (Product) (d ÷ 2)	(f) Raw Material Costs as a % of Total Manufacturer Sales Price (c ÷ e)
Roll Top Counter Organizer:						
Current	6	$900/mbf	$5.40	$60.00	$30.00	18%
Prior	6	$225/mbf	$1.35	$60.00	$30.00	5%
Wine Rack:						
Current	7	$900/mbf	$6.30	$56.00	$28.00	23%
Prior	7	$225/mbf	$1.57	$56.00	$28.00	6%
Knife Block:						
Current	2	$2,200/mbf	$4.40	$50.00	$25.00	18%
Prior	2	$400/mbf	$0.80	$50.00	$25.00	3%

Manufacturer Sales Price = 50% of Retail Sales Price

Source: Mater Engineering

Figure 5

About 185 in-depth interviews were conducted for the study, which generated over a 60% response rate. Those interviewed in the three-state region included:

- Primary (lumber) wood product producers.

- Value-added (component) wood product producers.

- Secondary (finished product ready for consumer use) wood product producers.

- Corporate wood product buyers for do-it-your-self (DIY) and home center retail stores.

Key survey findings that should be of interest to Parsons Pine are noted below.

DIY / Home Center Wood Product Buyers' Responses

- Over 80% of the respondents were aware of third-party certification for wood products. Of those respondents, 42% indicated they had received requests for certified wood products. Of those respondents, 42% also indicated they sold certified wood products in their stores. California and Oregon stores led in these categories.

- The survey indicated that over 50% of the respondents said they were willing to pay a 5% premium for certified wood products; almost 40% indicated a willingness to pay a 10% premium; and 13% indicated a willingness to pay a 15% premium for certified wood product. California buyers led in all three categories.

- Of all 38 retail buyers interviewed for this study, 100% indicated they would purchase certified wood product direct from the manufacturer.

- Almost 80% of those DIY/home center retail buyers requested additional information on certified wood product suppliers and marketing opportunities for certified wood products. Oregon producers, again, dominated this category with furniture and cabinet manufacturers indicating the highest level of interest.

Secondary and Value-Added Wood Product Manufacturers' Responses

- Of those secondary wood product producers surveyed, 16% had received requests from buyers for certified wood products. However, only 9%, or a little more than half of those receiving requests for certified wood products, were actually selling certified wood products. While Oregon producers received the highest number of requests for certified wood products, California producers were actually selling more certified wood products than Oregon producers. Furniture, cabinet, and door and box producers had the highest number of requests from buyers for certified wood products.

- Between 30%-59% of all surveyed secondary wood product manufacturers who work with hardwoods in the study region indicated a willingness to pay a 5-15% premium for certified hardwood.

- Furniture, cabinet, door, and box manufacturers led the affirmative respondents to this question. Producers in Oregon clearly dominated the 5% premium question; matched California respondents at 10% premium; but lagged behind California producers who lead the 15% premium mark. Washington producers consistently appeared to have the least interest in paying premiums for certified hardwood resources.

- Between 20% and 33% of all surveyed secondary wood product manufacturers who work with

softwoods in the study region indicated a willingness to pay a 5-15% premium for certified softwood. Furniture, cabinet, and door manufacturers were the lead affirmative respondents to this question. Producers in California clearly dominated the affirmative response in this category for all premium levels.

- On a state-by-state basis, cabinet and furniture producers dominated those respondents that indicated a willingness to pay at least a 5% premium for certified wood. For example, over 80% of all furniture producers surveyed *in each state* indicated a willingness to pay a premium for certified wood, as did over 60% of cabinet manufacturers surveyed in each state.

- Almost 80% of those secondary wood product producers surveyed in all three states requested additional information on certified wood resources and marketing opportunities for certified wood products. Oregon producers, again, dominated this category with furniture and cabinet manufacturers indicating the highest level of interest.

New Opportunity

These survey results suggest opportunities for Parsons Pine. The company is located within a market region containing retailers who sell product types similar to what Parson Pine produces. The survey results indicate that a meaningful percentage of retailers in that region have indicated both a preference for and a willingness to pay a premium for wood products manufactured from certified forests. By the same token, two of the four internationally-recognized certification organizations, SmartWood and Scientific Certification Systems (SCS), have corporate and field offices in the Oregon-California market region.

Meanwhile, more private, nonindustrial forestland owners in the region are trying to certify their forests, with species matching Parsons Pine product manufacturing requirements. Parsons Pine already understands the value of differentiating its products in the marketplace through offering wood products made from recycled wood and wood waste. With supplies of waste and recycled wood diminishing dramatically, a certified wood supply may afford Parsons Pine a new opportunity to continue to offer differentiated product, possibly at a premium price. In 1997 the company was evaluating the decision to offer certified products.

SOCIAL IMPLICATIONS: GIVING BACK

Parsons Pine has been a valued member of the small Oregon community of Ashland for over half a century, providing hundreds of full-time, family-wage jobs. With tourism a dominant, but seasonal, economic generator for the community, Parsons Pine is one of the few product manufacturing operations in the community that provides year-round employment.

In coordination with Oregon State University and the newly-created Sustainable Forestry Partnership, Parsons Pine president Jerry Sivin is actively working to develop new coursework and curriculum to be offered within the Oregon State University Forestry and Forest Products departments at both undergraduate and graduate levels. Curriculum development is focused on new SFM practices, and sustainable forest products development through innovative value-added manufacturing.

Lessons Learned

Parsons Pine Products' success as a sustainable forest products manufacturing operation provides useful insights into the role of value-added manufacturing in SFM. These insights include:

1) Value-added wood products manufacturing is a necessary and important component for achieving and maintaining SFM practices.

2) It can be profitable to convert wood waste to valuable consumer products. Yet, the conversion of wood waste to *highest and best use* applications continues to be an often overlooked industry option. Further, little if any network information exists to help value-added product manufacturers identify and gain access to wood waste supply for value-added conversion.

3) Both U.S. and Canadian lumber producers have been reluctant, even adverse, to recognizing the value of their wood waste. Internal purchasing policies such as limiting the purchase of high-grade lumber without the sale of lower-grade lumber and wood waste have been effective mechanisms for changing those mind-sets.

4) Product design and marketing acumen are critical skills for successful sustainable forest product development.

5) Shifts in business-as-usual practices are essential for success in a sustainable forest products business. Rethinking the production process, purchasing appropriate equipment, establishing a new customer base, and engaging new product distribution channels are critical tools for success.

6) Recent U.S. trade agreements with Canada to govern natural resource management and lumber flow, while intended to level the playing field for primary wood product producers in the United States, have had the opposite effect on smaller value-added wood product manufacturers, especially when combined with region-specific harvest restrictions imposed by legislative action. The results may be counter-productive to achieving sustainable forestry practices on a global scale. This is an important factor that is often overlooked or simply unrecognized within policy development surrounding the forest products industry. Future policy actions need to take into account the differences in impact that restrictions have on commodity wood flow, or lumber used in building construction, and industrial wood flow, lumber used in remanufactured and value-added wood products.

Portico, S.A.:
Strategic Decisions
1982–1997

CASE STUDY

PREPARED BY:
BETTY J. DIENER

A Case Study from "The Business of Sustainable Forestry"
A Project of The Sustainable Forestry Working Group

The Sustainable Forestry Working Group

Individuals from the following institutions participated in the preparation of this report.

Environmental Advantage, Inc.

Forest Stewardship Council

The John D. and Catherine T. MacArthur Foundation

Management Institute for Environment and Business

Mater Engineering, Ltd.

Oregon State University
Colleges of Business and Forestry

Pennsylvania State University
School of Forest Resources

University of California at Berkeley
College of Natural Resources

University of Michigan
Corporate Environmental Management Program

Weyerhaeuser Company

The World Bank
Environment Department

World Resources Institute

CCC 1-55963-626-2/98/page 12-1 through page 12-13
© 1998 John D. and Catherine T. MacArthur Foundation
All Rights Reserved

Portico, S.A.:
Strategic Decisions
1982-1997

CASE STUDY

PREPARED BY:
BETTY J. DIENER

A Case Study from "The Business of Sustainable Forestry"
A Project of The Sustainable Forestry Working Group

Contents

This case was prepared by Betty J. Diener, based on previous materials written about Portico by the author and Hunter Saklad, and information gathered by Sarah Cordero Pinchansky, as a basis for discussion rather than to illustrate either effective or ineffective handling of an administrative situation.

Portico, S.A.:
Strategic Decisions 1982-1997

Introduction

Increasingly, countries encounter a dilemma: How to protect their forests while maximizing their economic potential. In developing countries the dilemma is particularly acute. Poverty leads to multiple, short-term demands on forests, and governments in poorer countries are often unable to mediate economic demands on forests with the broader concerns of forest sustainability. In such an environment, wood products companies find it harder to secure reliable supplies of high quality tropical hardwoods, and to operate where governments are increasingly regulating forest management.

Portico, S.A.—Business Description

Portico, S.A., of Costa Rica, has confronted those challenges by adopting sustainable forest management (SFM) as a means to obtain a stable supply of high quality logs that it can use to manufacture high-end residential mahogany doors and accommodate increasingly stringent controls on the use of commercial forests. These doors command a premium price in the marketplace and can be exported worldwide without controversy, in an era of rising public concern about tropical deforestation. The case of Portico suggests that it is possible to combine SFM with successful businesses in a setting characterized by heightened government control of forests, and even benefit by increasing product quality and prevailing in competitive international markets.

In the fall of 1997, Leopoldo Torres, one of Portico's owners, reflected on the many strategic changes that Portico had undergone since 1982, including the latest decision to split the company's two main operations—forest management and wooden door manufacturing/export—that was to be implemented over the next year. He said the opportunity to successfully create these two different operations would have been impossible unless Portico had made a series of earlier key strategic decisions. These included acquiring and then vertically integrating forestry operations with the manufacture/export operations, making the commitment in 1987 to SFM, and attaining certification from Scientific Certification Services (SCS) in 1992 as a "state-of-the-art, well managed forest."

Certification enabled Portico to expand its business with major U.S. retail chains, and in 1997 was facilitating the company's efforts to raise capital to purchase and operate more forestlands. By conducting its forestry in an environmentally-certified, well-managed fashion, Portico has been able to produce materials for its high-end mahogany doors, avoid much of the controversy surrounding the logging industry in Costa Rica, become a leader in shaping new logging legislation, and preserve a fragile ecosystem.

EARLY COMPANY HISTORY

Portico was founded in 1982 when Torres and a small group of his business associates purchased a small Costa Rican company, *Puertas y Ventanas de Costa Rica* (Costa Rica Doors and Windows). The original company had been producing 70-80 wooden doors and window frames a month for the Costa Rican market since 1973 and for export to Panama and El Salvador since 1978. Torres and the other investors intended to immediately convert this company into one that manufactured and exported high-end, exterior-use, mahogany doors to the United States. They chose mahogany because of its availability in Costa Rica, as well as its strength, appearance, and popularity with U.S. customers.

In 1982, doors for the Costa Rican market had to be custom-made to fill nonstandardized spaces once the doorframe had been built. The U.S. and world markets, however, required doors of standardized dimensions and consistent quality. To serve these new markets, Portico needed to change completely its methods of production, factory layout, and product design. In addition, the company needed a stable supply of wood that was of consistently high quality so that the doors would not crack in transit or after installation.

Prior to 1982, several species of tropical hardwoods had been used, including oak and mahogany. Even though those woods were abundant in Costa Rica, their supply in the form of high-quality lumber at a consistently low price was uncertain. Logging was impossible during Costa Rica's six-month rainy season because heavy equipment could not be transported through muddy forestland. Logging practices were generally poor, with loggers often allowing the cut logs to receive either too much or too little humidity before they arrived at the mill. Costa Rica had no formal lumber market. Few standards for lumber quality existed, and prices had to be negotiated individually for every purchase. Moreover, lumber mills generally did not use high-quality equipment, such as drying kilns, circular saws, or band saws, and thus could not supply high-quality lumber with consistency.

Addressing the Problems

Beginning in 1986, Portico decided to address these price, supply, and quality problems, and to gain greater quality control over its raw materials by moving to full vertical integration. The company began to acquire forestland to supply its raw materials, trucks for transportation, and saw mills, kiln dryers, and other more sophisticated equipment for processing lumber. Vertical integration was intended to enable Portico to have "total quality control" over its products throughout the process, from the Costa Rican forest to U.S. retail outlets.

In 1987, Portico, S.A., acquired over 5000 hectares (12,355 acres) of additional natural forest in the northeast coastal region of Costa Rica, near

Nicaragua, as a result of a sizable infusion of capital from Norwest Corporation, a U.S. financial institution based in Minneapolis, Minnesota. Norwest's investment was part of a debt-for-nature swap agreement with the government of Costa Rica. In 1987, Norwest held Costa Rican government debt in its portfolio, debt that might never have been recovered. The market price of that debt had fallen on one occasion as low as $.16 on the dollar. The Central Bank of Costa Rica purchased this debt with bonds denominated in Costa Rican colones at a higher-than-market price, with the proviso that Norwest invest those funds in a Costa Rican company, preferably an export-oriented business. Norwest chose to buy 5000 ha of Costa Rican rain forest and exchanged these for a major share of Portico.

Portico then bought additional hectares of native Costa Rican rain forest, and, today manages slightly more than 20,000 acres of natural forest. A subsidiary, *Tecnoforest del Norte*, was established to oversee the management of the forestry operation.

Historical Deforestation and Logging Practices in Costa Rica

Costa Rica has a land area of 51,000 square kilometers with a population of three million. In Costa Rica, as in other developing nations, significant amounts of forest cover have been removed to facilitate agriculture or cattle grazing. At one time, an estimated 90% of Costa Rica was covered with wet and moist tropical forests. By 1992, 31% of Costa Rica's land remained forested. Of this remaining forest cover, only an estimated 5% has logging potential, while the remainder is in protected areas. The Forest Bureau estimates that between 1950 and 1990, the average annual rate of deforestation was 50,000 hectares.

During this same period, cattle and croplands increased from 15% of the land to over half of it. Government subsidies that promoted exports of beef, bananas, coffee, and other agricultural products supported this deforestation. Another contributing factor was the illegal deforestation practice known in Costa Rica as *socola,* a kind of gradual clear-cutting. With *socola,* the forest understory would be illegally cut, removed, and replanted with pasture grass. Then, the landowner would petition the government for permission to remove the large trees still standing because they were impeding the growth of his "pasture." Forestry law mandated that trees could be legally removed from pastureland, so the government had to grant the cutting permit. Most of the trees felled in this process were seen as merely a byproduct of shifting the land into a more productive use. They were not processed for timber, but were burned or left to rot. A 1990 World Resources Institute report stated that Costa Rica had the highest rate of deforestation in the world.

Clear-cutting, the process of removing all trees and other vegetation from a forest area, is a technique used in plantation forestry (when all trees are the same species and age), or when converting land use from forest to grazing or crops. Clear-cutting is seldom used in commercial logging in the tropics because too many trees have low commercial value in a natural forest stand.

In Costa Rica most traditional logging is selective, with cutting decisions made tree-by-tree. Traditional selective logging is a one-time operation, intended to reap maximum high-value lumber in one harvest. Because the logger never intends to return to log the site again, sustainable resource management is not considered.

In selective cutting, all valuable trees in the extraction area greater than 60 cm (24") in diameter at breast height (DBH) are marked and mapped for cutting or preservation. In tropical forests, however, ecosystem damage from selective cutting is much higher than in temperate forests because the net-

works of vines in the forest bring down trees other than the ones being cut. The trees being cut also destroy more trees and vegetation by falling on them. Once the trees are selectively felled, the logs need to be extracted from the forest. The extraction process further harms the forest because there is no economic incentive to minimize damage after the one-time harvest. For example, skid trails created by dragging the trees out of the forest on skidders damage seedlings needed for regeneration, cause erosion, dry out the soil in some places, and reduce drainage in others.

TECNOFOREST DEL NORTE AND NATURAL FOREST MANAGEMENT

Tecnoforest del Norte was the subsidiary established to oversee the management of the initial 7000 ha (17,300 acres) of natural forest that Portico purchased in northeast Costa Rica. This primary- and secondary-growth forest supplied the mahogany that Portico used for its doors. The mahogany that Portico obtained from its forests could not be grown on plantations because it depends on the surrounding biodiversity to protect it from massive infiltration by ants and other wood-destroying pests. Mahogany can be supplied only by a natural forest system in the lowland coastal areas of the humid tropics.

Tecnoforest was guided by two basic principles to ensure long-term natural forest management (a relatively new concept in the tropics) for its commercial purposes. First, it sought to guarantee the stability of the species used to manufacture the doors, so that it would have consistent long-term yields of high-quality wood. The species Portico was using were *carapa guianensis,* commonly called carapa and labeled by Portico as "royal mahogany."

Second, Tecnoforest tried to minimize disturbance to the surrounding ecosystem during the extraction process, so that its natural forest management would be considerably less damaging to the forest than traditional logging.

Traditional loggers sought to obtain a maximum amount of high-value wood from a single forest cutting, before moving on to another virgin area. Tecnoforest's natural forest management program intended to produce a consistent flow of valuable timber from the same land in perpetuity with low-impact, sustainable forestry practices. To accomplish that Tecnoforest used different practices for preharvest planning, harvesting, transportation, and maintenance under natural forest management from those used in traditional logging operations.

Tecnoforest first took a prelogging inventory. All trees greater than 70 cm (28") DBH were marked and mapped for the management plan that was submitted to the *Dirrecion General Forestal* (DGF). Carapa (royal mahogany) trees were marked in three ways. The best specimens were left as seed trees, others were left as shade trees to protect carapa seedlings from excessive sunlight during regeneration, and the remaining trees (about three to four trees per hectare) were designated for felling. Only trees greater than 70 cm DBH were marked for felling, even though trees greater than 60 cm DBH could legally be cut. Trees greater than 70 cm DBH were chosen because they were believed to be "overmature" and already deteriorating.

The average Tecnoforest crew was twice the size of a traditional crew and always included an accredited forester. All Tecnoforest loggers were educated by foresters in techniques of minimizing forest damage during the felling and extraction processes. Unlike other logging operations, Tecnoforest employed five full-time researchers to study its forest holdings and to oversee maintenance. Tecnoforest researchers were further helped by programs developed with U.S. universities. Preselected (marked) trees were directionally felled (cut to fall in a specific direction and land in a predetermined place) to minimize damage to other trees and allow the skid trails to be run along the same paths. Running the skid trails in confluence reduced damage to the forest floor by minimizing the area over which the heavy trees

were dragged. In this way, fewer saplings were destroyed, less forest undergrowth was damaged, and fewer animal dwellings were disturbed. This care was taken to ensure that the area, 15 years later, would be sufficiently regenerated to support another harvest.

An independent study conducted in 1990 found that Tecnoforest's natural forest management procedures were in fact much less damaging to the forest than traditional logging on a similar forest. The comparison of a traditional site with a Tecnoforest site showed that the residual stand (after logging) of the traditional site contained only 61% of the trees from before the cut, while the Tecnoforest residual stand contained 95% of the original trees. While forest composition (relative numbers of each species present) was relatively unchanged at the Tecnoforest site, it was significantly changed at the traditional site as new species became dominant. The forest canopy opening was computed to be 46% at the traditional site after logging but only 29% at the Tecnoforest site. (Large openings in the canopy can allow saplings to be damaged by excessive sunlight and can also interrupt important wildlife corridors that are crucial to maintaining biodiversity.) Also, while roads and skid trails encroached to cover 9% of the territory at the traditional site, they only covered 3% of the Tecnoforest site. The study concluded that "Portico's logging operation seems much more viable than traditional logging in terms of long-term sustainability."

History of Logging Regulations and Legislation

Prior to the passage of the 1977 Reforestation Law, the Costa Rican government freely handed out wood-harvesting and forestland grants. This was a period of rapid destruction of forests, much of it carried out illegally. The 1977 Reforestation Law set extraction limits on logging sites at 25% of the standing trees. Unfortunately, a stipulation that a tax be paid on the projected harvest before cutting served instead to stimulate illegal practices. In 1981, Costa Rica still had the highest rate of deforestation among tropical countries, at 4% of forested lands. By 1991, as government controls increased, this rate of deforestation had fallen to 2.6%. In 1987, and again in 1992, the government established new guidelines requiring forest management plans, and agreements to reduce logging to 60% of salable trees with diameters of 24" or wider. Despite such efforts, as of 1990, it was estimated that 60% of ongoing logging continued to be illegal.

INDUSTRY RESPONDS

In response, the country's forest producers founded the Costa Rican Chamber of Forest Producers—a group of more than 600 private businesses—to help pass forestry laws that would be enforced, and that would make forestry at least as profitable as was true for other uses of the land. Their initiative, led by Portico, had as its objectives the stabilization of government forest policy, the development of forest product export businesses, and the assurance of access to forests.

The new law, enacted as Forest Act #7575 in May 1994, contained several major provisions. First, it promoted the harvesting of wood from environmentally-certified sources in Costa Rica, a measure that would lower supply and raise wood prices. Second, the government agreed to compensate forest owners for certification costs and for the increased costs of maintaining a certified or sustainable forest. This would make the value of forested land competitive with other uses, with the added benefit of soil and water protection, carbon sequestration, and preservation of biodiversity. Third, the government would gradually open up the export market for raw logs and lumber, thereby generating higher world market prices for Costa Rican wood.

In addition, the forestry sector expected to eventually benefit from a new, international, joint implementation program. Under this program, industrialized countries would provide payments to countries such as Costa Rica that would use the funds to increase their sustainably-managed forests and plantations. Such payments would enable industrial countries to offset their increased emissions

Skepticism remained, however. Environmental advocates in Costa Rica generally agree that Costa Rica has had tough legislation to control logging and to keep forests intact. Unfortunately, the enforcement system has not always proven effective. The DGF was considered ineffective; the 1994 law, at the instigation of the Chamber of Forest Producers, removed DGF from issuing logging permits and collecting taxes.

One member of the Rainforest Alliance said in 1994:

> They have good laws, but they're easy to get around because there is a lack of effective enforcement by the DGF. Hardly anyone ever pays taxes. You have to file a management plan to get a cutting permit, but almost no one checks to see whether you follow your management plan.

Another new forestry law was adopted in April, 1996. It banned all further land-use changes in natural forests and deregulated forestry plantations. Municipalities and private forestry engineers were given the power to issue logging permits for farmlands and nonforested areas.

Environmentalists felt that too much control was being given to commercial loggers and to the Chamber of Forest Producers.

REGULATORY IMPACTS ON PORTICO

In 1994, delays in the permit process had caused Portico to buy an unusually large proportion of mahogany on the open market. In 1990, a normal harvest year, Portico had controlled less forestland but only had to buy 15% of its wood on the open market. Because of the uncertainty in the permit process, Portico, in 1994, did not plan to expand its Costa Rican forest operations. Management did not feel that the growth of its door business could be assured if it was dependent on Costa Rican holdings. Given the risks at the time of owning forestland in Costa Rica, Portico did not want to incur the risks associated with transforming its "small" forest management undertaking into a truly large-scale commitment.

In October, 1994, when Leopoldo Torres was interviewed about Portico's strategic options, he said:

> We can only log in a window of about four months from late January through May. This year [1994] we missed some of the season because of problems with getting all of the necessary permits from the DGF in time.

Nevertheless, vast stands of the lowland coastal forest still contained the type of carapa tree that Portico used and called royal mahogany. These forests extended into Panama to the south and Nicaragua to the north. There was adequate potential to supply whatever level of raw material Portico might need, as long as the company could gain access to it. However, many groups worldwide and within Costa Rica opposed any further cutting by anyone, regardless of their record as forest managers. Some local environmental groups and some legislators proposed banning all logging, even on private land.

By summer, 1997, however, the situation was radically different. Torres was so encouraged by the changes in the laws and by the changes in the attitudes that he had encountered in the environmental departments that he actively sought to increase the managed forestlands in Costa Rica and was seeking capital investment to do so.

> I believe that with the new legislation and the controls established by the government, everything that's being cut legally is from sustainably-managed, selectively-harvested lands. The government has brought in good people and

has removed many of the problems we used to experience. The previous government was very negative to our industry and was responsible for the destruction of the forests over many years. The government officials may not be as sophisticated as we are, but there are many more limits on logging practices, and the legal operators have plans that would qualify for certification. So, now we're trying to set up a certification process.

In addition the permitting processes had changed. In the past permits were issued each year, which had made them difficult to get. Under the new system logging permits no longer expired each year.

Portico's Market and Strategic Decisions, 1982-1997

U.S. DOOR MARKET

The size of the door market for high-end, exterior-use hardwood doors fluctuated from 500,000 to 1.5 million doors each year, depending on the health of the economy, housing starts, major home renovations, and general construction.

Residential homes and a select group of commercial properties constitute the market for exterior-use wood doors. The market can be further segmented by price, quality, and materials, ranging from low-end ($50) plywood composites, through mid-range ($150–500) softwoods and hardwoods, and up to premium ($500 and up) 100% hardwood doors made from mahogany, other hardwoods, and oak. Oak doors sold for anywhere from $50 to $150 more than mahogany.

In the United States, high-end hardwood doors were generally sold by home center chains such as Lowe's and Home Depot, and also in small independent housing supply and hardware stores that catered to high-end markets.

The market for wooden doors is split into two major categories—interior and exterior use—and two styles of construction—flush and stile and rail. Interior doors are usually flush style doors (flat one-piece surface on both sides), but are occasionally stile and rail types (four solid perimeter members and cross beam midsections that house a number of solid floating panels) if they are being used commercially. Exterior doors are generally stile and rail type construction with a floating panel design which makes them more resistant to weather damage from wind, rain, extreme temperature change, and humidity.

The National Wood Window and Door Association and buyers for national home center retailers like Home Depot agree that the highest quality wood doors available in appearance and performance are generally stile and rail, 1-3/4" thick, premium grade, hardwood doors employing mortise and tenon type joinery. Premium grade hardwood refers to a hardwood, mahogany, or oak, which is free of defects, such as knots, kiln burn, or bright sap spots that might affect its performance. Sometimes, this is also referred to as select grade, and is intended to receive a natural or stain finish. Portico manufactures this type of door.

FORESTRY AND MANUFACTURING OPERATIONS

From its beginnings in 1982 until 1997, Portico made a number of strategic decisions with regard to its forestry operations and its manufacturing operations that enabled it to obtain a year-round stable supply of raw materials. In forestry, these decisions included the acquisition of natural tropical hardwood forests, the commitment to manage forestland sustainably, and the decision to seek third-party certification of management practices. In the door manufacturing operations, Portico, concerned about the predictability of supply from the original natural forest, introduced new materials (American red oak and other tropical hardwoods), and new "engineered" construction methods. Each of these strategic decisions is discussed below.

Vertical Integration

As reported earlier, one of Portico's first key strategic decisions was to vertically integrate its operations, from Costa Rican forest to U.S. retailer, at its inception in 1982. Forest ownership ensured the company a stable source of supply, at a reasonable cost, for raw materials that were of consistently high quality.

Sustainable Forest Management

The next decision was to acquire, in 1987, a natural forest, under a debt-for-nature swap agreement, and then to manage it, even at considerable additional operational costs, under the principles of long-term SFM.

Third-Party Certification

The decision to seek certification was made to gain access to key distribution channels at a time of heightened public interest in tropical forest destruction. This decision was made fairly easily since the sustainable forestry practices used by Portico were already consistent with the standards necessary for certification, and the cost of the certification amounted to roughly $1/door.

In 1992, Tecnoforest received certification of its 7000 ha (17,300 acres) as a "state-of-the-art, well managed forest." This endorsement came from Scientific Certification Systems (SCS) of Oakland, California (formerly known as Green Cross), an independent environmental certification organization.

Portico chose to obtain certification from SCS rather than through a similar SmartWood program of the Rainforest Alliance because one of its retailers had a previous relationship with SCS and preferred this seal of approval.

The SCS inspection process took about six months from the time of the original application to final approval and cost Portico more than $40,000. SCS hired and sent a forester, a forest ecologist, a forest economist, and one of its own staff, all of whom were already familiar with the region, to inspect all of Tecnoforest's holdings. Their reports were then reviewed by others with expertise in the region.

After reviewing all reports, SCS scored the operation in three standard forestry categories: sustaining timber yields, maintaining the ecosystem, and providing socioeconomic benefits to the surrounding community. A score of 60 or better in each area won certification. Portico/Tecnoforest scored 82, 79, and 73, respectively, to become the first SCS-certified "forest manager" in Costa Rica and the third in the world.

As of October, 1994, SCS had approved six forests as being well managed; two of these, Menominee Tribal Enterprises and Seven Islands Land Management Company, had American red oak in their certified stands. SCS had also approved two retailers for selling wood from these six approved sources and eight manufacturers of wood products for using lumber from these sources. Portico was certified only as a forest manager, not as a manufacturer, because it was still buying a portion of its wood from uncertified sources.

SCS certification allowed Portico to advertise that it owned a forest that was managed in an environmentally-responsible manner and that this wood was used in its doors. It could display the SCS green cross of environmental approval in promotional literature and alongside its mahogany products. However, it could not place the seal directly on each door unless it had conducted a chain-of-custody audit documenting that every piece of wood in that door came from a forest that was certified as well managed. Portico felt that, unless all of the mahogany it used came from a certified source it would be far too labor-intensive to audit the source of each piece of wood used. At the time of certification, Portico was purchasing between 15-25% of its mahogany from other Costa Rican suppliers.

Single Piece Solid Mahogany

In 1982, Portico decided to concentrate on the manufacturing and export of high-end residential use mahogany doors. By 1987, Portico was selling 29,000 mahogany units per year in the U.S. door market. These doors were single-piece solid mahogany, generally 1-3/4" thick, with solid stile and rail construction and mortise and tenon joinery. The retail prices in the United States ranged from $500 to $2500, occasionally reaching $5000, depending on whether the door system included one or two doors, custom glass, and so on. These prices, which represented considerable value-added in manufacturing, enabled Portico to absorb the extra costs of SFM.

Engineered Construction

By the end of 1993, Portico was selling 60,000 units per year. That year Portico also introduced engineered construction into its doors. Torres noted that, increasingly, the industry was substituting engineered products for solid wood construction, and that, in a few years, the finest tropical hardwoods would be used solely for veneer. The cores of the engineered doors could either be mahogany (to retain the solid mahogany designation), other plantation species, or medium density fiberboards.

The engineered doors looked like their single-piece ancestors, but they were constructed by using the very best wood for a 2 mm veneer that covered the surfaces of the stiles and rails. Underneath the veneer were perfectly fitted and glued pieces of mahogany called butcher block. These smaller pieces had the dual advantages of improving the product, by reducing internal stresses that caused larger pieces of wood to warp, and by allowing more complete utilization of the company's raw material resources. Portico turned to engineered doors not only to increase efficiency and stability, but also because Portico's supply of mahogany was threatened by what it perceived to be the negative government policy toward logging natural forests.

Engineered doors cost 10-20% more to manufacture because of the additional capital expenditures needed for more technologically-advanced machinery and the extra labor required. Much of the wood used by Portico in the construction of these doors came from its own certified forest.

American Red Oak and Other Tropical Hardwoods

Portico began to sell high-end American red oak doors in early 1993. These doors were in the same range of styles and prices as the mahogany line. These oak doors cost the company approximately 15% more to produce than the mahogany doors, but it also received $20-50 more for each one. Final retail prices were $50-150 more than for mahogany. In 1994 oak accounted for 8% of Portico's sales; by 1997 30% of Portico's sales were in American red oak, with the proportion continuing to grow. This material was purchased from a U.S. oak plantation, then shipped to Costa Rica. Portico did not want to use tropical oaks, which had a less desirable appearance than American red oak, even though it could manage these wood sources in Costa Rica. The oak used by Portico was not environmentally certified.

According to Torres, Portico needed to offer high-end American red oak doors because buyers wanted to carry lines from premium door manufacturers that offered both oak and mahogany. Oak was a preferred wood in the Midwest and Northeast, whereas mahogany was preferred in the Southeast.

In addition to broadening its type of woods to satisfy its customers, Portico also introduced a third line of doors, targeted toward the lower end of the market, made with another species of tropical hardwood, priced at about 60% of the mahogany doors. These were designed to offer similar quality, but lighter molding, and a much lower retail price. This line of doors accounted for 20% of Portico's sales in 1997. The forests from which this wood was acquired were not environmentally certified although they were sustainably managed.

STRATEGIC OPTIONS

Portico had, by 1997, expanded its product line beyond royal mahogany (now 50% of sales) to include American red oak (30%) and a lower-priced tropical hardwood (20%). The doors had gone from solid construction to butcher block engineered products, not only making the products stronger but also conserving natural resources more efficiently.

Portico's sales of 60,000 units, however, were the same volume as in 1994 when it held a 60% share of the southeastern U.S. market. As the market increased in the Southeast, however, Portico's share of market declined even though its unit sales held steady and its prices had increased at roughly the same rate as inflation. Torres maintained that Portico's sales were flat primarily because many more companies had entered the high-end residential door market, from factories in lower-cost regions from Southeast Asia and from South American markets such as Bolivia, Peru, Brazil, Guatemala, and Chile.

> If a country has the forest resources, it is looking for wood products to make and export. Before, small companies couldn't meet the needs of the U.S. buyers. They shipped a few containers and failed. Now, their quality is up. They begin with a low price, selling to small customers. New entrants like that from South America are succeeding now.

Portico continued to export its high-end residential doors, with 10-15% going to countries other than the United States. Torres thought that environmental certification would be helpful as the company sought to increase sales in northern European countries such as Germany, "If we wanted to go for the main merchants in Germany, and supply them directly, certification would be almost indispensable."

The Dutch government had an initiative to import only sustainably-produced tropical woods. B&Q, the largest home center in the United Kingdom, had adopted a similar policy. The Austrian government had tried to launch (although it later abandoned) a policy labeling all tropical timbers it imported by country of origin and certification status. Southern European and Latin American markets seemed to be much less impressed by environmental certification than North American and northern European markets.

Portico had not, however, sought broader environmental certifications for its total operations or for its other types of wood. Portico's manufacturing process used only water soluble and nontoxic glues. Sawdust and particulate matter is vacuumed by dust extraction equipment, stored in two silos, and sold as mulch to farmers. And, using only wood from its own forests would allow the company to get certification as a manufacturer in addition to its forest certification.

According to Torres, however, while having SCS certification had advantages for Portico, he had no clear evidence from retail customers that they would pay more for a door simply because the wood used to make it was certified. He believed that customers still made their buying decisions largely based on the price-to-quality ratio.

However, Torres felt that the SCS certification had probably helped Portico by enhancing the company's image among its key accounts. A buyer for the Home Depot confirmed Torres' opinions and explained that:

> We have to buy out of managed rain forests because everything we buy has to be certified if it's imported. There's a lot of bad wood out there, but Portico does well-managed forestry. They've got the highest quality doors too.

Even here, Torres felt that Portico's customers were buying not because of the certification but for the package of services Portico offered, including price, quality, delivery, and consistency and service, as well as certification. Certification, he agreed, was one of

the criteria that Home Depot had used originally, but the services offered were more important in keeping Home Depot as a customer. He agreed that all things being equal, Portico has been in a better position with the certification and its proof of environmental claims. He just does not see any current need for expanding that certification.

This was particularly true in terms of Portico's next strategic challenge—to separate the forestry management and harvesting operations from the door manufacturing and distribution. By fall, 1997, Portico expected to separate the two operations, and then to offer capital investment in the forestry management operations through local stockbrokers.

The funds acquired would be used to increase forest reserves. Torres attributed this financing opportunity to Portico's record and certification as a sustainably-managed forest.

> Investors are interested in us because we are a proven forest manager. We can prove a reasonable long-term yield, not a high yield but not a high risk. It is very long term and very low risk. And, by separating the two areas, the investors in the forest don't have to worry about the door market and vice versa.

Lessons Learned

The challenge to Portico over the years had been to assure itself of a stable year-round supply of raw materials, within an environment of diminishing supplies of tropical wood, a high awareness of the issue of tropical forest destruction, and antagonistic government policies toward its industry. By 1997, it seemed to be a success after a series of strategic moves that included vertical integration, engineered wood technology, leadership in changing government policies, and certification of its forestry management. These actions had well positioned the company to move ahead, both in forestry management, and in door manufacturing and export. Portico's early adoption of SFM gave it access to high-quality wood, access to markets, and, now, increased access to capital.

Despite Portico's success, it is unclear whether certified, high value-added products actually command premium prices or whether the key benefit, at least in the United States, is simply gaining access to channels of distribution. If so, and should the environmental pressures on retailers moderate in the United States, there might be a concurrent reduction in any interest in certification. In addition, if the prices on early certified products include high margins, those products become vulnerable to undercutting by other countries or companies seeking to enter the markets with products produced from traditionally harvested wood.

Exterior—Panel Doors

Figure 1

Exterior—Glass Doors

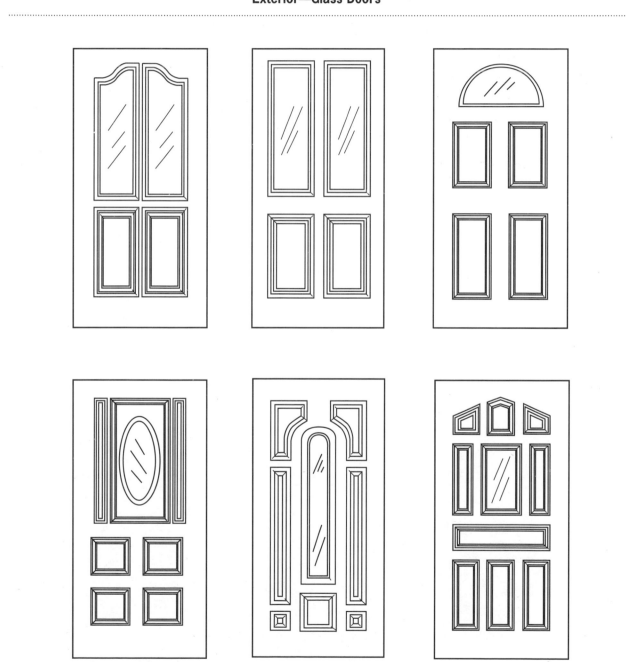

Figure 2

Precious Woods, Ltd.

CASE STUDY

PREPARED BY:
CHARLES A. WEBSTER
DIANA PROPPER DE CALLEJON

A Case Study from "The Business of Sustainable Forestry"
A Project of The Sustainable Forestry Working Group

The Sustainable Forestry Working Group

Individuals from the following institutions participated in the preparation of this report.

Environmental Advantage, Inc.

Forest Stewardship Council

The John D. and Catherine T. MacArthur Foundation

Management Institute for Environment and Business

Mater Engineering, Ltd.

Oregon State University
Colleges of Business and Forestry

Pennsylvania State University
School of Forest Resources

University of California at Berkeley
College of Natural Resources

University of Michigan
Corporate Environmental Management Program

Weyerhaeuser Company

The World Bank
Environment Department

World Resources Institute

CCC 1-55963-627-0/98/page 13-1 through page 13-24

Precious Woods, Ltd.

CASE STUDY

PREPARED BY:
CHARLES A. WEBSTER
DIANA PROPPER DE CALLEJON

A Case Study from "The Business of Sustainable Forestry"
A Project of The Sustainable Forestry Working Group

Contents

Precious Woods, Ltd.

Introduction

Destruction of the world's tropical forests remains a dramatic problem. Despite energetic public activism that has led to the creation of global treaties and the reform of national policies to address the issue, the deforestation of tropical forests not only continues, but in many cases has increased. In Brazil, for example, recent data show that deforestation actually rose 34% between 1991 and 1994, as the number of square miles disappearing annually climbed from 4,296 square miles in the 1990-91 burning season to 5,750 square miles a year by 1994.[1] In nearby Costa Rica, where deforestation has slowed, roughly 17,000 hectares were destroyed annually from 1988 through 1992, continuing a trend that has reduced the nation's primary natural forests from 4.5 million hectares at the turn of the century to less than 1.5 million today.[2]

In the midst of this destruction, sustainable forest management (SFM) has grown from a theoretical concept to a set of practical, procedural guidelines for harvesting natural forests in a way that minimizes damage to forest and ecosystem, while maximizing sustainable economic value. Still, the commercial viability of SFM has not been clearly demonstrated. Indeed, as of the early 1990s less than one percent of tropical forests around the world were managed even on a sustained-yield basis (an approach far less environmentally rigorous than SFM), and today most experts believe the actual figure is much lower.[3] Precious Woods, Ltd., a Swiss-founded corporation active in Costa Rica and Brazil, is one of the few companies in the world attempting to conduct SFM.

Precious Woods' achievements are many—the company raised over $40 million in equity capital, hired leading forestry engineers and experts to craft and implement a management plan, and launched production and sales of tropical hardwood to high-paying European markets. Nevertheless, Precious Woods experienced significant setbacks and financial crises during its start-up operations, leaving the company well short of meeting its revenue projections in 1996, its first year of full mill operations. In response, the company began to implement changes in operations at the end of 1996. While it is too soon to know the results from these changes, financial projections prepared by the case writers, using conservative assumptions, suggest that Precious Woods will be able to meet the expectations of its investors for returns in 1999.

An Industry's Challenges and Opportunities

This case examines Precious Woods' efforts to establish a sustainable tropical forestry business in Brazil, and identifies both the company's challenges and potential. In the process, the case evaluates how SFM can provide a viable alternative to uncontrolled logging in tropical forests—an alternative that, if widely implemented, could help slow tropical deforestation and preserve biodiversity of rainforests. The case begins by reviewing tropical deforestation, the global and Amazon tropical wood industries, and Brazilian government regulation, as issues that affect the competitiveness and commercial potential of the company. It then documents Precious Woods' formation, its forestry program and competitive strategy, and closes with projections and analysis of the company's profitability.[4]

The case, however, captures only an early snapshot in time of the company—the period between 1994 and December 1996. This period includes both Precious Woods' preoperating phase and the first few months of mill operation. An update of the company's operations in 1997 is provided in an addendum at the end of the case.

TROPICAL DEFORESTATION

Tropical deforestation destroyed an estimated 154,000 square km annually of the world's rainforests between 1981 and 1990.[5] In Latin America, some 27% of the region's 20 million square km of tropical forest had been destroyed by 1980, and a further 13% vanished over the next ten years.[6] The impact of tropical deforestation is far-reaching. Forests in the world's tropics—including the Amazon, the Congo Basin, and the myriad islands of Indonesia and MalaysiaCare home to an estimated 50% of animal and plant species on earth, containing great ecological, medicinal, aesthetic, and economic value. In addition, tropical forests play a key role in the absorption of atmosphere-threatening carbon dioxide, and when each hectare of tropical forest is burned, some 220 tons of carbon are released.[7] Tropical forest loss is estimated to have contributed one quarter of the carbon released into the atmosphere in the last decade.[8]

The burning of forests to clear land for livestock and agriculture has been one of the principal causes of Brazilian deforestation. However, with the growth of the timber industry, logging has played an increasingly important role.[9] The expansion of logging has come at great cost to the rainforest, not only because loggers extract diverse and exotic tropical timber, but also because logging operations crush and compact fragile soil, open the forest canopy, disturbing both flora and fauna, and increase the risk of fire. For each tree removed by loggers, an average of 27 others that are 10 cm or more in diameter are severely injured, 40 meters of road are created, and 600 square meters of canopy are opened.[10] Researchers estimate that it can take 70 years for logged forests to resemble their original states.[11]

TROPICAL TIMBER INDUSTRY

Although production of tropical roundwood increased steadily over the past few decades, production levels have slowed since 1990.[12] More than 80% of tropical timber production is consumed for fuelwood and non-industrial uses, and only about 6% of all tropical wood produced enters international trade.[13] Led by Malaysia and Indonesia, tropical Asia dominates the tropical forest products trade, representing approximately 80% of total exports of tropical forest products. Demand for timber in many of these same tropical countries, however, has grown sharply over the past ten years, and coupled with steady depletion of primary tropical forests, means that many large nations have become net timber importers. Bans on log exports to spur value-added domestic industry have also sharply reduced Asian log exports, creating a decline in overall exports that has only been partially offset by value-added products such as plywood and veneer. Brazil remains the dominant Latin American producer of tropical wood products, but still accounts for a small portion—roughly 8% in 1995—of international exports. Still, as evidenced by the growing presence of Asian firms in the Amazon, the perceived value of Brazil's forests is increasing.[14] Table 1 show trends among leading export countries since 1991.

In general, the tropical wood products industry is characterized by low barriers to entry, with little product differentiation among producers and, except in higher value-added activities, relatively low capital requirements.[15] Essentially a commodity product, tropical sawnwood is differentiated only by species and quality. Even sizes and cuts of sawnwood are largely standardized, although customers sometimes specify particular dimensions for sawnwood products. Quality is measured by the presence of defects evident in the wood (ranging from knots and holes to twisting or warping in dried wood), and wood is categorized in strict accordance with defect levels. Import markets such as those in Europe, Japan, and the United States typically require the highest-quality, lowest-defect category woods, including FAS, No. 1, No. 2, and Common, while domestic markets in Brazil and other developing countries are frequently less exacting.

Tropical Exports—Selected Countries and Regions

		(thousands of m³)						Percent Change YOY				
		1991	1992	1993	1994	1995	% Tot. (95)	92/91	93/92	94/93	95/94	Avg.
Indonesia	Logs	0	0	0	0	0	0.0%	n.a.	n.a.	n.a.	n.a.	n.a.
	Sawn	936	711	639	308	300	3.4%	-24.0%	-10.1%	-51.8%	-2.6%	-22.1%
	Veneer	31	30	13	10	10	0.1%	-3.2%	-56.7%	-23.1%	0.0%	-20.7%
	Plywood	8970	9761	9724	8852	8500	96.5%	8.8%	-0.4%	-9.0%	-4.0%	-1.1%
	Total	9937	10502	10376	9170	8810	100.0%	5.7%	-1.2%	-11.6%	-3.9%	-2.8%
Malaysia	Logs	19320	17797	9382	8561	8000	48.6%	-7.9%	-47.3%	-8.8%	-6.6%	-17.6%
	Sawn	4932	5417	5371	4560	3950	24.0%	9.8%	-0.8%	-15.1%	-13.4%	-4.9%
	Veneer	477	765	720	613	620	3.8%	60.4%	-5.9%	-14.9%	1.1%	10.2%
	Plywood	1186	1670	2421	3004	3900	23.7%	40.8%	45.0%	24.1%	29.8%	34.9%
	Total	25915	25649	17894	16738	16470	100.0%	-1.0%	-30.2%	-6.5%	-1.6%	-9.8%
PNG	Logs	1500	1929	2867	3100	2900	99.9%	28.6%	48.6%	8.1%	-6.5%	19.7%
	Sawn	3	5	3	3	4	0.1%	66.7%	-40.0%	0.0%	33.3%	15.0%
	Veneer	0	0	0	0	0	0.0%	n.a.	n.a.	n.a.	n.a.	n.a.
	Plywood	0	0	0	0	0	0.0%	n.a.	n.a.	n.a.	n.a.	n.a.
	Total	1503	1934	2870	3103	2904	100.0%	28.7%	48.4%	8.1%	-6.4%	19.7%
Myanmar	Logs	1500	1500	1029	600	700	94.0%	0.0%	-31.4%	-41.7%	16.7%	-14.1%
	Sawn	200	200	38	36	44	5.9%	0.0%	-81.0%	-5.3%	22.2%	-16.0%
	Veneer	0	0	0	0	0	0.0%	n.a.	n.a.	n.a.	n.a.	n.a.
	Plywood	0	0	1	1	1	0.1%	n.a.	n.a.	n.a.	n.a.	n.a.
	Total	1700	1700	1068	637	745	100.0%	0.0%	-37.2%	-40.4%	17.0%	-15.1%
Asia Pacific Total	Logs	22328	21229	13281	12266	11608	39.9%	-4.9%	-37.4%	-7.6%	-5.4%	-13.8%
	Sawn	6187	6434	6181	4999	4369	15.0%	4.0%	-3.9%	-19.1%	-12.6%	-7.9%
	Veneer	543	821	745	659	651	2.2%	51.2%	-9.3%	-11.5%	-1.2%	7.3%
	Plywood	10295	11544	12220	11899	12437	42.8%	12.1%	5.9%	-2.6%	4.5%	5.0%
	Total	39353	40028	32427	29823	29065	100.0%	1.7%	-19.0%	-8.0%	-2.5%	-7.0%
Brazil	Logs	100	71	200	1000	1000	33.3%	-29.0%	181.7%	400.0%	0.0%	138.2%
	Sawn	230	484	627	850	1000	33.3%	110.4%	29.5%	35.6%	17.6%	48.3%
	Veneer	40	109	188	85	100	3.3%	172.5%	72.5%	-54.8%	17.6%	52.0%
	Plywood	350	509	656	800	900	30.0%	45.4%	28.9%	22.0%	12.5%	27.2%
	Total	720	1173	1671	2735	3000	100.0%	62.9%	42.5%	63.7%	9.7%	44.7%
All Latin America	Logs	110	85	251	1052	1050	29.4%	-22.7%	195.3%	319.1%	-0.2%	122.9%
	Sawn	440	703	992	1273	1405	39.4%	59.8%	41.1%	28.3%	10.4%	34.9%
	Veneer	46	116	200	96	111	3.1%	152.2%	72.4%	-52.0%	15.6%	47.1%
	Plywood	381	550	714	881	1000	28.0%	44.4%	29.8%	23.4%	13.5%	27.8%
	Total	977	1454	2157	3302	3566	100.0%	48.8%	48.3%	53.1%	8.0%	39.6%
Africa Total	Logs	3460	2719	3250	4003	3476	70.3%	-21.4%	19.5%	23.2%	-13.2%	2.0%
	Sawn	1020	1081	1028	1211	1135	22.9%	6.0%	-4.9%	17.8%	-6.3%	3.2%
	Veneer	171	216	203	240	246	5.0%	26.3%	-6.0%	18.2%	2.5%	10.3%
	Plywood	53	66	70	82	91	1.8%	24.5%	6.1%	17.1%	11.0%	14.7%
	Total	4704	4082	4551	5536	4948	100.0%	-13.2%	11.5%	21.6%	-10.6%	2.3%

Source: ITTO, Annual Review 1995

Table 1

Influences on Demand

Demand for tropical wood varies considerably by species of wood, as each species has properties and characteristics that render it suitable for certain uses, and as wood characteristics vary greatly across species. A number of Brazilian species, for example, are very heavy and high in density, making them unsuitable as light construction woods, but desirable for their strength and resistance to decay as heavy, marine construction woods. Demand is influenced, however, not only by the species' suitability, but also by basic dynamics of market acceptance. In particular, market acceptance is a function of buyers' familiarity and confidence with the wood's suitability, which is mostly a product of time in use. Important characteristics by which wood is judged include the following:

- density (weight, hardness, ease of "workability" in shaping, cutting, joining, etc.)

- strength (structural rigidity, lateral and vertical load resistance, etc.)

- durability (resistance to fungus and insects, stability, shrinkage, etc.)

- appearance (color, grain, texture, luster, finish, etc.)

Tropical wood products firms are chiefly price takers, with little or no leverage to influence prices for their products either in domestic or international markets. Prices vary in direct relation to the quality of the wood (per the categories mentioned above) and, over the past thirty to forty years, have remained stable. International tropical timber prices rose an average of approximately 1.2% in real terms from 1950-1992 and, as shown in Figure 1, have remained fairly stable during the 1990s. The principal reason for price stability—despite the long-held impression that growing scarcity of tropical timber should result in rapidly-increasing prices—has been competition from high-quality temperate hardwoods, which provide potential substitutes for many tropical wood products.[16]

In sum, the global tropical wood industry is fragmented and competitive. Still, because the industry is characterized by numerous small firms—that in many cases produce with irregular access to supply, low-quality capital equipment, and, it is often alleged, low levels of professionalism—reliability of supply can have a very positive influence on producer-buyer relationships. In short, one of the more important ways in which firms can differentiate themselves is through producing high-quality product with consistency, paying meticulous attention to customer specifications, and providing professional service in orders and delivery.

AMAZON TIMBER INDUSTRY

Triggered by the development of roadways and the provision of government subsidies, logging in the Amazon increased dramatically over the past twenty years. In the early-1960s, fresh on the heels of establishing the country's new capital in Brasilia, the Brazilian government opened the Brasilia-Belem highway, suddenly making hundreds of miles of Amazon forest more accessible. In addition, the government began a program to promote cattle herding by offering subsidies as great as 75% of the costs of starting a ranch, and property to those who could demonstrate they had cleared land.[17]

For years cattle ranching grew rapidly, but it began to lose its appeal in the 1980s as nutrient-poor Amazon soil proved unable to provide sufficient grass for grazing, and as fiscal incentives dried up. Facing the collapse of their livelihood, ranchers turned to selling logging rights to their properties. Logging began selectively—focusing only a few high-value species such as mahogany—but grew steadily as hardwood stocks began to decline in southern Brazil, and prices rose in both domestic and international markets. From 1976 to 1988, hardwood production in the north Amazon increased to meet demand, rising from 7 to 25 million cubic meters.[18] Logging became increasingly mechanized and intensive, and the wood industry grew into a dominant economic force in the northeastern

Price of Latin American Sawnwood, 1990-1995

Bold lines show prices in constant 1990 US$ per cubic meter (deflated by the G-5 MUV Index used by the World Bank for deriving real commodity prices). Normal lines show nominal price trends.

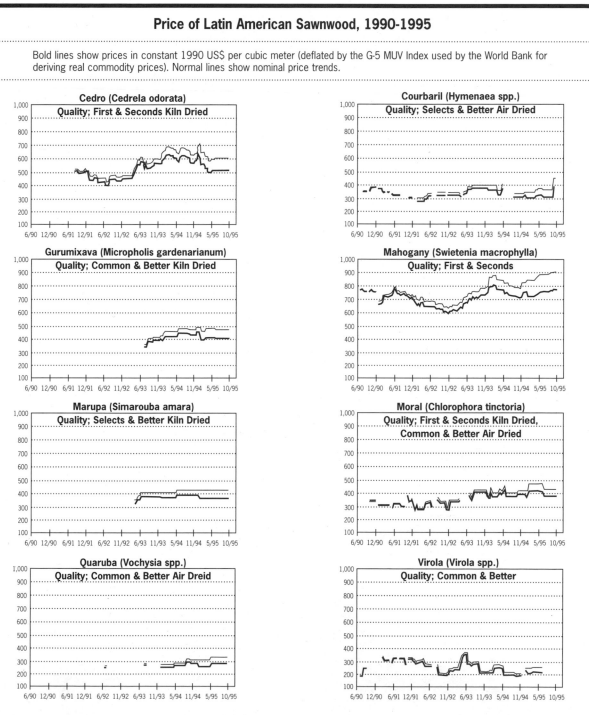

Source: ITTO, Annual Review 1995

Figure 1

Amazon. By 1990 more than 100 species of tropical woods were harvested, with bulldozers opening wide swaths of forest, and diesel trucks hauling timber to rapidly proliferating sawmills.

The Amazon timber industry developed most rapidly in Par·, a state of 22,000 square km in northeastern Brazil. Currently almost 90% of all roundwood production in the Amazon occurs in Pará.[19] As Figure 2 indicates, most of the growth in logging and sawmills occurred in the late 1970s and early 1980s.[20] By 1990, the industry's structure was composed of three principal types of firms: loggers, who harvested, or extracted woods from the forest; saw millers, or processors, who cut and processed the logs; and value-added firms including veneer and plywood producers.

Of the roughly 240 sawmills in Paragominas, about 80% were small, operating with one band-saw, and producing between 4,000-5,000 cubic meters per year, while 20% were large mills, with two or more band-saws and annual production of 10,000-14,000 cubic meters.[21] Almost all of these mills operated with used capital equipment of relatively low quality. Roughly 60% of the sawmills were vertically integrated, however, involved in both the harvesting and processing of logs. Almost all sawnwood was sold domestically, in Brazil, due to the strength of local demand, and the fact that most firms were too small, and did not produce wood products of sufficient quality for international markets.[22]

ECONOMIC TURMOIL AND INDUSTRY CHANGE

In general, economic conditions facing the logging industry have deteriorated since 1990. As land tenure has stabilized and areas of accessible forest have receded over the past five years, stumpage fees (fees charged for cutting logs) have risen more than 30% in real terms. Over the same period, high *Real Brazilian* interest rates have slowed civil construction, depressing a principal source of demand for wood products, and causing real domestic prices for end-products (principally sawnwood) to decline an average of 20% (see Table 2).

Wood Processing Firms in Paragominas (1970-1990)

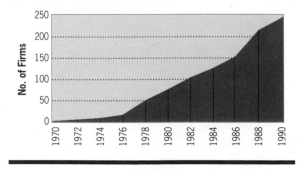

Figure 2

Wood Prices in Paragominas, 1990 and 1995
(in constant 1995 US$ per m³)

Grade	Product	Average Price		%Change
		1990	1995	
High	Log	$60	$82	37%
	Sawnwood	$336	$291	-13%
Medium	Log	$38	$43	13%
	Sawnwood	$216	$174	-19%
Low	Log	$24	$30	25%
	Sawnwood	$168	$98	-42%
Very Low	Log	$18	$27	50%
	Sawnwood	$96	$89	-7%
Average	**Log**	**$35**	**$46**	**30%**
	Sawnwood	**$204**	**$163**	**-20%**

Source: Steven Stone, "Economic Trends in the Timber Industry," p.22

Table 2

Although international prices for export-quality tropical woods have remained stable, Brazilian producers' margins have been reduced by inflation and the currency's appreciation.[23] In addition, the costs of electricity and diesel fuel have risen by almost 70% since 1990, while service quality has generally fallen. Finally, although the cost of labor has declined in real terms, firms continue to pay social security and other taxes on labor amounting to an additional 50% of salaries.[24]

For firms devoted exclusively to wood processing, profits have fallen sharply, compressed between higher log prices and energy costs, and lower market prices. Firms vertically integrated into land ownership and logging have suffered less, but have also felt the impact of lower wood prices. As a result, many firms have sought to integrate vertically and to enhance margins through economies of scale. Though statistics are sparse, industry consolidation has occurred, particularly among small firms. In addition, a number of firms have expanded into value-added steps of processing, including the manufacture of veneer and plywood. Over the past five years a small number of "export clearing houses" have also been established. Formed by consortia of six to seven of the largest processors, export clearing houses kiln dry, plane, and package wood for export (from various suppliers), each with an estimated throughput of 40,000-50,000 cubic meters per year.[25]

BRAZILIAN GOVERNMENT POLICY

Over the past thirty years the Brazilian government has established an institutional and legislative structure for regulating the country's forestry industry. The basic legal framework for regulation of the Brazilian forests was established in the Forest Code of 1965, which set up permanent preservation areas, created biological reserves, and led to the establishment of a public forestry agency responsible for oversights of the forest products industry.

Subsequent decrees and regulations have further specified conditions under which Amazon forest resources could be utilized, including the Normative Instruction of 1980, stipulating that the primary forest could only be exploited under a management plan approved by the public forestry service, and a 1987 decree requiring larger companies to secure approval for a reforestation plan before harvesting wood. Other regulations have prohibited the felling of trees smaller than 45 cm in diameter, the export of logs and sawnwood thicker than 76 cm, and the felling of certain species such as the Brazil nut tree, and the rubber tree. Most recently, temporary, two-year bans have been placed on the felling of mahogany and virola.

Despite these regulatory advances, the Brazilian government has been criticized for its ineffectiveness in slowing deforestation. The progress of IBAMA, the national environmental agency, has been hindered by a number of factors, including high management turnover (from 1989-1992 the agency had five different directors), understaffing (fewer than 1,300 employees monitor almost 5 million square km) and inadequate financial support.[26]

Over the last several years, however, the government's regulatory efforts appeared to be toughening. New federal protocols have strengthened IBAMA's reach, and officials reported that 70% of all forestry projects submitted for approval were declined in 1996 for inadequate forest management plans. Increased government control has created higher costs for logging firms that comply with regulations. (Forest management plans cost $10,000-$20,000 and can take as long as six months to process.) For those companies pursuing well-structured sustainable forestry projects, however, the likelihood of government intervention or shutdown is significantly reduced.

Precious Woods, Ltd.

Precious Woods, Ltd. was founded in 1990 by Roman Jann, a Swiss entrepreneur and former corporate lawyer. Jann's experience in tropical reforestation dated to 1988 when, as a financial advisor to Swiss investor Dr. Anton E. Schrafl, he began a several-year assignment evaluating a Costa Rican teak reforestation project. Encouraged by the economic potential and ecological benefits of reforestation, in late 1990 Jann developed a proposal for an independent company and, with important contributions from Dr. Schrafl, launched Precious Woods, Ltd. The company's stated business objective was to invest in tropical reforestation and hardwood plantation management in Costa Rica.

INITIAL DECISIONS

After various rounds of fund-raising, the company acquired a total of five properties in Costa Rica, ranging in size from 290 to 3,900 hectares, and located on the eastern coast and northern border of the country. From 1990 through the fall of 1996, the company had planted (or reforested) approximately 2,500 hectares, and expected first commercial harvests to commence roughly twenty-five years after planting.[27] Precious Woods' Costa Rican reforestation program has focused primarily on teak and pochote, though a mix of native species is also being planted (reaching 5.3% of total plantings by 1996) to help reestablish the natural flora and fauna, and to minimize forestry problems associated with monoculture plantations.

Shortly after launching operations in Costa Rica, Precious Woods considered expanding its activities into natural forest management in Brazil, both to widen the company's operations in tropical forestry and to enhance medium-term cash flow. In late 1991, the company commissioned Hans Peter Aeberhard, a Swiss economist with lengthy experience in business management in Latin America, to prepare a feasibility study for the project expansion. After extensive analysis and consultation with

leading tropical forestry experts including Dr. Ronnie de Camino of Costa Rica and Dr. Reitze de Graaf of the Netherlands, Precious Woods had both a forest management plan and a capital investment program for Brazil.

In January 1994, the Swiss board of directors of Precious Woods approved the Brazilian project, and began further fund-raising. The company then conducted due diligence for the purchase of appropriate forest land in the Brazilian Amazon, completed preparation of its forest management plan, and submitted the plan for approval to IBAMA. In May 1994, the company purchased 80,000 hectares of Amazon forest approximately 230 km east of Manaus along the Amazon River, roughly 90% of which was primary, undeveloped forest. The company paid a total of $4.3 million for the land, which included some road infrastructure, buildings, and sawmilling equipment.[28]

After securing purchase of the land, Precious Woods began to buy and import machinery and equipment in accordance with the capital investment program detailed in the company's feasibility study. The feasibility projected total start-up costs of approximately $15 million. In addition to land, this estimate covered the purchase of equipment for harvesting and transporting logs (skidders, tractors, and trucks), machinery for sawmilling (several European-quality bandsaws), working capital, and project development expenses. Given the company's inexperience in operating a forestry company in the tropics, the initial feasibility study was revised, and the company was forced to incur additional costs to establish its operations. After a number of delays, the first sawmill line was completely installed in June 1996. The second sawmill line and the first kiln-drying oven were completed in September 1996.

APPROACH TO SFM

From its founding in Costa Rica to its expansion into Brazil, Precious Woods' principal objective has been to introduce sustainable, market-economy management in primary tropical forests. The company based its work in Brazil on the practices of SFM, which is most clearly defined as, "controlled and regulated harvesting, combining silvicultural and protective measures, to sustain or increase the commercial value of forest stands, all relying on natural regeneration of native species."[29] By utilizing SFM, the company's objective has been to minimize tropical forest destruction by developing a system of harvesting that is not only sensitive to ecological impact, but that also provides incentives to protect and enhance the forest's natural value.

Precious Woods' forest management program was based chiefly on the Celos model, a polycyclic (or multiple-cycle) system developed in Surinam between 1965 and 1983 by Dr. Reitze de Graaf of the Wageningen Agricultural University in the Netherlands.[30] The Celos management system, combining both a harvesting program and a silvicultural program, differs from conventional logging methods in several important ways. First, through its meticulous organization of labor, its use of special, low-impact logging techniques, and its emphasis on forest inventory as a key planning tool, the Celos system places great emphasis on protecting and sustaining the forest ecosystem, while increasing the efficiency of extracting tropical timber.[31] Second, through its use of careful silvicultural practices, the Celos model stresses active management of the forest to enhance regeneration between cutting cycles, particularly of more desirable, commercial species.

The Celos SFM system was adapted by Precious Woods to Amazon conditions by scientists at INPA (a scientific research institute based in Manaus) and through the assistance of Dr. de Graaf, who was retained as a scientific advisor. The system began with a preliminary review of the topography and natural characteristics of the company's property, after which approximately 20-30% of the company's forest was set aside, to be held in reserve and not harvested at all. This land included areas considered ecologically sensitive, such as watersheds, as well as areas selected as buffer zones. The remaining 70-80%

of forest surface was divided into twenty-five parcels, or compartments, of roughly equal size (approximately 2,100 hectares), reflecting the company's decision to adopt a harvest rotation period (or time between each successive harvest) of 25 years.[32]

Inventory and "Prospecting"

As prescribed by the Celos system, Precious Wood's management plan calls for the completion of a comprehensive forest inventory, or preliminary survey of the forest's characteristics, followed by thorough prospecting of the entire forest. Conducted on each compartment of the company's land—ideally at least six months to one year before that parcel is harvested—the prospecting review identifies each tree over 50 cm in diameter by species and location. In addition, the process documents topographical factors such as gullies, steep hills, or other natural obstacles that can influence the harvesting process. All details from the inventory are recorded and entered into a computer, yielding a map of each compartment such as the one shown in Figure 3.

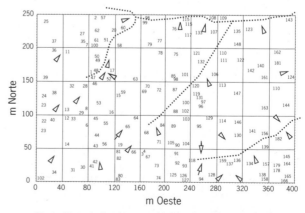

**Sample of Prospecting Printout
Precious Woods**

Chart of botonical and topographical data gathered in the field.

Figure 3

Low-impact Logging

The inventory is then used to locate all commercially-desirable trees, to determine which of these species and trees will be extracted (see also below), and to map out low-impact roads and skid trails that the provide the most direct, least damaging access to these trees. Further prescriptions of the SFM plan include directional felling and the cutting of vines to minimize the destruction of other, nearby trees as the desired trees are cut down, and winch skidding to drag logs across the forest floor with lower soil impact.

Lesser-known Species

Precious Woods' objective of generating attractive financial returns from its SFM program requires that the company utilize as great a variety of tree species as possible, including many species that are not well known on international markets. The principal reasons for using lesser-known species are: (1) to avoid over-harvesting and depleting the genetic stock of a handful of commercially-known species; and, (2) to generate sufficient harvest volumes (and thus sales) from the forest. As Table 3 indicates, the Brazilian rainforest is very heterogeneous, and a few better-known species such as Ipe, Amapa and Jatoba would not generate adequate financial returns.

Silvicultural Treatments

The SFM plan also includes silvicultural treatments that are designed to increase the representation of commercial species, and to enhance natural regeneration. Silvicultural steps range from cutting vines and other plants in the forest canopy to allow sunlight to reach desired species, to selectively killing undesirable species by cutting around the base of the trees and applying arboricide. The Surinam experience indicates that these post-harvest steps can increase growth in commercial timber volume by a factor of at least four, from about 0.5 to 2 cubic meters per hectare per year.[33]

SALES AND MARKETING STRATEGY

Precious Woods' strategy is to market two forms of high-quality, sustainably-harvested tropical wood products—sawnwood and semi-finished products—to high-paying international markets including Europe, the United States, and Asia. In 1997, the company expected to sell approximately 24,000 cubic meters of sawnwood, and approximately 2,500 cubic meters of semi-finished products, which it planned to produce utilizing roughly thirty different species (see Table 3). The company hired an international salesman in April 1996, who has lengthy international experience in tropical woods, and appears to have strong sales contacts in Europe. Importantly, Precious Woods' capital equipment is of sufficiently high quality to produce for European and other international market standards. The company's European sawmilling machines, for example, contain variable saw speeds, stelite-edged saw blades, and a hydraulic log-rotator—characteristics that place their equipment among the more technologically advanced tropical producers in Latin America.[34]

Precious Woods' strategy also rests heavily on expansion into semi-finished products. As mentioned above, the company's loyalty to its SFM plan—coupled with the economic need to harvest large volumes—requires that the firm harvest a very large number of species (Precious Woods markets roughly thirty species of a total of sixty species in their forests which have commercial potential), many of which are not well known in target markets. The company believes semi-finished goods will offer two important advantages in sales and marketing. First, semi-finished goods can be created from the scrap and odd-size pieces that result from the production of standard-size sawnwood. In many cases, this scrap is high-quality wood that, because of its dimensions, would otherwise be thrown away. Second, semi-finished products allow the company to pre-package lesser-known tropical woods for end-users, and in effect, to accelerate market acceptance by predetermining for consumers the woods' suitable uses.

Projected Precious Woods 1997 Production Volumes, by Species
(volumes in cubic meters)

No.	Species	Brazilian Name	Botanical Name	Total Vol.	Sawn Vol.*	Percent
1	CANA	Castanharana	Lecythis Pranccei	9,237	3,233	13.2%
2	LOPR	Louro preto	Ocotea Fragantissima	5,746	2,011	8.2%
3	MASS	Massaranduba	Manilkara Huberi	4,857	1,700	6.9%
4	CUPI	Cupiuba	Goupia Glabra	4,034	1,412	5.8%
5	LOGA	Louro Gamela	Nectandra (Ocotea) Rubra	3,817	1,336	5.5%
6	JARA	Jarana	Lecythis Poiteuai	3,386	1,185	4.8%
7	FAVI	Favinha	Entherolobium Schomburgkii	3,211	1,124	4.6%
8	PANA	Piquiarana	Caryocar Galbrum	2,866	1,003	4.1%
9	CDRI	Cardeiro/Cedrinho	Seleronema Micranthum	2,826	989	4.0%
10	BREV	Breu Vermelho	Protium Altosonii	2,437	853	3.5%
11	ABIU	Abiurana Vermelha	Pouteria Guianensis	2,206	772	3.2%
12	GUAR	Guariuba	Clarisia Racemosa	1,977	692	2.8%
13	AMDO	Amapa Doce	Brosium Potabile	1,814	635	2.6%
14	CUMA	Cumaru	Dipteryx Odorata	1,811	634	2.6%
15	PIQU	Piquia	Caryoca Villosum	1,743	610	2.5%
16	LOIT	Louro Itauba	Mezilarurus Sinandra	1,734	607	2.5%
17	ANVE	Angelim Vermelho	Dinizia Excelsa	1,477	517	2.1%
18	ANPE	Angelim Pedra	Hymenolobium Heterocarpum	1,463	512	2.1%
19	AMAP	Amapa	Brosium Parinarioides	1,391	487	2.0%
20	TAUA	Tauari Vermelho	Cariniana Micrantha	1,369	479	2.0%
21	ARVE	Arurua Vermelho	Iryanthera Grandis	1,260	441	1.8%
22	UCHI	Uchi Torrado	Sacoglottis Guianensis	1,246	436	1.8%
23	MAJU	Maparajuba	Manikara Cavalcantei	946	331	1.4%
24	SUVE	Sucpira Vermelha	Andira Unifolialata	943	330	1.3%
25	ANFA	Angelim Fava	Parkia Pendula	891	312	1.3%
26	MUIR	Muiracatiara	Astronium Lecontei	857	300	1.2%
27	PIMA	Piquia	Caryocar Villosum	749	262	1.1%
28	JATO	Jatoba	Hymenea Courbaril	729	255	1.0%
29	CAJU	Caqui	Anacardium Giganteum	563	197	0.8%
30	LOAM	Louro Amarelo	Ocotea Sp.	549	192	0.8%
31	SUAM	Sucupira Amarela	Qualea Paraensis	503	176	0.7%
32	TANI	Tanimbuca	Buchenavia Viridifolia	360	126	0.5%
33	VIOL	Violeta	Peltogyne Cantingae	311	109	0.4%
34	MARU	Marupa	Simaruba Amara	283	99	0.4%
35	IPE	Ipe	n.a.	209	73	0.3%
36	TABR	Tauari Branco	Couratari Stellata	180	63	0.3%
37	TAJU	Tatjuba	Begassa Guianensis	14	5	0.0%
38	LOAR	Louro Aritu	Licaria Chrysophylla	9	3	0.0%
TOTAL				**70,003**	**24,501**	**100.0%**

*Sawnwood volumes are shown as 35% of total volumes, as projected by the company. A conversion yield of 35% is what the company expects to produce, on average, for European-quality (FAS, I and 2) export product. Note that actual yields are likely to vary considerably from one species to another, depending on wood density, presence of defects, etc.

Table 3

Precious Woods expects that its semi-finished capabilities will include kiln-drying, custom-cutting, and manufacturing pieces for specific wood-based products such as molding, door parts, parquet flooring and certain furniture items. The company had six kilns installed and functioning in November 1996, with a total annual capacity of approximately 5,000 cubic meters.[35] In addition, the firm placed orders for approximately $350,000 in semi-finished production machinery, and expected to take delivery in February 1997.

Finally, in November of 1996 the company began discussions with a Brazilian firm manufacturing parquet flooring at a site 10 km from Precious Woods' operations. The possible joint venture is intended to combine Precious Woods' sustainably-harvested wood with the partnering firm's high-quality European parquet machines (one of which was already on-site and operational), and promote sales to European markets. Total output of the joint venture could reach as high as 500 cubic meters per month.

The company does not expect to secure a "green premium" for its products, but does anticipate that its distinctive status as a sustainable producer of high-quality tropical wood should assist it in gaining access to markets where buyers are more sensitive to tropical deforestation issues. (Precious Woods was certified by the SmartWood Program, which is accredited by the Forest Stewardship Council (FSC), as of June 1997.) Further, the company believes that markets for sustainable forest products will grow significantly. The company emphasizes that its mission and harvesting practices have not only been designed by several of the world's leading experts in tropical forestry, but they have also been widely endorsed by an impressive array of individuals and institutions including the president of Costa Rica, and a former Brazilian minister of the environment.[36]

EUROPEAN MARKETS

Precious Woods' competitive strategy depends heavily on successful penetration of European markets, where the company's sustainable forestry techniques are expected to help it gain competitive advantage, and where higher pricing levels are critical to its profitability. According to studies by the World Bank, certified sustainable producers could capture as much as 20% of the European tropical timber import market, up significantly from an estimated 1–3% today.[37] Although this figure varies considerably across markets, it is expected to grow over the next few years and, while unlikely to offer producers a green premium over market prices, may nevertheless represent an important tool in penetrating markets and developing strong buyer relationships.

A number of other global markets, however, also offer premium prices, and in many cases indicate more favorable demand trends. As Table 4 indicates, European markets imported 6.4 million cubic meters of tropical wood products in 1995, or roughly 21% of total global imports. Yet for a variety of reasons—including environmental activism resulting in consumer boycotts and purchasing bans—European demand for tropical timber has fallen steadily over the 1990s, declining at an average annual rate of 5.2%. In the United States, imports of tropical wood products have remained fairly stable over the 1990s. In addition, though reputedly lower-paying, Asian markets such as China and Taiwan exhibit much more dynamism. Overall tropical timber demand rose an average of 7% annually over the past five years in China; Taiwanese imports of tropical sawnwood increased an average of 18% over the same period.

By the standard of typical European buyers, Precious Woods' total production levels are low, presenting both an advantage and a challenge. Projected annual sales levels are 25,000–30,000 cubic meters, or roughly 3% of only one of Europe's smaller markets, the Netherlands. For most buyers of tropical hardwoods, a company's ability to

Tropical Imports—Selected Countries and Regions

		(thousands of m³)					% Tot. (95)	Percent Change YOY				
		1991	1992	1993	1994	1995	% Tot. (95)	92/91	93/92	94/93	95/94	Avg.
France	Logs	878	880	895	881	900	62.3%	0.2%	1.7%	-1.6%	2.2%	0.6%
	Sawn	464	354	360	324	325	22.5%	23.7%	1.7%	-10.0%	0.3%	-7.9%
	Veneer	20	19	18	20	20	1.4%	-5.0%	-5.3%	11.1%	0.0%	0.2%
	Plywood	213	189	180	195	200	13.8%	-11.3%	-4.8%	8.3%	2.6%	-1.3%
	Total	1575	1442	1453	1420	1445	100.0%	-8.4%	0.8%	-2.3%	1.8%	-2.0%
Germany	Logs	317	281	213	192	150	22.4%	-11.4%	-24.2%	-9.9%	-21.9%	-16.8%
	Sawn	416	353	240	259	250	37.3%	-15.1%	-32.0%	7.9%	-3.5%	-10.7%
	Veneer	85	85	77	72	70	10.4%	0.0%	-9.4%	-6.5%	-2.8%	-4.7%
	Plywood	202	218	193	195	200	29.9%	7.9%	-11.5%	1.0%	2.6%	0.0%
	Total	1020	937	723	718	670	100.0%	-8.1%	-22.8%	-0.7%	-6.7%	-9.6%
Italy	Logs	674	439	438	476	440	45.8%	-34.9%	-0.2%	8.7%	-7.6%	-8.5%
	Sawn	363	360	369	420	400	41.7%	-0.8%	2.5%	13.8%	-4.8%	2.7%
	Veneer	130	69	58	72	70	7.3%	-46.9%	-15.9%	24.1%	-2.8%	-10.4%
	Plywood	47	53	35	49	50	5.2%	12.8%	-34.0%	40.0%	2.0%	5.2%
	Total	1214	921	900	1017	960	100.0%	-24.1%	-2.3%	13.0%	-5.6%	-4.8%
Netherlands	Logs	117	123	111	120	111	14.4%	5.1%	-9.8%	8.1%	-7.5%	-1.0%
	Sawn	525	578	587	461	500	64.9%	10.1%	1.6%	-21.5%	8.5%	-0.3%
	Veneer	10	13	13	12	10	1.3%	30.0%	0.0%	-7.7%	-16.7%	1.4%
	Plywood	287	322	238	162	150	19.5%	12.2%	-26.1%	-31.9%	-7.4%	-13.3%
	Total	939	1036	949	755	771	100.0%	10.3%	-8.4%	-20.4%	2.1%	-4.1%
UK	Logs	24	19	19	23	17	1.8%	-20.8%	0.0%	21.1%	-26.1%	-6.5%
	Sawn	435	547	450	450	438	46.6%	25.7%	-17.7%	0.0%	-2.7%	1.3%
	Veneer	11	17	18	23	20	2.1%	54.5%	5.9%	27.8%	-13.0%	18.8%
	Plywood	536	634	478	485	464	49.4%	18.3%	-24.6%	1.5%	-4.3%	-2.3%
	Total	1006	1217	965	981	939	100.0%	21.0%	-20.7%	1.7%	-4.3%	-0.6%
All Euro. (EU)	Logs	3178	2768	2435	2527	2416	37.8%	-12.9%	-12.0%	3.8%	-4.4%	-6.4%
	Sawn	2955	2913	2599	2486	2479	38.8%	-1.4%	-10.8%	-4.3%	-0.3%	-4.2%
	Veneer	297	239	216	234	226	3.5%	-19.5%	-9.6%	8.3%	-3.4%	-6.1%
	Plywood	1505	1657	1370	1287	1271	19.9%	10.1%	-17.3%	-6.1%	-1.2%	-3.6%
	Total	7935	7577	6620	6534	6392	100.0%	-4.5%	-12.6%	-1.3%	-2.2%	-5.2%
China	Logs	1395	1776	1595	2030	2000	31.3%	27.3%	-10.2%	27.3%	-1.5%	10.7%
	Sawn	88	559	703	717	700	11.0%	535.2%	25.8%	2.0%	-2.4%	140.2%
	Veneer	27	217	287	193	200	3.1%	703.7%	32.3%	-32.8%	3.6%	176.7%
	Plywood	1417	1424	1371	1094	750	11.7%	0.5%	-3.7%	-20.2%	-31.4%	-13.7%
	Total	2927	3976	3956	4034	3650	57.1%	35.8%	-0.5%	2.0%	-9.5%	6.9%
Taiwan	Logs	4218	3961	2180	2000	2000	31.3%	-6.1%	-45.0%	-8.3%	0.0%	-14.8%
	Sawn	529	709	1052	900	950	14.9%	34.0%	48.4%	-14.4%	5.6%	18.4%
	Veneer	146	226	204	194	220	3.4%	54.8%	-9.7%	-4.9%	13.4%	13.4%
	Plywood	432	741	788	800	830	13.0%	71.5%	6.3%	1.5%	3.8%	20.8%
	Total	5325	5637	4224	3894	4000	62.6%	5.9%	-25.1%	-7.8%	2.7%	-6.1%
Japan	Logs	10402	10990	8324	7632	6550	102.5%	5.7%	-24.3%	-8.3%	-14.2%	-10.3%
	Sawn	1013	1248	1805	1642	1500	23.5%	23.2%	44.6%	-9.0%	-8.6%	12.5%
	Veneer	250	192	239	160	132	2.1%	-23.2%	24.5%	-33.1%	-17.5%	-12.3%
	Plywood	2941	2882	3864	3791	4074	63.7%	-2.0%	34.1%	-1.9%	7.5%	9.4%
	Total	14606	15312	14232	13225	12256	191.7%	4.8%	-7.1%	-7.1%	-7.3%	-4.2%
USA	Logs	6	4	5	4	1	0.0%	-33.3%	25.0%	-20.0%	-75.0%	-25.8%
	Sawn	196	193	175	222	188	2.9%	-1.5%	-9.3%	26.9%	-15.3%	0.2%
	Veneer	20	20	15	13	11	0.2%	0.0%	-25.0%	-13.3%	-15.4%	-13.4%
	Plywood	934	1053	919	742	957	15.0%	12.7%	-12.7%	-19.3%	29.0%	2.4%
	Total	1156	1270	1114	981	1157	18.1%	9.9%	-12.3%	-11.9%	17.9%	0.9%

Source: ITTO, Annual Review 1995

Table 4

supply minimum, regular volumes of a wood species is as important as the quality of the wood itself. Constant, ongoing supply is particularly important in promoting market acceptance of lesser-known species. The company's central challenge, therefore, is to provide end-markets with sufficient volumes of its diverse, tropical species.

FINANCIAL PERFORMANCE

Precious Woods registered significant progress during its initial start-up phase. As of mid-1996, an inventory of the first 4,000 hectares was completed, and the company had logged an area of roughly 1,500 ha. in Compartment A, its first parcel of land scheduled for harvest. In addition, the extraction of timber from Compartment B was well underway, creating inventory of over 35 tropical wood species in sufficient volumes to begin sales, and ongoing monthly harvest of roughly 5,500 cubic meters. Further, the first sawmill line was operational as of midyear, and the second line became operational in early October 1996. Finally, six kiln ovens had been installed, orders for semi-finished machinery had been signed, and a possible joint venture to produce highly priced parquet flooring was under consideration.

Despite these solid achievements, the company suffered setbacks and delays from 1995-96 that created financial strain and slowed the establishment of an operational track record. Chief among these setbacks were: (1) inadequate administration and management of the capital investment program that resulted in start-up delays and considerable loss of wood inventory; (2) significantly higher than initially expected preoperational expenses, which contributed to cost overruns; (3) excessive operational costs, caused by the difficulties inherent in new business development, particularly in the Amazon, and by inadequate financial control systems and oversight; and (4) a weak sales record that resulted in excess production (and further at-risk inventory) and ongoing cash shortages. High inflation and an appreciating exchange rate further increased costs.

As a result of cost overruns and slower-than-expected sales in the summer and fall of 1996, Precious Woods' board of directors demanded stricter financial controls and greater discipline over expenditures. The board's concerns triggered operational and management changes, as well as the beginning of a program to cut costs and improve efficiency in production. The board, however, was not satisfied with the pace and level of change, which in turn led it to reconsider its funding of the company. Due to urgent cash needs, in late November 1996, the company was on the brink of financial crisis, and its future was in question.

Although Precious Woods' recent financial troubles and nascent sales record make an accurate evaluation of its future profitability extremely difficult, Table 5 shows that, under the conservative assumptions of the case writers, the company's income statement is likely to improve considerably through 1998. As shown in Table 6, these projections assume basic levels of production and, most importantly, increasing sales in both sawnwood and semi-finished products, growing to an overall conversion rate of 47% by 1998. In addition, the projections assume significant progress in the company's cost-cutting efforts—with operational cost reductions of 20% in 1997, and a further 10% in 1998—and important reductions in the company's investment program, reflecting Precious Woods' internal budgeting plans as of November 1996.

Under these assumptions, the company would break even in the first quarter of 1998, and by the end of the year generate a net margin of 28%. Total equity capital allocated to Brazil was approximately $23 million at the time of this writing (Figure 4 shows the firm's balance sheet at year-end 1996). Assuming that an additional $5-6 million in shareholders' funds would be required through breakeven, the company's net margin of 28% would produce a return on equity (ROE) of approximately 7% in 1998, improving to 11-12% in 1999.[38] What is significant about these numbers is that the pro-

Precious Woods Income Statement, 1996-1998
(in thousands of constant US dollars)

Item	1996				1996*	1997*				1997*	1998*				1998*
	1st Qtr.	2nd Qtr.	3rd Qtr.	4th Qtr.*	Total	1st Qtr.	2nd Qtr.	3rd Qtr.	4th Qtr.	Total	1st Qtr.	2nd Qtr.	3rd Qtr.	4th Qtr.	Total
Total Net Sales Revenue	0	0	0	263	**263**	488	1,000	1,338	1,663	**4,488**	2,000	2,000	2,350	2,700	**9,050**
Operational Expenses															
Harvesting	150	150	150	150	**600**	120	120	120	120	**480**	108	108	108	108	**432**
Sawmill	190	190	190	190	**760**	152	152	152	152	**608**	137	137	137	137	**547**
Semi-Finished	0	0	0	0	**0**	0	105	158	158	**420**	210	210	315	420	**1,155**
Administrative Expenses	380	380	380	380	**1,520**	304	304	304	304	**1,216**	274	274	274	274	**1,094**
Mngmnt. Expenses*	114	114	114	114	**456**	91	91	91	91	**365**	82	82	82	82	**328**
Maintenance	305	305	305	305	**1,220**	244	244	244	244	**976**	220	220	220	220	**878**
Restaurant	138	138	138	138	**552**	110	110	110	110	**442**	99	99	99	99	**397**
Prospection	74	74	74	74	**296**	59	59	59	59	**237**	53	53	53	53	**213**
Manejo Florestal	0	0	20	20	**40**	16	16	16	16	**64**	14	14	14	14	**58**
Sales	0	0	1	5	**6**	4	4	4	4	**16**	4	4	4	4	**14**
Agro/Madama	20	20	20	20	**80**	16	16	16	16	**64**	14	14	14	14	**58**
Total Oper. Expenses	1,371	1,371	1,392	1,396	**5,530**	1,117	1,222	1,274	1,274	**4,887**	1,215	1,215	1,320	1,425	**5,175**
Corporate Overhead (Swiss, BVI)	253	253	253	253	**1,012**	253	200	180	170	**803**	170	170	170	170	**680**
Investment Program Expenses															
Forest	162	162	162	162	**648**	153	153	153	153	**612**	84	84	84	84	**336**
Sawmill	328	328	328	328	**1,312**	50	50	50	50	**200**	40	40	40	40	**160**
Semi-finished	250	250	250	250	**1,000**	75	75	75	75	**300**	90	90	90	90	**360**
Infrastructure & Logging	150	150	150	150	**600**	45	45	45	45	**180**	40	40	40	40	**160**
Vehicles, Comm.s, Misc.	50	50	50	50	**200**	45	45	45	45	**180**	30	30	30	30	**120**
Total Invest. Program	940	940	940	940	**3,760**	368	368	368	368	**1,472**	284	284	284	284	**1,136**
Total Oper., O/H & Investment	2,564	2,564	2,585	2,589	**10,302**	1,738	1,790	1,822	1,812	**7,162**	1,669	1,669	1,774	1,879	**6,991**
Interest	0	0	0	0	**0**	0	0	0	0	**0**	0	0	0	0	**0**
Tax	0	2	7	42	**51**	46	53	56	56	**210**	60	60	60	60	**238**
Net Earnings	**-2,564**	**-2,566**	**-2,592**	**-2,368**	**-10,090**	**-1,296**	**-842**	**-541**	**-206**	**-2,885**	**271**	**271**	**516**	**761**	**1,821**
Net Margin (percent of sales)	n.a.	n.a.	n.a.	n.a.	**n.a.**	n.a.	n.a.	n.a.	n.a.	**n.a.**	13.6%	13.6%	22.0%	28.2%	**20.1%**

*Numbers are estimates of the author, not the company.

Assumptions: Operational costs decline 20 percent in 1997, and an additional 10 percent in 1998, both in constant US dollar tems.

Investment program expenses fall from $3.76 m to $1.472 m in 1997 per company budgets completed November, 1996; and decline to 22 percent in 1998. Costs of semi-finished production assumed $210 per m³ for kiln dried and semi-finished per company's own estimates. Corporate overhead in Switzerland and the British Virgin Islands falls 20 percent in 1997, and another 15 percent in 1998. The company is assumed to have no debt. Taxes are assumed to average $3.50 per cubic meter of gross logs harvested (ICMS tax), the average in October, 1996; the company believes it may not have to pay this tax at all.

Table 5

Precious Woods Sales Projections, 1996-1998
(actuals through 1996; estimates beyond; in constant US dollars)

Item	1996 1st Qtr.	2nd Qtr.	3rd Qtr.	4th Qtr.*	1996* Total	1997* 1st Qtr.	2nd Qtr.	3rd Qtr.	4th Qtr.	1997* Total	1998* 1st Qtr.	2nd Qtr.	3rd Qtr.	4th Qtr.	1998* Total
Total Net Sales Revenue	0	0	0	263	**263**	488	1,000	1,338	1,663	**4,488**	2,000	2,000	2,350	2,700	**9,050**
Gross Vol. Harvested ('000 m3)	0.0	0.5	2.0	12.0	**14.5**	13.0	15.0	16.0	16.0	**60.0**	17.0	17.0	17.0	17.0	**68.0**
Sawnwood Conversion Rate	35%	35%	35%	35%	**35%**	35%	35%	35%	35%	**35%**	35%	35%	35%	35%	**35%**
Net Export Sawnwd Produced ('000 m3)	0.0	0.2	0.7	4.2	**5.1**	4.6	5.3	5.6	5.6	**21.0**	6.0	6.0	6.0	6.0	**23.8**
Net Export Sawnwood Sold ('000 m3)	0.0	0.0	0.0	0.8	**0.8**	1.5	2.0	2.5	3.5	**9.5**	4.0	4.0	4.0	4.0	**16.0**
Percent of sawnwood produced	0%	0%	0%	19%	**16%**	33%	38%	45%	63%	**45%**	67%	67%	67%	67%	**67%**
Average Price ($US)	0	0	0	325	**325**	325	325	325	325	**325**	325	325	325	325	**325**
Net Sales from Sawnwood ('000 $US)	0.0	0.0	0.0	263.3	**263.3**	487.5	650.0	812.5	1,137.5	**3,087.5**	1,300.0	1,300.0	1,300.0	1,300.0	**5,200.0**
Net Semi-Finished Sold ('000 m3)	0.0	0.0	0.0	0.0	**0.0**	0.0	0.5	0.8	0.8	**2.0**	1.0	1.0	1.5	2.0	**5.5**
Average Price ($US)	0	0	0	0	**0**	0	700	700	700	**700**	700	700	700	700	**700**
Net Sales from Semi-Fin. ('000 $US)	0.0	0.0	0.0	0.0	**0.0**	0.0	350.0	525.0	525.0	**1,400.0**	700.0	700.0	1,050.0	1,400.0	**3,850.0**
Total Conversion Rate	0.0	0.0	35%	35%	**35%**	35%	38%	40%	40%	**38%**	41%	41%	44%	47%	**43%**
Total Net Sales Revenue	0	0	0	263	**263**	488	1,000	1,338	1,663	**4,488**	2,000	2,000	2,350	2,700	**9,050**

* Numbers are estimates of the author, not the company.

Assumptions: Annual harvest roughly 15 percent lower than company forecast of approximately 70,000 cubic meters for 1997; 10 percent lower for 1998.

Sawnwood conversion rate for European-quality exports remains stable at 35 percent, per the company's own projections.

Sawnwood sold to international markets begins slowly; grows to no more than 68 percent of production, or the representation of the 25 species identified by the company as Categories I and II (most marketable).

Effective conversion rate assumed to rise from 35% to 47% in 1998 with growth in sales of semi-finished goods.

Source: Harvest data for first 9 mo.s of 1996 are estimates taken from discussions with management, other figures are author's projections.

Table 6

Precious Woods Balance Sheet

Consolidated balance sheets December 31, 1996 and 1995 (in U.S. Dollars)

ASSETS	NOTE	1996 US $	1995 US $
CURRENT ASSETS:			
Cash and marketable securities	1b and 2	2,224,247	7,329,807
Accounts receivable	3	400,802	151,656
Inventories	1c and 4	1,264,639	450,909
Prepaid expenses		87,095	62,647
Total current assets		3,976,783	7,995,019
PROPERTY, PLANT AND EQUIPMENT - Net:	1d, 1e and 5		
Costa Rica		566,328	1,122,795
Brazil		9,586,726	6,252,595
Others		27,136	20,174
FOREST INVESTMENTS - COSTA RICA	1f,1g and 6	12,530,941	10,851,176
FOREST INVESTMENTS - BRAZIL	1f, 1g and 7	13,357,978	9,430,770
OTHER PROJECTS	1f, and 1g		220,373
OTHER ASSETS	8	651, 324	10,218
TOTAL		40,697,216	35,903,120
LIABILITIES AND STOCKHOLDERS' EQUITY:			
CURRENT LIABILITIES:			
Notes Payable	9		268,070
Accounts payable	10	512,408	401,482
Accured expenses	1h and 11	648,110	398,641
Total current liabilities		1,160,518	1,068,193
CONTINGENCIES	15		
STOCKHOLDERS' EQUITY:			
Common stocks A	12 and 13	7,696,860	6,105,860
Common stocks B	12 and 13	20,272,500	18,532,500
Additional paid-in capital		17,637,127	13,754,316
Promotional and capital acquisition costs	1i	(3,652,920)	(3,113,690)
Net capital proceeds		41,953,567	35,278,986
Translation effect	1l	(26,209)	(8,233)
Accumulated losses		(2,390,660)	(435,826)
Total stockholders' equity		39,536,698	34,834,927
TOTAL		40,697,216	35,903,120

Figure 4

jected returns of the company would meet the range of returns (11–16%) that the investors expected at the time they made their long-term investment in Precious Woods.

RISKS AND OPPORTUNITIES

The principal risk to the company's profitable development is inadequate sales. The company's profitability will be affected most powerfully by two variables—the *volume* of product it is able to sell, and the *price* at which it sells this product. At present, Precious Woods expects that its conversion rate of natural logs to European-quality sawnwood will average 35%—a fairly low level of conversion efficiency[39]—but that this yield will improve markedly with the addition of semi-finished products, to as high as 50–55%. To illustrate the importance of these variables, Table 7 shows sensitivity analysis of profitability under different volume and price assumptions (but excluding semi-finished production, for which precise data was not available). The company's semi-finished products strategy, however, has yet to be implemented. While opportunities such as the partnership with the Brazilian semi-finished products company testify to its promise, the program's success hinges not only on adequate operational management of semi-finished goods, but also on the market acceptance of these products.

A second principal risk to Precious Woods' success is inadequate management in Brazil. While a number of the company's line managers are very capable, many of the company's financial and operational problems stemmed from its corporate structure—leaving the firm's CEO split between activities in Costa Rica, Switzerland, and Brazil—and inadequate Brazilian management. Responding to the board's

---

Net Profit Sensitivity Analysis

Vol. Sold (m³)	$200	$250	$300	$350	$400
15,000	-2,994,283	-2,354,158	-1,714,033	-1,073,908	-433,783
20,000	-2,140,783	-1,287,283	-433,783	419,717	1,273,217
25,000	-1,287,283	-220,408	846,467	1,913,342	2,980,217
30,000	-433,783	846,467	2,126,717	3,406,967	4,687,217
35,000	419,717	1,913,342	3,406,967	4,900,592	6,394,217
40,000	1,273,217	2,980,217	4,687,217	6,394,217	8,101,217

Average Sales Price

Net Margin Sensitivity Analysis

Vol. Sold (m³)	$200	$250	$300	$350	$400
15,000	-100%	-63%	-38%	-20%	-7%
20,000	-54%	-26%	-7%	6%	16%
25,000	-26%	-4%	11%	22%	30%
30,000	-7%	11%	24%	32%	39%
35,000	6%	22%	32%	40%	46%
40,000	16%	30%	39%	46%	51%

Average Sales Price

Assumptions:
Analysis assumes total cost levels per projections in 1998 Income Statement, less semi-finished products. Gross volume harvested assumed 70,000 m³. Taxes assumed equal to 14.65% of sales.

Source: Author's projections and analysis; ovals indicate most likely scenario.

Table 7

---

concerns, the November 1996 management change signals that steps are being taken to rectify the situation.

A third risk is that the company will not be able to cut costs sufficiently to ensure profitability. As mentioned above, Precious Woods' cost structure is high, leaving the firm little flexibility in the face of a difficult sales environment. (Table 8 compares Precious Woods' cost structure with those of other Amazon wood processing firms.) Further, cost projections shown in Table 5 assume important

Precious Woods, Ltd.

Integrated Logging and Processing Operational Cost Analysis
(common-size analysis—per m³)

	Sm. 1995 Stone	%	Lg. 1995 Stone	%	Sm. 1990 Verissimo	%	P.Woods 1996*	%
Amount logged (m³)	16,464		40,572		9,200		70,000	
Costs of Logging								
Logging rights	5.00	7.44%	5.00	8.44%	1.84	3.09%		
Salaries & benefits	2.21	3.29%	1.91	3.22%	2.55	4.28%		
Food	0.00	0.00%	0.00	0.00%	0.65	1.10%		
Fuel	1.34	2.00%	1.09	1.84%	1.12	1.87%		
Equip. Maintenance	2.04	3.04%	1.66	2.79%	2.07	3.47%		
Depreciation	2.35	3.50%	1.91	3.22%	2.65	4.45%		
Forest tax	0.00	0.00%	0.00	0.00%	1.54	2.59%		
Forest Mngmnt. Costs	0.61	0.90%	0.49	0.83%	0.65	1.10%		
Total Costs of Logging	13.55	20.17%	12.05	20.33%	13.06	21.94%	n.a	
Costs of transport								
Salaries & benefits	1.51	2.24%	1.22	2.06%	1.23	2.06%		
Fuel	2.02	3.01%	1.84	3.10%	1.99	3.34%		
Maintenance	2.04	3.04%	1.66	2.79%	1.63	2.74%		
Depreciation	5.88	8.75%	5.68	9.58%	3.39	5.70%		
Insurance	2.88	4.29%	2.56	4.32%	4.89	8.22%		
Total Transport Costs	14.33	21.33%	12.96	21.86%	13.13	22.05%	n.a	
Volume to Processing	5,600		13,800		4,300		24,500	
Conversion Rate	34%		34%		47%		35%	
Processing Costs								
Labor	25.68	38.23%	24.95	42.10%	17.48	29.36%		
Electricity	4.90	7.29%	4.11	6.94%	2.28	3.84%		
Fuel	0.70	1.04%	0.28	0.48%	0.70	1.17%		
Administration	1.90	2.82%	1.54	2.60%	6.24	10.49%		
Maintenance	2.79	4.15%	1.57	2.65%	3.03	5.09%		
Depreciation	3.33	4.95%	1.81	3.05%	3.61	6.06%		
Total Costs of Processing	39.29	58.50%	34.26	57.80%	33.34	56.01%	n.a	
Total Costs	67	100.00%	59	100.00%	60	100.00%	224	
Avg. Price Obtained	188		188		156		350	
Value of Production	1,051,680		2,591,640		670,800		9,600,000	
Taxes	154,071		379,675		98,272		1,406,400	
Net Profit	160		160		41		594	
Margin	20.8%		28.0%		28.1%		28.2%	

* Volume of 70,000 m³ used as rough estimate of company's expected annual production, October 1996 through September 1997. For Precious Woods, total costs per m³ are estimated based on projections contained in Income Statement for 1998. In specific, Precious Woods total operational and investment expenses (less semi-finished) are assumed to total $5.537 million per year. More precise, actual figures were not available from the company, and a breakdown across cost categories was not possible.

Sources: See Steven Stone, "Economic Trends in the Timber Industry," and Adalberto Verissimo, "Logging Impacts," as noted in bibliography.

Table 8

expense reductions in real terms. Inflationary pressures as well as continued appreciation of the Brazilian currency represent additional risks. In sum, the success of the firm's cost-cutting program will be critical to its future profitability.

Mitigating these risks are a number of opportunities stemming from Precious Woods' distinctive position as a sustainable harvester of high-quality, Brazilian tropical wood products. Key areas offering significant potential for improvement include:

- **Leveraging competitive advantages.** Precious Woods has a number of important competitive advantages that it can leverage toward profitable growth:

  - An industry-leading SFM program, which limits the company's exposure to environmental liabilities, and creates the opportunity for product differentiation.

  - High-quality productive machinery, which enables production quality to meet high-paying European and international standards.

  - Ownership of land, which creates vertically-integrated operations with no vulnerability to rising costs of principal raw material.

  - European and international connections, which increases opportunities to generate sales in high-paying markets.

- **Improving species utilization.** As the company becomes more familiar with the characteristics and properties of its many species, it will become more efficient in both harvest and production. Precious Woods has already discovered, for example, that certain species have extremely high levels of defects, and yield low output in sawnwood; harvests of these species will be stopped in favor of lower-defect woods. Other wood species, however, have proven surprisingly suitable in company tests for certain end-products such as parquet flooring; the use of these species will be increased.

- **Minimizing waste.** As mentioned above, the success of the company's expansion into semi-finished products is likely to sharply reduce wood waste, while boosting revenues. Other opportunities also exist, however, including using wood for electricity generation (offering significant savings from high diesel fuel costs).

- **Improving cost structure and operational efficiency.** As the company's investment program nears completion, strengthened senior management will be able to greatly improve operational efficiency in a number of areas ranging from divesting unnecessary assets and regulating production to minimize wood damage in inventory, to reducing administrative headcount and bolstering sales and order procedures.

Conclusions

Precious Woods has made tremendous strides from the business concept stage to becoming fully operational. As the above analysis indicates, Precious Woods' start-up difficulties and inadequate sales record prevent more precise analysis of the company's profitability, and broader comments and conclusions about the viability of its SFM model. The company has not yet demonstrated that its SFM-driven program of selling lesser-known species will be successful. In addition, the company's financial control systems need further improvements to accurately measure extra costs due either to its sustainable forestry program and high sensitivity to environmental issues, or to the expense of establishing high-quality operations that meet European standards. Despite these start-up difficulties, Precious Woods, if it is able to successfully turn its operations around, has the capability of meeting its investors return expectations.

Precious Woods pioneering efforts do yield several important lessons learned. First, the great bulk of the company's problems to date—and its challenges going forward—relate to basic issues of business management, not sustainable forestry per se. Facing far greater business challenges as a start-up in the Amazon, the company has learned the importance of stable, qualified management, keenly oriented to business efficiency and shareholder value.

Second, if Precious Woods succeeds in navigating this difficult start-up period, financial analysis suggests that its operational profitability will be attractive. A projected net margin of roughly 28% compares favorably with other wood processing businesses in the Amazon. Further, followers of the Precious Woods model should be able to save considerable investment and even operational expense, creating a leaner capital structure, and more attractive returns on shareholder funds.

Third, though it is too early to determine to what extent the company's SFM program will represent added costs, it appears probable that Precious Woods could capture economies of scale in wood processing that would enhance profitability. The company has an option to purchase an additional 270,000 hectares of neighboring primary forest and, while in no position financially to consider expansion at this time, it could obtain higher efficiencies through larger volumes in processing. In addition, larger volumes of lesser-known species could be particularly helpful in securing their market acceptance and promoting sales.

On a final, cautionary note, there are a number of aspects of Precious Woods' experience that are highly unusual, and may prove difficult to replicate. Most important of these has been the company's ability to raise a total of almost $40 million in private equity capital. Though future SFM projects should not require a funding base of this magnitude, the company's fund-raising achievements will be hard to repeat. Further, the firm's international profile and contacts are likely to prove important to its sales efforts.

Addendum: 1997 Update

The financial difficulties Precious Woods experienced in the end of 1996, led to the board's aggressive intervention to turn around the company's operations and performance. In December 1996, the board announced several sweeping changes, including the following actions:

- The President and CEO of Precious Woods stepped down from both his positions.

- The functions of the CEO and President were assumed by the Executive Committee of the Board of Directors.

- Friedrich Bruegger, a Swiss-Brazilian dual national with forest operations experience in the Amazon, was hired to take over responsibilities as General Manager of the Brazilian operations.

- A former CFO of Champion International Brazil, was hired to revamp the financial systems and controls of Precious Woods Brazil. This move was designed to improve financial reporting, as well as increase operational efficiency, cut costs, and increase productivity.

- Compensation for management was to be revised to link pay to the company's performance.

With these changes in place, the company reported several improvements and successes in its letter to shareholders in June 1997. Swiss investors registered their confidence in the board of director's intervention by investing an additional $3 million in Precious Woods in January 1997. These funds were critical to see the company through the beginning of 1997 until such time that sales and revenues picked up. In Brazil, the new management team put into service a third sawmill line which got overall production at the mill close to its 100 cubic meters a day (the level needed to operate profitably). Sales for the first two quarters were projected to be $1.1 million, with total sales projected to reach between $4.5 to $5.8 million by year's end.

Forty percent of the production was being sold into the local market (for lower grade quality lumber) and 60% of production was being exported. The average price for sales made in early 1997 was slightly below $300/m³, while in May, the sales contracts secured prices of between $350-$400/m³ which the company hoped to maintain during the remainder of 1997. The company stated that sales contracts included thirty-five species, suggesting that Precious Woods was gaining ground in marketing its lesser-known species. Clearly this news is grounds for cautious optimism, and the performance of the company will have to be evaluated at year end to determine how successful the turnaround plan has been.

Endnotes

1. Amazon Deforestation Has Accelerated. *New York Times,* September 12, 1996.

2. Kishor, Nalin M., and Luis F. Constantino. 1993. Forest Management and Competing Land Uses: An Economic Analysis for Costa Rica. LATEN Dissemination Note #7, the World Bank, October, p. 2.

3. Statistic taken from Dr. Richard Rice, Raymond E. Gullison, and John W. Reid. 1996. Can Sustainable Management Save Tropical Forests, unpublished manuscript, September 24, p. 4; current situation from author's personal research.

4. Information on Precious Woods was gathered both from a number of public sources including the company's annual reports and shareholder mailings, and from independent research. In specific, the authors made two on-site visits to the company's operations in Brazil, where they reviewed the company's forestry practices, sawmilling operations, and financial and administrative management. The authors also conducted interviews with senior managers of all principal departments, and held lengthy discussions with the president and CEO, Roman Jann. Throughout, the company was extremely cooperative and forthcoming with information about Precious Woods' experience, and the authors would like to express their gratitude to Jann and to the firm's other senior managers for their kind assistance, without which our study and analysis would not have been possible.

5. Statistics cited from the World Resources Institute and the FAO

6. Ibid., p. 7.

7. Statistics are cited in Holloway, Margeurite. 1993. Sustaining the Amazon. *Scientific American,* July, pp. 91-99.

8. Ibid., pp. 91-99.

9. In many parts of Asia and Africa, where tropical forests often tend to be more homogeneous in terms of wood species, logging has played a far more important role in forest destruction. Ironically, the heterogeneity of Brazilian forests has helped stave off Asian-style destruction, as this diversity of species has made logging far more difficult and costly.

10. Holloway, op cit., p. 95; statistics are cited from researchers of IMAZON, a highly respected tropical forestry research center in Belem, Brazil.

11. Ibid., p. 95.

12. See Joanne C. Burgess. 1993. Timber Production, Timber Trade and Tropical Deforestation. Ambio, vol. 22, May, pp. 136-137. See also, International Tropical Timber Organization (ITTO). *1995 Annual Review,* production tables, pp. 46-53.

13. Burgess, ibid., p. 137.

14. Several large Asian wood products firms have recently purchased land and begun logging and processing operations in the Amazon, as most of the high-quality, easily-accessible hardwood forest in their home countries have been depleted. See the *Wall Street Journal,* November 11, 1996, p. 1.

15. Capital investment requirements rise with the level of value-added of products. In developing countries such as Brazil, where interest rates have been extremely high by industrialized standards, the cost of capital represents an important barrier to entry.

16. See Vincent, Jeffrey. 1992. The Tropical Timber Trade and Sustainable Development. *Science,* June, pp. 1652-1653.

17. Total estimated tax-funded subsidies granted to encourage economic activity in the Amazon summed to $1.9 billion between 1965 and 1989. See M.A.F. Ros-Tonen, *Tropical Hardwood from the Brazilian Amazon: A Study of the Timber Industry in Western Pará,* p. 38.

18. See Verissimo, Adalberto, et. al. 1992. Logging Impacts and Prospects for Sustainable Forest Management in an old Amazonian Frontier: The Case of Paragominas. *Forest Ecology and Management,* 55, p. 170.

19. In 1992, logging and sawmill companies in the state generated sales that totaled roughly $190 million, and earned approximately $62 million in profits (or a impressive net margin of almost 33%). In addition, the industry generated almost 6,000 jobs, and in Paragominas, a city located in the heart of Pará's forestry activity, roughly 56% of the urban population depended directly on the wood industry for their livelihood. The next biggest industry, ranching, was only 10-20% of this size. See Verissimo, ibid.

20. Ibid., p. 177

21. See Steve Stone and Verissimo, et. al., op. cit.

22. According to the United Nations Food and Agriculture Organization (FAO), 94.6% of Brazilian production was consumed domestically up through 1989, and Brazil's total production, despite having the largest reserve of tropical rain forest in the world, was less than 1% in the mid-1980s. See M.A.F. Ros-Tonen, op cit., p. 34 and 173.

23. Particularly recently, Brazilian macroeconomic policy has been characterized by monetary tightening to control inflation, resulting in high real interest rates. In addition, the government has utilized the real as an additional tool to control inflationary pressure, allowing the currency's value to adjust, but at a slower rate than that of inflation. As a result, the currency has appreciated in real terms, further squeezing profit margins for export firms whose costs are principally denominated in reals, and whose revenue is denominated in hard currencies such as the U.S. dollar.

24. Real interest rates in Brazil have remained above 25% per year for most of the 1990s, falling only briefly in 1994, but rising sharply again under the Plano Real.

25. Steve Stone. Economic Trends, op. cit., p. 11.

26. See M.A.F. Ros-Tonen, op. cit.

27. The company plans to plant another 3,000 hectares by the year 2000, at a rate of approximately of 500 ha. per year. The company's long-term plan is to plant at least 12,000 ha., corresponding to a total surface area of 20,000 ha., and believes this amount of land is necessary to ensure commercially-viable quantities of supply.

28. A total of $4.3 million was paid for 80,900 ha. of land. This land, however, contained roughly $565,000 in houses, buildings, sawmill equipment, and other infrastructure. In addition, approximately 5,500 ha. of the land area were already deforested, which is more expensive in the Brazilian land market. Thus the company asserts it paid an equivalent of $35 per ha. for 75,400 ha. of primary tropical forest. The company considered this a fair price.

29. See Schmidt, R. Tropical Rain Forest Management: A Status Report. *Unasylva,* 39, pp. 2-17, as cited in Reid and Rice. Natural Forest Management as a Tool for Tropical Forest Conservation: Does it Work? Forthcoming in Ambio, submitted September 1996.

30. There are four principal SFM systems, including the Malayan Uniform System, the Tropical Shelterwood System, the Strip-Clearcut System, and the Surinam Celos System. For further detail, see Reid and Rice, Natural Forest Management, op. cit., pp. 3-6.

31. For additional detail see, "The Celos Management System: A Polycyclic Method for Sustained Timber Production in South American Rain Forest," by N.R. de Graaf, and R.L.H. Poels.

32. This time period was derived from an estimate of the average volume of wood that could be harvested per hectare of natural rainforest (approximately 35 cubic meters), and from the estimated growth rate of the forest (1.5 to 1.8 cubic meters per ha. per year). Thus over twenty-five years, the forest would regenerate at least 37.5 cubic meters of wood (25 x 1.5), more than replacing what had been harvested. For further detail on the scientific studies contributing to these statistics, see writings by Silva, et. al., in *Forest Ecology and Management,* 1995.

33. See de Graaf and Poels, op. cit., p. 120.

34. From authors' research.

35. The company believed annual capacity could be as high as 7,000-8,000 cubic meters per year, depending on the type of wood being dried (less-dense wood generally dries faster), and the dimensions of the wood (smaller pieces meant greater overall surface area, and faster drying).

36. Precious Woods has received letters of support from Jose Maria Figueras, president of Costa Rica, Dr. Jose Lutzenberger, former Brazilian minister of the environment, and Amazonino Armando Mendes, governor of the Amazon state.

37. See the ITTO, *Annual Review 1995,* p. 4, citing the results of a series of World Bank studies completed in 1995. Current assessments of the sustainable import market in Europe are very difficult to obtain, though it is widely believed that the market at present is very small.

38. Total equity capital allocated to Brazil to date sums to approximately $23 million (with the remainder of monies raised allocated to Costa Rica). Adding $5 million in equity capital gives a base of shareholders funds of $28 million. Earnings of roughly $2 million are projected for 1998, producing an ROE of 7%. Note that original profitability projections used by management to raise financing for Precious Woods cited a constant dollar-based IRR for the project of 11-12%, indicating that investor expectations for profitability have been quite modest (particularly in light of the many risks of the project).

39. The forestry industry is renowned for its low efficiency rates in converting raw material into product, with U.S. manufacturers using principally homogeneous, reforested logs showing yields of 40-60%. Brazilian tropical wood processors average 30-35% conversion rates; see Stone and Verissimo, op. cit.

Bibliography

Almeida, Oriana Trindade de, and Christopher Uhl. 1995. Brazil's Rural Land Tax. *Land Use Policy,* vol. 12, no. 2, pp. 105-114.

Barreto, Paulo, Paulo Amoral, Edson Vidal and Christopher Uhl. 1996. Impacts of Forest Management on the Economics of Timber Extraction in Eastern Amazonia. Unpublished manuscript.

Barros, Ana Cristina and Christopher Uhl. 1995. Logging Along the Amazon River and Estuary: Patterns, Problems and Potential. *Forest Ecology and Management,* April.

Browder, John O., Eraldo Aparecido Trondoli Matricardi, and Wilson Soares Abdala. 1996. Is Sustainable Tropical Timber Financially Viable? *Ecological Economics,* 16, pp. 147-159.

Burgess, Joanne C. 1993. Timber Production, Timber Trade and Tropical Deforestation. *Ambio,* vol. 22, no. 2-3, May.

de Graaf, N.R., and R.L.H. Poels. The Celos Management System: A Polycyclic Method for Sustained Timber Production in South American Rain Forest.

Hardner, Jared J., and Richard E. Rice. 1994. Financial Constraints to Sustainable Selective Harvesting of Forests in the Eastern Amazon. DESFIL paper prepared under AID contract, June.

Holloway, Margeurite. 1993. Sustaining the Amazon. *Scientific American,* July, pp. 91-99.

Howard, Andrew F., and Juvenal Valerio. 1996. Financial Returns from Sustainable Forest Management and Selected Agricultural Land-use Options in Costa Rica. *Forest Ecology and Management,* 81, pp. 35-49.

Kishor, Nalin M., and Luis F. Constantino. 1993. Forest Management and Competing Land Uses: An Economic Analysis for Costa Rica. LATEN Dissemination Note #7, the World Bank, October.

Pancel, L., ed. 1993. *The Tropical Forestry Handbook,* vol. 2. Germany: Springer-Verlag.

Reid, John W., and Richard E. Rice. Natural Forest Management as a Tool for Tropical Forest Conservation: Does it Work? Forthcoming in *Ambio,* submitted September, 1996.

Rice, Richard E., Raymond E. Gullison, and John W. Reid. 1996. Can Sustainable Management Save Tropical Forests. Unpublished manuscript, September 24.

Stone, Steven W. 1996. Economic Trends in the Timber Industry of the Brazilian Amazon: Evidence from Paragominas. CREED Working Paper Series No. 6, July.

Verissimo, Adalberto, et. al. 1992. Logging Impacts and Prospects for Sustainable Forest Management in an old Amazonian Frontier: The Case of Paragominas. *Forest Ecology and Management,* 55, pp. 169-199.

Verissimo, Adalberto, Paulo Barreto, Ricardo Tarifa and Christopher Uhl. 1995. Extraction of a High-value Natural Resource in Amazonia: The Case of Mahogany. Forest Ecology and Management, 72, pp.39-60.

Vincent, Jeffrey R. 1992. The Tropical Timber Trade and Sustainable Development. *Science,* vol. 256, June 19, pp. 1651-1656.

STORA:
The Road
to Certification

CASE STUDY

PREPARED BY:
RICHARD A. FLETCHER
JAMES MCALEXANDER
ERIC HANSEN

A Case Study from "The Business of Sustainable Forestry"
A Project of The Sustainable Forestry Working Group

The Sustainable Forestry Working Group

Individuals from the following institutions participated in the preparation of this report.

Environmental Advantage, Inc.

Forest Stewardship Council

The John D. and Catherine T. MacArthur Foundation

Management Institute for Environment and Business

Mater Engineering, Ltd.

Oregon State University
Colleges of Business and Forestry

Pennsylvania State University
School of Forest Resources

University of California at Berkeley
College of Natural Resources

University of Michigan
Corporate Environmental Management Program

Weyerhaeuser Company

The World Bank
Environment Department

World Resources Institute

STORA:
The Road
to Certification

CASE STUDY

PREPARED BY:
RICHARD A. FLETCHER
JAMES MCALEXANDER
ERIC HANSEN

A Case Study from "The Business of Sustainable Forestry"
A Project of The Sustainable Forestry Working Group

Contents

STORA:
The Road to Certification

Introduction

"We changed our attitudes, we listened, we learned, we cooperated, and we took the initiative."

–Åke Granqvist, supervising forester, STORA

Over the past ten years, Swedish forest products giant STORA has transformed its forest management to implement and verify a commitment to sustainable forestry. The company has hired a staff ecologist, implemented ecological landscape planning, brought local environmentalists into its management planning, retrained its workforce, and adopted new forest conservation measures. Most recently, STORA became Europe's first major timber company to have a large block of its forests certified by a third party as sustainably managed.

Headquartered in Falun, Sweden, STORA is one of the largest forest products companies in the world with 1996 sales of $6.0 billion. The company ranks fifth worldwide in paper and board production, producing 1.9% of the world's production compared to 3.2% for industry leader, International Paper Co. STORA sells primarily paper products, but also runs four sawmills and is involved in power production, banking, and associated financial operations. The company owns a total of 2.3 million hectares of forest, primarily in Sweden, but it has holdings in Portugal and Canada, as well.

In 1996 STORA became one of the first large commercial forestry operations in the world to attain third-party certification. Scientific Certification Systems used principles of the Forest Stewardship Council (FSC), the oldest and most credible certification system with environmentalists, to certify STORA's Ludvika district. STORA's size and its importance in the global forest products industry makes its actions a milestone in the development of sustainable forestry. As STORA's evolution toward sustainable forestry indicates, certification has already become a strategic consideration for some forward-looking companies.

Sustainable Forestry for Competitive Advantage

STORA decided to gain FSC certification for its forests after senior management carefully considered the company's competitive position, market dynamics, and strategic opportunities that they believe will leverage its unique sources of competitive advantage.

COMPETITIVE CHALLENGES

As a Swedish company, STORA faces a number of challenges. Swedish forests produce relatively low wood yields, an average of 4–5 cubic meters of wood per hectare each year, with a high reaching 14 cubic meters per hectare annually. That is well below yields in the Pacific Northwest that can range between 20–30 cubic meters, or those of the best plantations in New Zealand, Chile, or Brazil where yields rise to 30–40 cubic meters per hectare annually. Although Swedish forests produce limited quantities of fiber, these slow growing trees yield fibers of unmatched quality that are used in such products as high quality magazine paper and lumber for windows. In these markets STORA competes primarily with Canada, Scandinavia, and other northern forest regions.

STORA also has some of the highest labor costs in the worldwide forest products industry, according to senior managers. Such things as social programs and workforce regulations push labor costs high enough to place STORA at a significant cost disadvantage. Sweden is also experiencing double digit unemployment and wage increases in excess of productivity gains, according to the 1997 Economic Survey of Sweden by the Organization of Economic Cooperation and Development. These and other factors increase production costs making it difficult for STORA and other Swedish companies to compete in the global market.

Consequently, STORA is vulnerable to price competition from lower cost regions of the world, especially Asian companies that are trying to expand their market share. In the past, Sweden has devalued the Krona to lower the costs of its products to export customers. Since Sweden has joined the European Union (EU), however, this tactic is no longer an option.

STORA Sales by Product Category, 1996

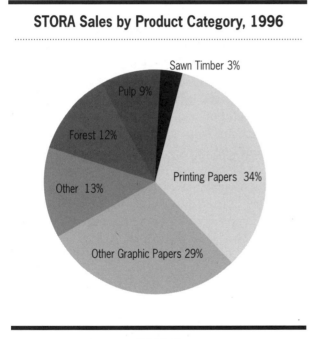

Exhibit 1

The market for forest products is highly cyclical. Recently, production of paper, which accounts for 63% of STORA's sales (see Exhibit 1), has outpaced demand. As a result, average prices for paper pulp dipped 12% in 1996 from those in 1995. Accordingly, STORA's sales revenue declined in 1996 to $6.0 billion (1US$=7.477 Swedish Krona) from $7.6 billion in 1995, while net income dropped to $2.1 million from $7.2 million during the same period. In response to the weak paper

market STORA curtailed production at several facilities. The divestment of Stora Building Products also contributed to declining revenues.

Despite the lower productivity of its forests, higher labor costs, and a soft market for its most important products, STORA earned an average annual return of 16% between 1986 and 1995, according to the 1995 annual report. Its U.S. and European counterparts averaged 12% and 11% for the same period, respectively. STORA's performance suggests that management has responded well to its economic, political, and competitive challenges.

CHANGING MARKET DYNAMICS

The competitive dynamics of the global forest products market is also changing. Production capacity in Southeast Asia and South America is rising. A number of substitutes threaten important wood markets—in particular, steel for lumber and plastic for paper. Environmental pressures are rising as environmental groups, consumers, and regulators register rising interest in the management of natural resources. The preservation of tropical rain forests and biodiversity have become politically charged issues. For STORA, environmental concerns are priorities due to the economic and social considerations in Sweden and the reality that export markets in Europe, which account for most of its sales, have become sensitive to environmental issues.

The Swedish Milieu

The Swedes, with a population of nearly nine million living in an area slightly smaller than California, have a history of sensitivity to the protection of their forests. Although Sweden has several large urban areas that contain a significant portion of the nation's population, Swedish citizens spend a good deal of time at outdoor pastimes and many still live a farm and forest lifestyle. This closeness to the natural environment is manifest in a feeling among the Swedes that the people own the forests—a feeling codified in Swedish law, which guarantees citizens the right of access to all lands. The Swedes frequently take advantage of their right to hike, bike, or ski across privately owned forest and farm property.

Citizen interest and involvement with the forests have limited the latitude that industrial owners have to extract economic benefits from their land. In the late 1800s and early 1900s logging operations across Sweden badly degraded significant portions of the forest. The Forest Act in 1903 was Sweden's first attempt to address growing concerns over the potential of these cutover forests to produce wood in the future. By 1923, Swedish forests were given additional protection by a law that prohibited clear-cutting young forests. In 1950 the General Director of the state-owned forest lands, Eric Höjer, issued Letter Circular #151, which concluded that Sweden needed to reconstruct its forests.

Clearing the land to establish plantations of soft-wood, particularly spruce and pine, was the most common rehabilitation method. This model of forestry, based in Germany, and the economics of Adam Smith, had as its goal growing as much fiber as quickly as possible. When growth slowed, the trees were cleared, so the time between cuttings, or rotations, was short. Soon companies and small landowners followed suit and plantations became the Swedish norm. These plantations are successful timber producers—Swedish wood growth has exceeded harvest for several decades, and currently is running about 20% above harvest. The visual effects of clear-cuts, combined with practices like burning and herbicide spraying, however, eventually sparked public concern and debate over the health effects of burning and using chemicals.

That debate led to the adoption of a new national forest management code in 1993. Then, the priority in forest management shifted from wood production to the shared priorities of producing wood, and preserving ecosystems. Public support in Sweden for forest conservation and the adoption of

the 1993 forest management code were the primary reasons STORA decided to adopt environmentally-based management strategies for its forests.

A Green Market in Europe

The emergence of market forces creating demand for certified forest products, especially in Europe, was also instrumental in leading STORA to seek certification. European markets account for nearly 90% of STORA's total sales (see Exhibit 2). In recent years, German publishers, important customers for STORA's high quality paper products, have become particularly demanding in the environmental realm. At one point they called for suppliers to provide paper drawn from forests that did not use clear-cutting practices. For the time being, the German publishers have backed off such demands, and turned their attention to the broader concept of forest certification, without publicly endorsing a specific certification system.

"Buyers Groups" are also an important catalyst in Europe for creating demand for certified forest

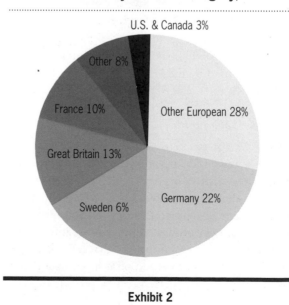

STORA Sales by Market Category, 1996

U.S. & Canada 3%
Other 8%
France 10%
Great Britain 13%
Sweden 6%
Germany 22%
Other European 28%

Exhibit 2

products. These groups, especially the 1995 and following Plus Group in the United Kingdom, made up of large wood products customers want to purchase products certified by third parties, such as the FSC system. The products affected cover everything from sawn wood, all manner of paper products, including napkins, newspapers, diapers, furniture and doors.

The actions of the 1995-Plus Group members, which include the largest and most influential retailers in the United Kingdom, have had a high profile with STORA and other Swedish companies. As membership in the 1995-Plus Group has continued to grow, so has the scope of affected products. For instance, Tetra Pak, a maker of liquid packaging, recently joined, which will put pressure on suppliers like STORA to produce packaging that is third-party certified.

The 1995-Plus Group members have pressed suppliers in a variety of ways. Their managers have made special visits to producers, including STORA, to stress their commitment to purchasing FSC-certified products. Group members have exerted pressure by participating in the FSC Working Group process, which is setting standards for forest management in Sweden, and through questionnaires that 1995-Plus Group companies send out to try and document their sources of supply. For example, TimberTracker, the supply tracking system used by the U.K.'s J Sainsbury plc, is well known to STORA personnel.

Strategic Opportunities

Given that comparatively high labor costs, less productive forests, and soft demand put STORA at a competitive disadvantage, STORA management needs to identify innovative ways to differentiate its products and/or seek preferential relationships with specific customer groups. Third-party certification provides the sort of opportunity that STORA might capitalize on to accomplish its competitive goals. Since few suppliers currently provide third-

party certified products, STORA has a strategic opportunity to sell a differentiated product that, at least in the immediate future, may have an edge in price-sensitive markets.

By moving quickly to fill this market demand, STORA may forestall competitors and maintain strong relations with current customers, many of which belong to buyers groups. Sweden's geographic and political links with Europe, including EU membership, give STORA preferential access to European customers that is unavailable to U.S. and Asian producers. By promoting the value and importance of certification to a healthy environment, STORA might also attract new customers to these products.

STORA appears to be particularly well suited to accomplish third-party certification for its forests. Swedish culture fits well with the requirements of third-party certification. The Swedish government's social and environmental policies relating to company responsibility for social welfare and worker treatment are consistent with the underlying philosophies of FSC-accredited certification. Swedish forestry's record of cutting less than annual growth is also an advantage in earning certification.

After 1,500 years of logging, Sweden has little of the old growth forest left that has been so contentious with environmentalists in other regions of the world. Any expectations for growing new, ancient forest on these cutover lands seem lower than in other places in the world where large stands of ancient forest remain to compare with plantations. What ancient forest there is in Sweden grows largely on non-company lands. Logging those lands, however, is a controversial national issue.

The Swedish Society for Nature Conservation recently launched a campaign to preserve all remaining old growth forest in Sweden. Complete with the signatures of more than eighty prominent forest experts, the literature claims that logging has endangered more than 1,500 species in those forests. A telltale hint of what may be coming is contained in this highlighted statement from the group's publication *Stop Logging Sweden's Last Old Growth Forests Now*, "That [stopping harvests of old growth] will not be sufficient to ensure the long-term survival of all endangered species; but it is a vital first step that can be taken immediately, without additional inventories or studies." In the future STORA and other Swedish wood producers may well face demands to regrow old forests to restore lost habitat for species that depend on ancient trees.

As part of its strategy, Swedish managers have adopted comparatively innovative tactics to address public pressures on environmental issues. Rather than take a defensive, hard-line approach to environmentalists as have some U.S. companies, STORA has found a way to collaborate with the environmental nongovernmental organizations (ENGOs) that is mutually advantageous. As a member of the FSC Working Group, for instance, STORA is working with ENGOs such as the World Wide Fund for Nature (WWF) and Greenpeace to develop national standards for forest management. If these efforts prove fruitful, STORA will have helped to shape a constructive solution to land-use conflicts that meets the company's needs and addresses the concerns of the ENGOs, instead

of engaging in a drawn out, destructive battle.

Throughout most of STORA's long history (see box), forests were used to the supply the timber for smelting ore, supports in the mine, and in the extraction process for its copper mining operations. This single-minded, extractionary view of forests prevailed throughout Sweden at the time. Two developments altered the course of STORA's forest management: During the 1970s and 1980s corporate restructuring and acquisitions took the company away from mining toward forests products production, and STORA decided to emphasize paper over other forest products.

As a major papermaker, STORA naturally focused on the most valuable species for paper production. Spruce is clearly the winner in yield of high value newsprint and other papers. It will almost double the yield of other native tree species on the best sites. As do other major papermakers worldwide, STORA adopted a forest management strategy that maximized pulp production from intensively-managed plantations of spruce, or pine where spruce was not suitable. Through the 1970s this meant clear-felling, burning, use of herbicides for weed control, and liquidation of other tree species. Public pressure mounted during the 1970s to discontinue

---

STORA Forest Management: The Road to Third-Party Certification

Timeline of Events—Stora's Evolution to Ecology-Based Land Management

1988 Test of ecological landscape planning	**1991** Hire staff ecologist	**1993** New forest management	**1996** Certify Ludvika

From Copper Mining to Forestry

STORA's history has shaped its current business philosophies, and accounts in large measure for its importance to Sweden. STORA claims to be the world's oldest company. This modern forest products company has its historic roots in copper mining at the Copper Mountain in the town of Falun. A Deed of Exchange belonging to Bishop Peter of Västerås is the oldest existing document dealing with the mine at the Copper Mountain. It is dated June of 1288. The mine and the wealth it generated has played a significant role in the development of the Swedish economy: That wealth helped Sweden ascend to world leadership during the 1600s. Although mining ceased at Falun in 1992, significant volumes of Falun Red paint are still produced from the Copper Mountain. This paint, which has been produced for generations, is a tradition in Sweden. Many buildings throughout Scandinavia are still painted with the distinctive red color.

Early mining methods required huge volumes of wood, which was used not only to smelt ore, but also for support timbers in the mine and the extraction process. Large fires, built against the rock walls in the mine, heated the rock. After the rock cooled, it became brittle and easier to work with. Because of the extensive use of wood, the mine controlled large tracts of forest land, which eventually led STORA to a transition into the forest products industry.

The company's mining experience also got it involved in iron ore and steel production. By the late 1970s, however, this part of the business was losing large sums of money. Through reorganization and help of the Swedish government, STORA divested itself of the steel component of its business. From these spin-offs, the state-owned Swedish Steel Company, SSAB was formed.

Large-scale development of the forest products component of the business started in the late 1800s with investments in saw and paper mills. After the company eliminated the steel sector, it funneled significant capital investments into forest products and quickly grew that business. During the late 1980s and early 1990s STORA purchased a number of companies and subsequently spun off pieces of those. Billerud was purchased in 1984, followed by Papyrus in 1986, Swedish Match in 1988, and Feldmühle in 1990. These purchases added a total of 78,800 employees. However, a variety of sales took place between 1990 and 1995 which eliminated segments of each of these acquisitions and reduced the number of employees by 48,750. In 1996 the building products segment (primarily window and door production), was sold, taking an additional 3,900 employees. By 1996 total employment had dropped to 22,716.

Source: Sven Rydberg, The Great Copper Mountain, 1988, Värnamo, Sweden and Tommy Forss and Kurt Netzler, Chronicles, The Annals of STORA.

burning and use of herbicides, so STORA and other Swedish companies abandoned both practices by adopting a higher removal standard, which reduced logging slash, and manually cutting, rather than spraying, weeds.

ECOLOGICAL LANDSCAPE PLANNING

In response to public concerns, in the late 1980s STORA forest and timber group decided to develop a new style of forest management on their Ludvika District lands. Under the direction of assistant district manager, Åke Granqvist, this pilot project, called ecological landscape planning (ELP), eventually led to forest certification for the district. In 1991, Stora Forest president Dr. Bjorn Hägglund hired Börje Pettersson as staff ecologist to define a corporate forest management strategy that would maintain the company's economic viability while improving its ecological performance. Dr. Pettersson came with solid credentials as a forest scientist and credibility with Swedish ENGOs.

STORA used an 8,500 hectare area five miles west of Grangärde in southern Dalarna as its test site for ELP. Maps, aerial photos, GIS technology, and field inventories were used to identify key ecological features and the presence of endangered species. Fire resistant areas along streams and bogs were set aside to develop naturally. Those areas with tree species that depend on fire to regenerate are being restored with a variety of management methods, including controlled burning, increasing the proportion of hardwoods in the stands, retaining green trees in harvested units, and leaving more rotting wood to maintain the overall health of the ecosystem.

Listening To Neighbors and Environmentalists

"The environmental nongovernmental organizations are trustworthy, but they won't tell that we are behaving well if we aren't in their eyes. They are the only ones, really, that can reach the end consumers."

–Åke Granqvist, supervising forester, STORA, for ELP development at Grangärde

Incorporating the views of local citizens and ENGOs in the planning process is perhaps the most innovative feature of the ELP process. STORA established local "reference groups" to learn how local people view and use its forests. STORA personnel found that by listening to people and groups who use their forests—hunters, fishermen, bird watchers, berry pickers, skiers, hikers, educators—they not only increased their own knowledge about the resources that exist on STORA lands, but also gained credibility in the Grangärde area. They are no longer treated as part of an impersonal corporation, and they have become more integrated with the local community.

Since getting locals involved in ELP, STORA's relationship with the local media has evolved from confrontation to a point where "no negative press has come from local environmentalists during the last two years," according to STORA forester Granqvist. STORA personnel find it easier to work with local environmental groups than national ENGO officials. Local members, they said, seem to be motivated more by ecological values, than political considerations, and tend to be more concerned with scientific and conservation issues.

STORA's Grangärde ELP test area has now been expanded to the entire Ludvika District of 300,000 hectares. The real test of ELP's viability, however, will come as STORA expands the program to its entire ownership. One Swedish forestry expert pointed out that the test unit near Grangärde was unique: It was less expensive to implement ELP there than it will be on other company lands. A key concern is whether STORA will be able to afford the internal costs of implementing ELP across 2.3 million hectares long term. The areas set aside for conservation reduce harvest levels 10% annually. The company has also assigned fifty employees to ELP management to date, which, combined with other ELP expenses, costs $10.5 million a year.

The Goals of Ecological Landscape Planning

STORA has designed and implemented its ELP to achieve multiple goals on adjacent tracts of forest:

1. Exclude key ecological areas, especially those containing endangered species, from normal management.

2. Create possibilities for populations of endangered species to travel across the landscape so that chances of reproducing are improved.

3. Maintain adequate proportions of old forest, hardwoods, and other important stages of forest growth.

4. Balance the use of intensive management practices against the goal of maintaining a diversity of species in the landscape, which may affect the use of exotic plants, ditching, and fertilizers.

1993—A NEW FOREST MANAGEMENT STRATEGY

"The problem is a lack of old trees. Many of the most demanding species, they need old trees to survive."

–Per Skoog, silviculturalist, Ludvika Forest District

As the ELP experiment was under way at Grangärde, STORA forestry executives designed a new forest strategy for all company lands, that has as its overall goal to "contribute to the achievement of STORA's financial goals, as well as maintain a high, valuable, and sustainable forest production, while preserving the biological diversity." In keeping with Swedish protocol, the strategy was announced at the same time as a new national for-

est policy revision, which covered all types of lands in Sweden. The changes made in both sets of standards are comparable. One could conclude that STORA was just keeping pace with Swedish forestry, not leading the way.

The new strategy led to a number of changes. Harvest levels will be increased 0.5% to 1.0% per year over the next fifty years from spruce and pine plantations, which contain a high proportion of young trees. The company is also seeking to maintain the productive capacity of the land, and preserve biodiversity, so in addition to traditional methods for managing spruce and pine plantations, STORA has extensive sections on nature conservation in its new strategy. They include discussions on ELP, matching management techniques to the characteristics of individual sites, principles for day-to-day nature conservation, and how to address special forest management questions, such as the use of fertilizers and ditches.

The strategy does not address the backlog of ecosystem restoration activities built up by centuries of intensive harvesting—such as the many hectares of wetlands that have been ditched and drained to establish spruce plantations. STORA's strategy seems to assume that restoration will occur as part of the ELP process, or under day-to-day nature conservation measures. Whether it actually will, and how quickly, is unknown.

Forest Management Practices

"In the end, even the machine operator has to have quite a good knowledge of what it is all about."

–Åke Granqvist, supervising forester, STORA, on the need for ecosystem training for personnel

STORA's approach to implementing its new forestry strategy is centered on educating forest workers. They are trained to understand ecological principles and forestry techniques, and are responsible, as well, for carrying out day-to-day nature conservation measures. Harvesting crews typically work in teams of four. Two workers operate harvesting equipment during the first half of the work shift, while the other two mark and plan the next field work to be done. Halfway through the day, the two crews change places. This cuts down on monotony, equipment-generated stress, and involves forest workers closely with the implementation of conservation measures.

The new practices are visible across the STORA forest. More green trees are left standing within clear-cut areas. A higher percentage of deciduous trees dot the landscape. Controlled burning is used to regenerate harvested areas. The reintroduction of fire has been a challenge for the company. Much of the experience and knowledge of how to use fire effectively resides with long-retired STORA foresters. Current staff have gone back for help from these retired professionals to reestablish the practice. Fire also poses risks, raises air pollution questions for STORA foresters, and after a generation convincing the public of the evils of fire, requires reeducating the public about the benefits of fire.

Maintaining Ecosystems

"If you think I am focused on biodiversity, it has its reason. The whole nature conservation strategy of the company is also focused on biodiversity. It is not that we have forgotten the other things, but they are not big problems in Sweden."

—Börje Pettersson, chief ecologist, STORA

Restoring and maintaining healthy ecosystems is an integral part of STORA's new forestry strategy. From an inventory standpoint this involves identifying unique ecological areas, threatened or endangered species, and areas that need restoration. As do other large Swedish forest companies, STORA conducts an inventory of its lands every ten years, as part of a national project. A range of methods, including computerized GIS technology, are used to document and track the various ecosystem components. The real dilemma here, as elsewhere in the

world, is how much can the company afford to measure and monitor. Inventory methods are heavily oriented towards tree growth, but are being adapted to track other ecosystem components.

STORA currently targets three types of areas for restoration and set asides. Critical habitats for species such as the white-backed woodpecker are reserved and protected. Wetland sites are prominent among these areas. Difficult or risky landscape features, such as rock outcrops, bogs, steep slopes, and other low productivity sites make up the second category. Areas that can serve as corridors between populations of animals, enhancing the flow of genetic diversity between nonadjacent forest tracts are the third target. These tend to follow the courses of streams. Management expects to meet requirements for preserving old forest habitat this way, but outside pressure is mounting to retain additional older, commercial forest stands on STORA's land.

Overall, STORA expects to lose no more than 10% of its fiber productivity by setting these lands aside for general ecosystem maintenance. STORA forest managers estimate that meeting the requirements of Swedish law account for about a 5% set aside, while ELP requirements make up the additional 5%. Since most Swedish sites already produce less than 50% of the fiber that high-yield forest plantations elsewhere produce, these ecological setasides are financially riskier in the long term for STORA managers.

CERTIFICATION OF THE LUDVIKA DISTRICT

"We can't say 'the law says this.' We are such a major player, we need to make our own decisions in this matter, because it's important to have them for our own structure."

—Börje Pettersson, chief ecologist, STORA

During 1996, STORA contracted Scientific Certification Systems (SCS) of Oakland, California, to conduct a certification of its Ludvika Forest District, in south-central Sweden. STORA considered the certification a pilot effort, in part because forest certification standards for Sweden are currently being drawn up by the Swedish FSC Working Group. The company wanted to gain experience with the certification process, and to see how Swedish forest management would be assessed from an international perspective. STORA management stated that it did not intend to use the results of certification to market products. Information from the certification project was integrated into Sweden's FSC working group.

The team conducting the certification project for SCS consisted of Dr. Robert Hrubes, a U.S. forest economist, Dr. Per Angelstam, research leader at Sweden's Grimso Research Station, and Lars–Olof Österström, director of the Swedish Forestry Association. They customized the certification standards used by SCS to fit Swedish forestry, and in June 1996 conducted an eight-day evaluation of the 300,000 hectare Ludvika Forest District.

Under FSC guidelines, forest certifications examine three management elements—sustainable wood production, the maintenance of ecosystems, and the socioeconomic benefits to the human communities involved. A "passing score" is judged as 80 out of 100. STORA received a score of 92.7 for sustainable wood production, 80.4 for ecosystem maintenance, and 88.9 for socioeconomic benefits. The team listed a number of commendations, observations, and requirements that STORA needed to meet for continued certification. The company was commended for efficiency in harvesting, investments in timber stand improvement, and treatment of workers. The company also received high marks for self control in managing forest inventories and cutting less than it was growing.

However, the team had a number of observations. It pointed out that the Ludvika forest over-relied on conifer species, and had inadequate numbers of deciduous trees. The team also said the company needed a better inventory of nontree resources, more training for forest workers in ecosystem man-

agement techniques, and to pay more attention to the economic development of communities surrounding the Ludvika District.

The FSC team imposed six conditions on STORA for continued certification. STORA must inventory all its stands of trees older than the lowest age for cutting, and determine which ones might be held to achieve a goal of 5% old forests in the district. Within one year of certification, STORA must incorporate methods into its strategic plans to assess biological diversity over the long term. New policies and methods to reduce damage to deciduous trees by moose and deer, and the restoration of wetland spruce habitats were required within eighteen months of certification. The FSC team also directed the company to draw up plans to minimize and counteract the effects of air pollution, namely acid rain, on the forests, and to help a local community draw up strategies for more diversified economic development.

STORA personnel were neither surprised, nor disappointed with the results of the certification project. They were satisfied with a passing score, and felt that the company was already addressing the deficiencies pointed out in the project. The only condition that seemed to cause major concern was the one directing the company to address local economic development. STORA personnel do not consider economic development outside of forest products to be the company's responsibility. They think the communities themselves need to provide that leadership.

If STORA decides to certify its entire ownership, real challenges may lie ahead. One Swedish forestry expert, knowledgeable of the STORA ownership and forest certification, does not think that other STORA districts would pass the certification test. STORA officials, however, do not anticipate significant difficulties bringing these other areas into compliance with certification standards. Whether current initiatives will accomplish this goal is clearly the subject of some disagreement.

Non-company Lands—A Key to Sustainability Claims

STORA appears to be successfully implementing a range of new environmentally-sensitive policies and forest management practices on its own lands. STORA has just begun, however, the difficult task of responding to 1995-Plus Group members by documenting its sources of wood supplies. The raw materials for STORA's products come from five basic sources: company lands (25%), independently owned land that is harvested by company crews (15%), small woodland owner associations (25%), imports (15%), and chips from sawmills (20%). STORA's vertical integration gives it some advantage in keeping track of the chain of custody of its products, which is required by certification. In the production of solid wood, the chain is fairly intact. Nearly all the wood used in the company's sawmills comes from its own lands.

The challenge lies in paper production since the majority of the wood used for pulp comes from non-company lands. Maintaining a clear chain of custody when mixing many different sources of fiber to make paper, is much trickier. STORA is pushing the FSC to allow producers to claim a percentage of certified content, similar to recycled paper, which would help the company manage the chain of custody difficulties. If unsuccessful, STORA will find it impossible to market certified paper without expensive changes in production.

The greatest obstacle to STORA's ability to provide sustainably produced products is getting the landowners from whom it buys timber to come into environmental compliance. If other owners refuse to adopt and maintain these new forest practices, it greatly weakens STORA's claim as a sustainable operation. Where STORA personnel are supervising or conducting harvest operations, they implement these techniques, but even then, they admit that the landowners involved may not maintain the practices. This may be one reason STORA is part of

the Swedish Forest Stewardship Council Working Group that is trying to establish nationwide standards for forest certification. If adopted, these standards would apply to all ownerships. STORA has a clear vested interest in ensuring that all forest lands in Sweden are certified by a third party.

Whether Sweden will successfully develop national standards is uncertain. As of April 1997, the Swedish FSC Working Group had failed to reach a consensus on standards for certification. The small woodland owner cooperatives are no longer participating, but the group continues to work toward consensus. If the group does reach consensus, it will be much easier for STORA's paper making operation to certify its product.

Implications of Sustainable Forestry

"Commitment to sustainable forestry must be real. A media campaign to change attitudes will not work. They will find you out."

–Ragnar Friberg, chief forester, STORA

STORA recognizes that its relationship with the general public has often been difficult, and that the public had come to distrust claims made by business. By finding collaborative opportunities with diverse STORA stakeholders, the company has bolstered its public image and reduced conflict.

Evolution in Philosophy

Several Swedish foresters describe this collaborative approach as an outgrowth of an evolution in the Swedish philosophy of forest management that took place in four stages.

1. 1960s—*Ignore the Critics*

Similar to the U.S. forest industry, the Swedish foresters tried to marginalize the growing criticism from environmental groups by asserting that their critics were just a bunch of "whackos," who were not to be taken seriously.

2. 1970s—*We Are the Professionals, You Don't Understand This*

As environmentalism became widespread and forestry practices were questioned in the media, foresters stood on their credentials: They were the trained professionals, had the actual experience with the forest, and had everything under control, particularly in regard to the herbicide issue. Foresters presented a litany of scientific evidence that showed that the chemicals they used did not cause health problems. Still, environmentalists and a growing segment of the public were unconvinced that the science was complete, or on the side of herbicide use. Despite their efforts, Swedish foresters lost herbicides as a tool, because of public fear.

3. 1980s—*Listening, Learning, Cooperation*

After two decades of conflict, Swedish foresters took a new tack. Instead of trying to discredit and marginalize environmental groups, they began to sit down and listen to what the critics said. This process eventually led to cooperative ventures. When STORA and Assi Domän began ecological landscape planning in the late 1980s, local environmentalists were included in the discussions and planning.

4. 1990s—*Joint Development of Ecological Engineering*

The cooperation of the late 1980s led to collaboration in the 1990s. Local environmentalists are still involved in forest management planning in the field. At the national level, a coalition of industry, small private landowners, indigenous peoples, labor unions, and environmentalists have joined to form the Swedish FSC Working Group. This is the fifteen-member team currently trying to reach consensus on a set of forest certification standards for all of Swedish forestry. STORA is represented in this process by chief forester, Ragnar Friberg, who heads the economics subgroup of the Working Group.

CONVERT CRITICS TO SUPPORTERS

STORA officials recognize that the company needs both the message and the messenger to effectively garner public support for forestry practices. By hiring Börje Pettersson, and adopting ELP, STORA developed an Earth-friendly forest management method. By involving local environmentalists in planning, the company added voices to carry its message: STORA actually has the endorsement of local environmentalists. In STORA's estimation, local ENGOs are credible messengers for their corporate environmental performance message. Many of these same groups have been forest products industry critics in the past, and also participate in the FSC Working Group. From the company's perspective there is no stronger image enhancement than a former critic supporting company activities. Indeed, the endorsement from environmental groups is the planned end result of their current strategy—to convert critics into supporters. The strategy is played out, in part, by involving local ENGO representatives on "reference groups" to help implement ELP. Attaining third-party certification not only helps the company keep pace with Swedish forest industry standards, it also allows STORA to continue to improve its relationships with the public and environmentalists.

INTERNATIONAL AND COMPETITIVE CONSIDERATIONS

The international challenges confronting STORA are daunting: industry overcapacity, cost and productivity disadvantages, and new, low cost competitors in traditional markets. The commitment of their customers to purchase certified wood products is the one opportunity on the horizon that promises STORA the potential to hold on to its lead European markets. Fortuitously for STORA, a handful of its strongest European markets have chosen to publicly align themselves with FSC-accredited certification. Some members of the 1995-Plus Group are so eager to put FSC-certified products on their shelves that they are willing, at least in the short term, to pay a price premium for these goods.

STORA's management, however, is waiting for the establishment of an FSC national standard before using third-party certification as part of its marketing. Environmental performance is clearly an area of concentration for Stora Fors AB, a cartonboard mill, as well as other parts of the company. In 1995, Stora Fors became the first cartonboard mill in Europe to receive Ecomanagement and Audit Scheme (EMAS) Registration, an environmental management system similar to ISO 14000. The company has created promotional pieces for Stora Fors, including a video and several print pieces, that talk about "The Green Champion," "The Ecologically Balanced Mill," and their environmental statement. There is no question the company sees this positioning as a source of competitive value in its markets.

Management at the certified Ludvika forest district and the sawmill is frustrated at being unable to market certified products. Forest district managers would like to see a payoff from the significant investment in money and time they have made over the last several years. Sawmill managers recognize pent up demand in specific market segments and would like to capitalize on what they perceive as a competitive advantage if they could sell certified material. Some customers are apparently even irritated because they cannot claim that the products they buy from STORA are certified. Company personnel worry that if they do not capitalize on certification soon, STORA will lose any potential competitive advantage.

Third-party certification could be used to create market barriers to protect STORA's European markets. STORA managers have investigated the practices of a number of Southeast Asian competitors and concluded that these producers would have great difficulty obtaining third-party certification. If STORA can win FSC certification and supply sufficient product to maintain momentum among the assorted buyers groups and similar customers, management thinks it could effectively close the market to these competitors.

Certification may also help slow down the adoption of substitutes for wood products. Third-party certification may help provide ammunition to demonstrate that wood products are environmentally sound through life cycle analysis, which assesses the environmental impact of products from design to disposal. The documentation required for certification will help generate the detailed information needed to make these life cycle computations. The upshot should be a demonstrably stronger environmental position for forest products compared to many substitutes.

Third-party certification may, however, provide a strictly short-term competitive advantage. Certainly, the environmental groups that have pressed for sustainable forestry hope that such practices as third-party certification become the industry norm. However, STORA may reap greater longevity to any first mover advantage by establishing good relationships with customers such as the 1995-Plus Group members. Strong relationships between buyers and sellers can cultivate a form of brand equity that is difficult for competitors to replicate. The trust that grows from experience with supportive suppliers has value to commercial customers and creates bonds that are difficult to break.

Relying on third-party certification as a competitive tool does have risks. The success of certification as a competitive strategy for STORA depends on the validity of management's assumption that competitors will not be able to quickly mimic STORA's practices. A potentially even greater risk is that European customers will not have a lasting commitment to FSC accredited certification. Although these buyers have expressed a willingness to pay a small premium for these products in the short-term, they will be unwilling to do so indefinitely. These buyers have taken the position that third-party certification comes at little cost to the suppliers and merely reflects practices that any credible supplier should follow anyway. STORA may also fail in its efforts to provide FSC-certified products to its customers, unless other Swedish landowners that also supply raw material to STORA adopt similar practices. If STORA fails to supply certified products, it risks unfavorable publicity, which might place the company at an even greater competitive disadvantage.

The Future

It appears that STORA is headed toward enhanced sustainability of its forest lands. The eventual costs to fully implement these changes are still unknown, although STORA's management thinks they know the level of total implementation costs. While certification was not a driving factor behind STORA's forest management decisions, the changes in management practices are positioning the company to efficiently certify and reap whatever rewards may result from marketing certified products. Management considers certification an efficient way to communicate to a range of stakeholders its commitment to sustainable forestry and business practices. The impact and success of the Swedish FSC Working Group's attempts to complete its work, and the initial marketing efforts of Swedish companies, including STORA, will be a critical predictor of the future of certification worldwide.

CASE LIMITATIONS

The conclusions drawn from this case need to be tempered by the observation that the data from which this analysis was made came from a single visit with STORA management and one visit to STORA's forests that was supervised by STORA managers. The researchers have had little opportunity to triangulate their findings by interviews with competitors, NGOs, or lower level employees. Nor did the researchers have time to visit additional STORA-managed forest sites, other than Ludvika, to observe practices there.

LESSONS LEARNED

1. The forest industry and ENGOs can collaborate productively. Coming from a situation of heated conflict in the 1970s, the Swedish forest industry has forged working relationships with ENGOs over the past fifteen years by listening, learning, and collaborating on forest issues. The relationships have led to forest management actions that have achieved mutually beneficial environmental, social, and economic objectives. In the most optimistic scenario, environmentalists have become a messenger to help industry rebuild its somewhat tarnished image with the public.

2. A large pulp and paper company can achieve certification for its forest operations under FSC guidelines. The forest management practices that pulp and paper producers typically use to promote rapid growth, including shorter rotations than solidwood producers, are at odds with the multiple benefits and long-term orientation needed for ecologically-based forest management. That should make it more difficult for pulp producers to gain certification. However, STORA passed certification on 300,000 hectares of its land by having a strong internal commitment to change its practices and allocate the resources needed to implement a new forest management strategy. This course of action has included hiring new staff, retraining its workforce and reducing harvests by a potential 10%.

3. Environmental certification is not necessarily driven by final consumer purchasing decisions. STORA's decision to certify its forests is based on the demand for third-party certification by its European customers, buyers groups, and large paper customers, which indicates that these groups will play a decisive role in the spread of certification.

4. For the present time, the European forest products markets appear tied strongly to FSC-certified products. This reality is driving forest product producers to align themselves with this system. STORA managers see early certification as an opportunity to gain competitive advantage over other producing regions. Under this scenario, producers are using third-party certification as a nontariff barrier to producers outside the region.

5. Paper making presents difficult chain of custody challenges. Unlike some wood products, paper production can involve raw material from multiple sources that is fed simultaneously into a continuous process. In STORA's case less than 50% of the fiber it uses in paper comes from its own lands. To alleviate the problem STORA is pressing the FSC to allow claims of a percentage of certified content. It is also a member of the FCS Working Group that is trying to develop certification standards for Swedish forestry.

6. Companies set limits to environmental concessions to maintain economic viability. In STORA's case the board of directors has placed that limit at a 10% loss in potential harvest. Management considers any losses beyond that to have the potential to endanger the company's ability to compete in the worldwide pulp and paper markets.

7. In seeking certification, STORA is trying to capture a competitive advantage open to it due to current management practices. The company decided to seek certification well after it had decided to adopt ecologically-based forest management. In this case, certification is a reward for behavior the company was already committed to.

Appendix A: STORA's Environmental Policy[*]

In its environmental work, STORA supports the sections of the international agreements adopted at the Rio Conference that are applicable to the forest products industry, and the ICC's 16 Principles. Thus, STORA as a whole, as well as each STORA company, commits itself to continuous improvement efforts aimed at:

- Practicing a long-term, sustainable forestry policy that preserves biological diversity.

- Producing, developing and marketing high-quality products that are safe for people and the environment, with the emphasis on minimal overall environmental impact.

- Using production methods that are resource-efficient and safeguard both the internal and the external environment.

- Selecting raw materials and chemical additives, and substituting different ones when necessary, to reduce the total impact on the environment.

- Introducing environmental management systems that are compatible with EU Regulation No. 1836/93 relating to voluntary adoption of the Community's ecomanagement and audit scheme.

- Ensuring that employees are informed, receive supplementary training, and feel actively involved in regard to environmental matters, while also conducting an active dialog with suppliers and other interested parties regarding relevant environmental issues.

[*] STORA's Environmental Policy, *STORA Annual Report*, 1995, pg.10.

- Establishing the goal of exceeding the demands contained in various environmental laws and regulations, while also working to ensure that laws and norms in the environmental area are harmonized at an international level.

- Encouraging a balanced view in respect to recycling and energy recovery from recycled fiber.

- Encouraging transportation systems that will contribute to environmentally-sound overall solutions.

- Preventing the occurrence of environmental accidents and making careful consequence assessments in advance of any new operations being implemented, or old ones being phased out.

- Establishing as a long-term goal the development of all operations to achieve sustainable production.

STORA's ambition is to be a leader in the environmental area. Since the mid-1980s, STORA has had a common, Group-wide environmental policy to bring about a unified approach to environmental policy that will, in turn, bring about a unified approach to environmental work. The present environmental policy was ratified by the Board of Directors in August 1994.

This environmental policy summarizes in eleven points the guidelines for environmental work that apply to all operations within the Group.

BACKGROUND AND MOTIVATION

STORA's environmental policy is based on available information relating to environmental issues, the assessments and considerations expressed in what has come to be known as the Rio Declaration, the environmental program for commerce and industry sponsored by the ICC, and other national and international statements of purpose in this area.

The Group's environmental work shall always be based on the most recently available scientific information. STORA shall also work actively to encourage a more favorable climate for environmental work in society as a whole. This means that the Group's own environmental policy will need to be reviewed so that it continues to reflect advances in knowledge and changes in the requirements imposed in regard to the environment.

BASIC PRINCIPLES

STORA supports the applicable parts of the accords from the Rio conference. The three basic documents are:

- The Rio Declaration on Environment and Development, containing twenty-seven principles that will be followed in all long-term environmental and development work, nationally and internationally.

- Agenda 21, which is a practical action program for achieving the objectives expressed in the Rio Declaration.

- Forest Principles, setting out guidelines for utilizing the world's forests without creating a negative impact on the global environment.

The Conference also witnessed the signing of two conventions that are significant for the environmental work of the forest products industry:

- The aim of the Convention on Biological Diversity is to improve the preservation of species and ecosystems.

- The aim of the Convention on Climatic Change is to limit the concentration of greenhouse gases in the atmosphere.

Appendix B: Five-Year Financial History, STORA

(In Millions of Swedish Krona)

	1992	1993	1994	1995	1996
Income Statements					
Sales	47,623	50,435	48,894	57,106	45,161
Operating Expenses	-42,482	-44,564	41,490	-44,785	-38,777
Restructuring costs and capital gains/losses	-561	58	498	301	-37
Share in income of associated companies	58	53	31	–	–
Depreciation according to plan	-3,437	-3,785	-3,566	-3,648	-3,488
Operating income	1,201	2,197	4,367	8,974	2,859
Net financial items	-2,560	-1,668	-1,150	-954	-510
Income/loss after net financial items	-1,359	529	3,217	8,020	2,349
Taxes	288	116	-1,126	-2,605	-770
Minority interests	-44	6	-53	-48	-19
Net income/loss	-1,115	651	2,038	5,367	1,560
Personnel					
Average number of employees	38,881	33,64	26,858	25,619	22,716
Key financial ratios					
Return on shareholders' equity [1]	-6%	3%	8%	20%	5%
Return on capital employed [1]	2%	4%	9%	19%	6%
Ditto after deduction for tax liabilities [1]	3%	5%	10%	22%	7%

[1] Average capital has been calculated as the average value of the opening and closing balance adjusted for the effects of major corporate acquisitions and divestments.

Vernon Forestry:
Log Sorting for Profit

CASE STUDY

PREPARED BY:
CATHERINE M. MATER
SCOTT M. MATER

A Case Study from "The Business of Sustainable Forestry"
A Project of The Sustainable Forestry Working Group

The Sustainable Forestry Working Group

Individuals from the following institutions participated in the preparation of this report.

Environmental Advantage, Inc.

Forest Stewardship Council

The John D. and Catherine T. MacArthur Foundation

Management Institute for Environment and Business

Mater Engineering, Ltd.

Oregon State University
Colleges of Business and Forestry

Pennsylvania State University
School of Forest Resources

University of California at Berkeley
College of Natural Resources

University of Michigan
Corporate Environmental Management Program

Weyerhaeuser Company

The World Bank
Environment Department

World Resources Institute

CCC 1-55963-629-7/98/page 15-1 through 15-12
© 1998 John D. and Catherine T. MacArthur Foundation
All Rights Reserved

Vernon Forestry:
Log Sorting for Profit

CASE STUDY

PREPARED BY:
CATHERINE M. MATER
SCOTT M. MATER

A Case Study from "The Business of Sustainable Forestry"
A Project of The Sustainable Forestry Working Group

Contents

Vernon Forestry: Log Sorting for Profit

Introduction

Canada's Province of British Columbia, like other countries where private citizens and government agencies have clashed over the management and use of *public* forestlands, has faced conflict over logging in its public forests. Amidst confrontations between environmental and industry organizations over clearcutting practices, the arrest of more than one thousand environmental protesters, and major political upheaval within the Provincial Government, in the early 1990s the Vernon Forest District in the south-central region of the province undertook a small, innovative project that has the potential to define new solutions to sustainable forest management (SFM) for not only British Columbia, but Canada itself.

AN INNOVATIVE PROJECT

In 1990, with the forests of British Columbia the focus of economic, environmental, and social conflict over resource management, British Columbia's Ministry of Forests established the British Columbia Forest Resources Commission. At the time the public was vocal in its concern over the visual impacts of clearcuts. Forest communities were dissatisfied over the loss of jobs because small operators were unable to gain access to timber and with the processing of harvested trees outside the region; and in those communities demand was rising to develop smaller value–added wood product manufacturing. The commission was mandated to examine the state of the province's land base, recommend ways to improve its management, and address the economic and social issues.

In 1993, the commission recommended that the B.C. government conduct a pilot project to evaluate new forest management techniques that would embrace an ethic of enhanced stewardship. Its goal was to balance the old values of forest economics with new values that support the preservation of wilderness, environmental protection, water quality, recreation, and community stability. The project was accomplished under the Small Business Forest Enterprise Program (SBFEP) of the Ministry of Forestry (MOF) and carried out in the Vernon District of the Kamloops Forest Region.

Project Concerns & Objectives

Concerns

- Visual impacts of clearcuts

- Loss of Jobs

- Lack of assistance in value-added manufacturing

Objectives

- Evaluate alternative log harvesting practices (vs. large clearcutting)

- Evaluate alternative log sales methods

The project, initiated in 1993, addressed several key objectives. First, it was to technically and financially evaluate alternative silviculture systems in contrast to the clearcutting techniques traditionally used in the B.C. forests. Second, the project was to evaluate alternative methods of selling logs that would allow smaller wood product manufacturers in a region easier access to the resource *and* increase financial returns on the sale of timber harvested under alternative logging practices.

High Stakes Undertaking

At the beginning, the project was correctly viewed by many in industry, the environment, and government as a high stakes venture. If alternative silviculture and log sales practices were determined to be technically and economically feasible, the potential impact on existing industry operations in British Columbia and throughout other Canadian provinces might prove significant.

Some government and industry representatives had a major concern that the project would produce the high probability of *industry-wide expectations* generated from *site-specific results*. Any cost differences, for instance, between clearcutting methods and alternative logging practices documented in the Vernon project would be site-specific and simply might not apply to other forest sites. Price differentials for log sales that might occur at a designated log yard for this project might not be reproducible in other log yards where geographic conditions are different—where access to local value-added product manufacturers and local log buyers might be limited, as an example. The participants in the project were correctly concerned that, in the highly-charged political environment, site specific results could quickly convert to unrealistic expectations for generic solutions.

When the international environmental organization, Greenpeace, elected to undertake an environmental assessment of the alternative silviculture practices used in the project that only raised the stakes and the profile of the experiment. As an independent, third-party evaluator, the Greenpeace assessment was to be Canada's first forest certification process to evaluate whether alternative silviculture practices that didn't use clearcutting techniques could protect the integrity of the forest at the landscape level and among stands of trees.

The certification process was an equally a risky venture for Greenpeace, which was considered an effective adversary of most of the logging practices used by the forest products industry. Greenpeace's objective for the certification was to show that society could protect forests while providing timber in socially and economically successful ways. The forest ecocertification process represented the first time that Greenpeace tried to evaluate whether certain alternative logging practices could be supported while still protecting the overall health of forests.

ALTERNATIVE LOGGING PRACTICES—SETTING THE RECORD STRAIGHT

The forest management techniques used in the project and (called alternative silviculture systems) were not all different from those used in traditional logging. Although the term "alternative" was applied to the five different logging practices used in the project, four of the five practices evaluated are not new and have actually been used for many years in existing forestry operations in the Vernon Forest District and some parts of British Columbia. "Alternative" was a term coined to provide a simplified distinction between large-area clearcutting and other silviculture options where large-area clearcutting is not used. The five alternative logging practices evaluated in the Vernon Project were:

1) clearcuts with reserves

2) selection systems (group, strip, and single-tree)

3) seed tree systems

4) shelterwood systems (uniform and group)

5) small clearcuts

Of those five, only the strip shelterwood silviculture option had not been used in the past.

What differed about the four alternative logging options already used in B.C. forestry was the way in which they were applied. In most cases, the *amount* of timber that would have been removed from any one site would have been greater under typical industry practices.

Greenpeace evaluated only two sites that used group and single tree selection logging options for certification. The group did not consider any sites that used clearcutting as a harvest method, regardless of size, as eligible for certification. Other independent, third-party certifiers endorsed by the internationally recognized Forest Stewardship Council (FSC), however, do not require a clearcut-free zone for certification. Silva Forest Foundation, the Greenpeace certifier for this project, is not an FSC-approved certifier. Due to limits from private funding sources for the Greenpeace certification process, the other acreage using non-clearcut harvest methods were not evaluated. Of the total 161,000 cubic meters of wood harvested during the project period (1993-96), Greenpeace certified less than 2.5% , or 4,000 cubic meters.

The five alternative silviculture methods used in the project each had specific objectives as noted in the sections that follow.

Clearcuts with Reserves (New Forestry)

Often referred to as "New Forestry," clearcut with reserves is a harvesting method used in mature stands that maintains the biodiversity of the stand. The primary differences between this method and standard clearcuts is in what is left on the site. In this method, loggers leave a more diverse stand of trees in age and species, more woody debris to help maintain the long-term soil productivity and other plant and animal life, and snags for wildlife. These actions help soften the visual impact of the clearcut.

This method was used on approximately 31.8 hectares and accounted for approximately 29.2% of total project harvest. Approximately 457 cubic meters of timber was harvested per hectare. The species harvested on this site included Englemann spruce, subalpine fir, and lodgepole pine. In these areas the techniques were chosen to accomplish a number of goals:

• Increase the quality of remaining standing timber.

• Create snags (15 snags/ha).

• Protect soils from degradation.

• Cushion negative visual effect of traditional clearcutting.

• Regenerate the preferred commercial timber species.

Selection System

Selection method provides for the cutting of specific trees or small areas with strict rules and controls over how much damage can be done to the remaining trees. Three different selection systems were used—group, strip, and single-tree vs. clearcut methods of harvesting. One of the primary goals of these systems is to create or maintain uneven-aged stands of timber. Reducing the visual impacts of harvesting is another goal. When group selection was used the stumps were removed after harvest. Greenpeace certified this method as "Ecologically Responsible Wood."

This method was used on approximately 45.1 hectares and accounted for approximately 22.5% of total project harvest. Between 172 cubic meters and 350 cubic meters of timber was harvested per hectare. The species harvested on this site included Douglas fir, lodgepole pine, cedar, larch, and white pine. On these sites, foresters wanted to salvage wood infected with root rot, protect soil, enhance deer range, create uneven-aged stands, and maintain partial visual quality of the forest.

Seed Tree System

This method was used in an area heavily infested with mountain pine beetle (MPB) and root rot. The seed tree method was chosen so that the diseased trees would be removed, larch trees would regenerate naturally, and the visual quality of the forest and the quality of the soil would be protected. The area was steep sloped and required cable logging. This method was used on approximately 21 hectares and accounted for about 14.6% of total project harvest. Approximately 546 cubic meters of timber was harvested per hectare. Species types harvested on this site included Douglas fir, lodgepole pine, and larch.

Shelterwood Systems

Two types of shelterwood systems were used to create even-aged stands—uniform shelterwood and group shelterwood. Uniform shelterwood involves leaving selected species and sizes of trees uniformly throughout the remaining stand. Group shelterwood results in leaving groups of selected species and size of trees, which creates small clearcut openings throughout the stand. These methods were used primarily to maintain visual quality of the forest to satisfy a nearby resort, control root rot, regenerate the stands of trees, and, in the case of the uniform system, to leave about 130 of the best quality trees per hectare standing.

These methods were used on approximately 18.8 hectares and accounted for approximately 3% of total project harvest. Approximately 285 cubic meters of timber was harvested per hectare. Species types harvested on this site included hemlock, Douglas fir, lodgepole pine, larch, white pine, and cedar.

Small Clearcuts

This method uses clearcuts of less than 5 hectares in size versus traditional clearcuts of 20 to 30 hectares. The method was chosen to salvage white pine infected with white pine blister rust, protect visual quality, and to regenerate preferred species. Other benefits included controlling root rot, enhancing deer range, and protecting the soil.

These methods were used on about 9.2 hectares and accounted for approximately 6.4% of total project harvest. Approximately 375 cubic meters of timber was harvested per hectare. Species types harvested on this site included hemlock, Douglas fir, white pine, and cedar.

FROM THE FOREST

Between 1993 and 1996 about 161,000 cubic meters of wood were harvested from the targeted sites by approximately 65 different cutting contracts. Annual harvest volumes were fairly consistent from year to year: 1993-94: 53,000 cubic meters; 1994-95: 52,265 cubic meters; and, 1995-96: 56,479 cubic meters. The 4,000 cubic meters of certified wood volume was harvested in 1995-96.

During the Vernon project MOF retained ownership of the timber until it was sold at the log yard, which was unusual. Typically, the MOF would sell the timber on the stump and the purchaser would assume all costs for logging and transportation of the logs to a designated mill. The MOF would receive payment for an appraised value of the standing timber plus any additional amount (bonus bid) the purchaser would elect to offer to secure the award of the timber contract from the MOF.

The Vernon project was designed to address the results of a typical bidding process:

- The process usually involves larger volume sales, which restricts access to the purchase of needed timber for many smaller wood product manufacturers in the region.

- Contract loggers are typically required to contract with the large wood product producers in the region to secure work because when the bid is awarded, up-front log purchase is required and the volume of logs to be harvested is so large in any one contract.

- The total dollar value of the timber sold on the stump is assumed to be less than what could potentially be generated from a log yard if it sorted logs and sold them according to product applications that met market demand.

Most of the logging for the project has taken place in the Vernon Forest District located in the Okanagan Valley of south-central British Columbia.

ALTERNATIVE SILVICULTURE METHODS PRODUCE INTENDED RESULTS AT A COST

The environmental and economic impacts of the alternative silviculture systems employed in the project were evaluated for the first year of harvest (1993-94: 53,030 cubic meters). The results of the alternative methods illustrate both significant opportunities and constraints for future SFM projects.

Environmental Viewpoint

From an environmental viewpoint, the alternative logging practices used by the project produced tangible benefits. Management objectives such as maintaining visual quality of the forests for surrounding communities, protecting and enhancing deer ranges, salvaging diseased species, creating habitat snags, protecting soils from degradation, and providing forest landscapes that encourage desired regeneration and preserve higher quality tree species were all met.

The project provided direct opportunity to expand new forestry concepts in the area and created new partnerships among forestry officials, community stakeholders, loggers, worker safety officials, and environmentalists. The project was especially beneficial in fine tuning methods and procedures that can be used in future forestry projects in the region, thus providing a solid framework for resolving future conflicts between industrial use of the forest and environmental issues.

Perhaps most important, the overall sensitivity of the MOF foresters involved in the project to the environment and surrounding communities when laying out timber sales created a new level of trust and credibility for MOF forestry in the Vernon District. As an example, the harvest block selected for strip selection was located near the community of Cherryville in the Currie Creek area. The Vernon MOF involved the community in designing the harvest area, which resulted in maintaining a visual quality in the forest that not only pleased, but clearly surprised some Cherryville residents. When asked how they felt about the harvest project on their picturesque mountainside, some residents indicated they did not know that it had already happened.

In attempting to achieve a balance between industrial use of the forest and environmental values, again the project ranked high. Many involved in the project felt that in some of the project units logging most probably would not have been allowed at all without the use of the project's alternative harvesting systems and the sensitivity shown to the concerns of the visual impact of logging on neighboring communities.

Economic Viewpoint

From an economic viewpoint, however, there is little question that alternative logging practices can cost more to implement in the field when compared to traditional large-scale clearcutting methods (see Figure 1). Of the eleven harvest tracts where equivalent costs for conducting large-scale clearcutting on the sites were calculated by the MOF, costs increased an average of 20% to implement alterna-

Harvest Cost Comparison: Clearcutting vs. Alternative Practices

Track Species	Harvest Cost/m³ Alternative	Clearcut	% Difference
Fir Cedar	Group Selection $24.15	$20.38	+18%
Fir Larch	Group Selection $29.95	$24.36	+23%
Spruce Lodgepole Pine Balsam	New Forest $32.64	$24.44	+34%
Lodgepole Pine Fir Larch	Strip Shelterwood $39.95	$28.06	+42%
Fir Lodgepole Pine	Selection $34.55	$26.60	+30%

Source: Vernon Forest District

Figure 1

tive logging practices over clearcutting methods. For clearcut-free alternative logging practices that may have met Greenpeace certification standards, the average cost increase over clearcutting methods was approximately 30%. Even so, as the experience with the community of Cherryville illustrates, the use of alternative logging practices in the project allowed for access to between 10-20% more land base without citizen protest than what would have been allowed under traditional harvesting practices.

With the MOF retaining ownership of the timber through log yard sales, logging contracts for the project sites were opened up to smaller independent operators who normally would have been excluded from contract opportunity due to the requirement for purchasing the timber up front. Further, no bonding was required on the project, although a 10% security on the bid was required.

Finally, alternative logging systems required higher employment levels due to lower productivity caused by the logging and site protection restrictions of the project.

To the Log Yard

One of the most significant aspects of this project which, in effect, financed the increased costs of the alternative logging practices employed in the forest, was the development of alternative methods for selling timber from the targeted project sites. For this project, a log sort and sales yard was set up on approximately 16 hectares of land in Lumby, British Columbia. Unlike traditional MOF practices of selling timber on the stump, for this project all logs from the targeted sites were brought into the yard where they were scaled, sorted, and sold in lots based on species and grade factors to the highest bidder.

The log sales yard was established to allow smaller wood product manufacturers in the region easier access to the timber *and* to increase financial returns on the sale of the region's wood harvested under alternative logging practices.

If located in a geographic region with access to value-added wood product manufacturers and log brokers, log sales yards can maximize returns on the sales of logs by sorting logs according to product and manufacturer needs. As an example, under conventional log sales systems, it is standard practice to separate veneer-quality logs from logs of lower grade that may be manufactured into commodity construction lumber. However, it is not standard practice to recognize the value of lower-grade logs with defect or "character wood" that are often used in the development of higher-end products, such as furniture and log home construction. Sort yards provide greater returns on log sales by allowing for additional log sorts that separate those character wood logs, sold at higher prices per unit over lower-grade logs, to interested manufacturers.

Log sort yards can have additional economic benefits to surrounding communities by allowing small wood product manufacturers located in the region to purchase small volumes of wood. Typically, these small-scale producers do not get access to traditional MOF timber sales due to the heavy competition from large wood product producers in the area that can finance the large-volume timber sales traditionally offered by the MOF. By buying through a log sort yard, the smaller-scale product producer is able to use his full purchase volume, with no waste or log residual needing to be sold at a loss.

For this project, the log yard is run by a contractor who is paid a flat fee per cubic meter for logs brought to the yard. The MOF provides one full-time supervisor for revenue and cost control. The logging contractor provides all personnel and equipment to unload, sort, and load. The MOF retains direct control of the scaling and merchandising operations. The scaling crew personnel are independent contractors, contracting directly with MOF on the basis of a competitive bid.

The log yard directly employs ten people from the local area. All employees are independent contractors and bid for their jobs, which offer wages comparable to other jobs in the area.

Between 1992-93 and 1994-95 the average volume through the yard was approximately 54,000 m³/yr. In 1995-96, however, volume reached 57,500 m³/yr. This is approximately 5% of the district harvest volume. The total harvest volume in the Vernon District off of Crown land is approximately 1,200,000 m³/yr based on a sustained yield basis with approximately another 120,000 m³/yr coming from private and salvage sales.

Logs are delivered to the yard, then unloaded and spread on the ground for sort coding and scaling. The number of log sorts are based on market demand, log buyers' requests for desired log characteristics for multiple product applications, and on species, length, and top and butt diameter. Scalers also determine if a log should be bucked or cut to shorter lengths and may cut serious defects from the log at that time, which increases its value. The marked and scaled logs are then placed in appropriate sort bins. Bins are closed off once a week, if an adequate number of logs has accumulated. The volume in each bin is then advertised and competitively sold in lump sums through sealed bids. No minimum price is set, although the MOF reserves the right to refuse any or all bids.

Once the bins have been sold, the buyer has seven days to pay for and remove the logs from the yard.

LOG SORT YARD RESPONDS TO MARKET AND ENVIRONMENTAL DEMANDS

When the log yard started in 1993, only eleven sorts were planned, but that number was elevated to twenty-three log sorts within two months of the start of operations. Within the next twelve months, the log sorts had again doubled to forty-two. In 1997 there were forty-seven regular sorts plus sorts of certified wood.

Additional log sorts at the yard benefited the project in three key areas as noted below.

Market Considerations

Sorting is the primary method of marketing for the log yard. Since all logs must be sold to the highest bidder, direct marketing cannot be done. Based on customer requests, log sorts are established which meet particular industries needs. Sorts such as balsam peelers; firewood, dry white pine, spruce, and fir building logs, character logs, larch and pine shake logs, "acoustic" logs (used in the manufacture of musical instruments), and logs for log home production were all additional log sorts based on customer requests.

Environmental Considerations

The sort yard has allowed the number of log sorts in the forest to be reduced from seven to two, which has a number of benefits:

- Smaller log sort areas (landings) are required in the forest which means less forestland is taken out of production, and the risk of associated erosion and siltation is reduced.

- Increased safety on the site.

- Lower costs for loggers, which are reflected in the bids.

- Improved quality control since many of the decisions are made in the log yard by trained log graders rather than in the woods by cutters who often work under adverse conditions, and have limited market expertise.

Certified Log Handling Considerations

To maintain the identification of certified logs, they were separated in the log yard using the same categories of sorts as the noncertified logs. Sort numbers were distinguished by adding 50 to the standard sort code numbers, i.e. certified SPF sawlog sort would be 55 (Sort 5 + 50), a spruce O/S would be 63 (Sort 13 + 50), etc.. All certified sorts were clearly identified and all certified logs had a certification brand on their ends. They were sold in separate lots from the non-certified logs. Workers in the yard commonly referred to these sorts as "holy wood."

LOCAL VALUE-ADDED PRODUCT PRODUCERS BENEFIT FROM LOG SALES

Part of the success of the project is based on location. The Vernon District has numerous primary and secondary manufacturers within a reasonable distance of the log yard and the log yard is centrally located in the harvest areas. The number of log home manufacturers, specialty products manufacturers, and other small business operations form the competitive base for log sales. Prior to this project, smaller manufacturers had difficulty in obtaining logs for their operations.

In the first eight months of operation (fiscal year 1993-94), logs were purchased by twenty-nine different customers.

While the largest volume (73%) of logs that go through the sort yard are purchased by large operations manufacturing commodity lumber, the added number of sorts has increased participation and availability of logs to small value-added manufacturers as indicated in Figure 2. Most of the special sorts go to small companies.

The numbers of buyers has actually increased since the first logs sales in 1993, with over 63% of fiscal year 1995-96 buyers being repeat customers.

Some buyers were able to buy at the sort yard when they were unable to buy elsewhere.

- Two guitar manufacturers buy Englemann spruce for manufacture of guitars.

- Two log home producers that needed a few logs were able to buy them from the yard when they needed them to complete orders. Initially they only wanted cedar building logs. Now they are using dry spruce, pine, and fir logs. Currently log home builders are regular customers and buy the

Log Buyers by Product Type

Product	% of total	
	1993-94	1995-96
Log Homes	14%	20%
Lumber	18%	30%
Guitars	7%	7%
Chopsticks	4%	3%
Flooring	4%	3%
Millwork	4%	3%
Veneer	0%	3%
Post/Poles	18%	10%
Furniture	0%	3%
Pulp/Firewood	4%	3%
Shakes/Shingles	4%	7%
Brokers	18%	7%

Source: Mater Engineering

Figure 2

building log sorts. One of the manufacturers estimates that there are approximately thirty to forty log home companies are in the area. Each of them has trouble convincing other log buyers to sort long straight timber for them. The building log sorts at the Lumby yard fit their needs.

- Other sorts which specifically serve small business include character logs for custom construction and furniture (logs which would otherwise end up as fire wood or chips), larch shake logs, and pine logs for post and rail.

Not even the bark slabs from the trees are wasted. Thick fir bark is removed and given to the local Boy Scouts and Naturalist Clubs to construct bat houses that are then placed in the forest to assist bat populations.

CERTIFICATION NOT A FACTOR

Greenpeace certification of targeted harvest sites proved not to be a factor in the project's success. With less than 3% of the total volume of wood coming from Greenpeace-certified sites, sales of logs based on certification were clearly constrained due to the small, one-time volume offering and, according to project participants, due to the lack of marketing facilitation expected from Greenpeace. With only 4,000 cubic meters total of certified logs offered for sale at the log yard, only 50% of that certified volume has currently sold. Of that 50%, only 1% of that volume was purchased based on its certification status.

While initial interest in certified wood was high from around the world, the only load of logs specifically bought for certified wood was by Rouck Brothers sawmill for custom flooring. When interviewed for this case study, representatives of Rouck Brothers indicated they purchased the certified logs not as a result of customer demand, but in response to a Greenpeace request to purchase. The company did not pay a premium for the logs.

Although the log sort yard received international inquiries for certified wood supply, many of those inquiries were for certified logs that cannot be sold under current B.C. law, which requires that logs be initially processed in the province.

Had the project included a certified wood supply that could have satisfied potential buyers' concerns for consistency of certified supply over time in the species and grades required, results might have been different. Even so, from all indications interest in the project still remains high. As an example, Greenpeace is in the fourth printing of its report on the Vernon project, having received over 20,000 requests for the report from interested parties around the world in 1996. According to Greenpeace, 50% of the requests came from Canada, 30% from the United States, 15% from Europe, and the remainder from Australia, New

Zealand, and Malaysia. Of those requesting copies of the report, only a small percentage are government representatives or representatives from environmental organizations. The majority of requests come from private wood manufacturing concerns.

Despite the interest in the project in the industry, Greenpeace noted that media coverage of the project did not reach its normal threshold. The group cited reporters' comments that the project was not considered "conflict-oriented" enough to merit coverage.

The Bottom Line—Financial Success

Both an internal report produced by the MOF in September 1994 and an independent evaluation conducted by Price Waterhouse in May 1995, concur that the Vernon project was a financial success. The information gathered on alternative forest practices and the operation of the log yard provide a model for future operations. While both reports concluded that the project was at least as profitable as conventional logging, the MOF and Price Waterhouse disagreed on the comparison of the overall profitability of the project with what it would have been under conventional logging.

Based on costs and revenues analyzed by the MOF and Price Waterhouse, both determined that the project achieved a $2 million net profit after paying all stumpage fees. However, when calculating the estimated net profit that would have been realized from conventional harvesting methods, Price Waterhouse projected a $1.9 million profit; the MOF estimated a $1 million profit. Regardless of calculated net profit differences for conventional logging, the project proved a financial success for the Crown, realizing between $100,000 and $1,000,000 more net profit than conventional log sales and harvesting methods would have.

EXAMINING THE DIFFERENCE

The primary difference in the estimated profit from conventional logging methods was the estimate of what bidders would have bid as a "bonus bid" for the timber if it had been offered under the conventional method of bidding for timber on the stump. The bonus bid is the amount bid over the "upset" (minimum) bid price set for stumpage based on MOF appraisals. The MOF figure of $15 to $20 per cubic meter was based on estimates prepared for the area by the SBFEP forester who took into consideration the limited experience of the bidders in the area. The Price Waterhouse analysis used a weighted average of bonus bids from throughout the Kamloops region to arrive at a figure of $36.55 per cubic meter for the estimated bonus bid. This difference in estimated bonus bid accounts for the difference in conclusions of the two reports.

While this difference may affect the spread of additional log yards operated by the MOF, it does not affect the conclusion that alternative forest practices, when combined with a log sort yard, can be at least as profitable as conventional methods. And while some may argue that a less concerned forestland manager would simply maximize profits by both clearcutting *and* employing a log sort and sales yard, current political realities dictate that being allowed to conduct a business-as-usual option in the forest simply no longer exists.

SORTING IS KEY

Much of the economic success of the project was due to sorting at the log yard. In almost all cases, these special sorts yielded a higher price than if left in the original sort (typically sawlogs or chip logs). The original sort is monitored to determine if removing a specific category of log from that sort has an impact on the price of the original sort. In most cases the impact has been negligible. The impressive price differentials received from log sales due to added yard sorts that cater to customer product interests are shown in Figure 3.

Price Differentials Due to Added Yard Sorts

New Sort	Old Sort Price	New Sort Price	Differential (%)
Balsam Fir Peeler	$74.03	$105.15	(+42%)
Spruce O/S (Dry)	$73.92	$93.09	(+26%)
Spruce Peeler (Dry)	$89.13	$111.19	(+25%)
Pine Peeler (Dry)	$79.17	$88.14	(+11%)
Spruce Bldg. Log (Dry)	$79.17	$111.75	(+41%)

Notes:
1. Old sort is where the logs were placed when the log yard first started up.
2. Old sort price is the price of the old sort at the time the first logs were sold in the new sort.

Source: Vernon Forest District

Figure 3

One of the largest increases occurred in the sales of lower-grade logs, which included a large percentage of dry logs and logs that had died on the stump. The log sort yard manager, who was knowledgeable of the markets and the value of dry sorts, found that many of these logs were useful for more than just sawlogs or chips. They could be used as peelers, for shake manufacturing and house logs, for instance. Dry spruce logs could be used for chopsticks and window and door stock production, as well. A typical dry sort was sold at $110 per cubic meter. After deducting an average cost of $55/m³ for logging and sort yard and stumpage costs, these sorts provided a return of $54.75/m³ more than if sold for traditional lower-grade uses.

All costs were included in the up-front capitalization cost of the operation with the exception of some of the planning costs prior to start of the project. The lease costs are included in the operational costs. Lease costs have been increasing annually since the start of the operation; from $40,000 the first year to $83,000 in 1997. Costs have increased to the point that other sites closer to Vernon are now being considered for the log yard.

PROJECT CONTRIBUTES TO LONG-TERM CHANGES

The Vernon project, in its fifth year in 1997, has contributed to long-term changes in the way forestry is conducted in the region. The overall sensitivity to the environment and the surrounding communities exhibited by government foresters in laying out the alternative logging timber sales for this project has created a new level of trust and creditability of MOF foresters in the Vernon District. This has had a positive impact towards solving conflicts between the industrial use of the forest and environmental concerns.

Many of the alternative logging practices employed in the Vernon project are already in forest management use in British Columbia. But the success of the project in applying those alternative practices to *lower-volume harvest levels,* and the ability to offset the increased costs of implementing alternative silviculture practices through effective alternative log sales strategies has had a tangible impact and will serve as a benchmark for future activity in this arena. Aside from stimulating discussions on alternative logging practices, the success of the log sort yard has been the catalyst for other log yards development in the province:

• A new log yard is now operated by the MOF in Prince George.

• In 1996 Weyerhaeuser set up a log yard next to the Lumby log yard to assist in merchandising logs.

• It is estimated that similar log yards could be justified in other B.C. locations, including the East and West Kootenays as well as the Cariboo Region.

• Many visitors have come to see the log yard and hundreds of inquiries have been received from outside individuals and organizations that are considering developing similar log yards in other parts of Canada, the United States, and other regions.

The bidding on logs at the Lumby Log Yard has developed information on the log market for the interior of British Columbia. This important information was not available before the project and is now currently published in the *Forest Industry Trader,* a weekly industry publication from Vancouver, to help forest resource managers understand and gain the maximum value for their forest resource upon harvest.

Because of the project's success, a number of new economic development strategies are now being considered to increase the access of small value-added wood products businesses to sustainably-harvested timber supply in the area.

While not currently pursuing additional certification of logging operations in their district, the MOF will continue to separate certified wood in the sort yard and is willing to work with any certifying program, although they are not actively pursuing any of them.

Lessons Learned

With the management of public forestlands continuing to be a source of conflict in many regions, the results of the Vernon project suggests several useful lessons for the management of public forest lands.

1) Even in areas where the conflict between appropriate management of forestlands is highly charged, options do exist for forestland managers that can deliver both *environmental* and *economic* performance. Although the employment of alternative logging methods did increase logging costs over conventional harvesting methods in

this project, the ability even to harvest on some lands or harvest without protest proved a significant equalizing factor. In effect, the alternative logging practices used in the Vernon project actually expanded the timber base by being environmentally conscientious. The project graphically illustrated that alternative logging techniques made harvesting in some sensitive areas possible. In those areas the public simply would not have allowed harvesting at all using conventional techniques.

2) To make SFM practices in the field transfer to successful bottom-line results, working smarter *upstream* is a mandate. For the Vernon project, the creation of a log sort and sales yard after harvest proved a vital key to the project's overall economic success.

3) Although it can be an effective tool in SFM techniques, certification of forestlands is not a mandate for success. The certification component played a minor, if less successful, role in the

Vernon project, further illustrating the importance of consistency of supply in relation to market demand for certified product—all factors not evident in this project. Even so, the key project participants within the MOF felt that demand for certified wood does exist and that markets will develop with the advent of more certified supply coming on line.

Finally, much of what was learned in the Vernon project is transferable to other regions. Activities in the Province of British Columbia—new agreements between environmentalists and industry, the creation of new log sort and sales yards, and government reviews to reevaluate stumpage payment prices and practices for harvesting on Crown-owned lands—bear this out.

The Vernon project may have been small, but the results are significant.

Weyerhaeuser
Forestry:
The Wall of Wood

CASE STUDY

PREPARED BY:
ROBERT DAY
STUART HART
MARK MILSTEIN

A Case Study from "The Business of Sustainable Forestry"
A Project of The Sustainable Forestry Working Group

The Sustainable Forestry Working Group

Individuals from the following institutions participated in the preparation of this report.

Environmental Advantage, Inc.

Forest Stewardship Council

The John D. and Catherine T. MacArthur Foundation

Management Institute for Environment and Business

Mater Engineering, Ltd.

Oregon State University
Colleges of Business and Forestry

Pennsylvania State University
School of Forest Resources

University of California at Berkeley
College of Natural Resources

University of Michigan
Corporate Environmental Management Program

Weyerhaeuser Company

The World Bank
Environment Department

World Resources Institute

CCC 1-55963-630-0/98/page 16-1 through page 16-16
© 1998 John D. and Catherine T. MacArthur Foundation
All Rights Reserved

Weyerhaeuser Forestry: The Wall of Wood

CASE STUDY

PREPARED BY:
ROBERT DAY
STUART HART
MARK MILSTEIN

A Case Study from "The Business of Sustainable Forestry"
A Project of The Sustainable Forestry Working Group

Contents

Weyerhaeuser Forestry:
The Wall of Wood

Introduction

Born during the "cut and run" days of early twentieth century America, Weyerhaeuser defied conventional industry logic by holding onto timberlands after they were cut rather than walking away. By the late 1930s, the company was faced with a decision: What to do with previously logged land on which natural regeneration had been ineffective. It decided to regenerate forests and grow timber as a crop, first by seeding harvested areas (1940s) and later by planting seedlings (1950s to present). Beginning in the 1960s, Weyerhaeuser began producing seedlings in nurseries and integrated replanting into its plantation operations. Following this strategy, Weyerhaeuser, headquartered in Federal Way, Washington, has become the world's largest private owner of standing softwood timber, North America's largest producer of softwood lumber, and the world's largest supplier of softwood pulp.

Weyerhaeuser initiated sustained yield forestry to provide a guaranteed and consistent supply of wood, not out of direct concern for the environment. However, the company has come to realize that by investing in a long-term strategy, their decisions have positive ecological and economic consequences that will amplify into the future. Over the past thirty years, Weyerhaeuser has developed a form of sustainable forestry based upon high-yield plantations that are among the most productive in the world. This high-yield model provides higher returns while simultaneously minimizing overall environmental impacts by producing high-quality wood and fiber on substantially fewer, continuously regenerated, acres. In this sense, the Weyerhaeuser Forestry model may facilitate both environmental and economic sustainability.

Business Description

Weyerhaeuser was an $11.1 billion company in 1996 with the forest products segments accounting for slightly more than $10 billion in sales. The company provides products and services through three different business sectors: Timberlands & Wood Products, Pulp, Paper & Packaging, and the Weyerhaeuser Real Estate Company. Exhibit 1 shows a five-year trend of the approximate contributions to operating earnings made by each sector.

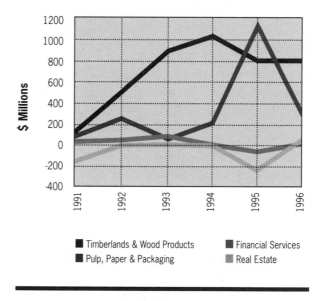

Approximate Contributions to Operating Earnings 1991-1996

Legend:
- Timberlands & Wood Products
- Pulp, Paper & Packaging
- Financial Services
- Real Estate

Exhibit 1

Weyerhaeuser's corporate strategy is based on the premise that the highest returns from timber growing are achieved through an emphasis on high-value, softwood timber and wood products over fiber for pulp and paper. As a result, Timberlands is run as a profit center, rather than as a cost center (internal supplier) for the product sectors. This arrangement does on occasion create friction with the downstream manufacturing operations, since company lands are not solely managed to produce raw material targeted to the needs of the wood products, pulp, paper, and packaging businesses.

TIMBERLANDS AND WOOD PRODUCTS

In 1996, operating earnings for the Timberlands and Wood Product sector together totaled $805 million on $5.2 billion in net sales. The Timberlands businesses produce primarily high-grade timber and sawnwood products, specifically, Douglas fir *(Pseudotsuga menziesii)* and western hemlock *(Tsuga heterophylla)* in the U.S. Northwest, and loblolly pine *(Pinus taeda)* in the U.S. Southeast.

Timberlands, in addition to operating company tree farms, manages genetics and selective breeding programs, as well as seed and seedling operations at the company's orchards and nurseries. These nurseries grow 257 million seedlings per year, of which 50 million are used on the company's 5.3 million acres of timberland. The balance, 207 million, are sold to third parties. Finally, they provide raw material in the form of timber to Weyerhaeuser's manufacturing facilities and to other forest products customers.

The Wood Product businesses manufacture structural and appearance-grade lumber, along with plywood and engineered products. In addition, Weyerhaeuser operates a wholesale building materials distribution business in the United States. Products consist of softwood lumber, plywood, veneer, composite panels, oriented strand board, doors, treated products, logs, and chips. In addition to Douglas fir and loblolly pine, the Wood Products business also extensively uses white spruce *(Picea glauca)* aspen *(Populus spp.)*, sourced from Crown concessions in Canada, and other hardwood species from U.S. locations.

Most tree species used by the company, with the exception of aspen and other hardwoods, are evergreens, or softwoods.[1] Despite the nomenclature, softwood species are favored over hardwoods in structural applications for their superior functional characteristics of strength, durability, and flexibility.

All softwood species, however, are not created equal: Douglas fir has strength and quality characteristics that are unmatched by other species. Douglas fir is particularly competitive in Japanese post and beam housing construction where customers are willing to pay premium prices for high structural performance and quality. Monterey or radiata pine *(Pinus radiata)*, grown in the temperate tropics, and loblolly pine are less expensive to grow, but are of lower structural quality.

Western Timberlands manages approximately 2.1 million acres located in Oregon and Washington. Southern Timberlands manages another 3.2 million acres in the states of Arkansas, Oklahoma, Louisiana, Mississippi, Alabama, Georgia, and North Carolina. In addition, Weyerhaeuser Canada oversees the company's long-term Crown license arrangements in the provinces of British Columbia, Alberta, and Saskatchewan which cover approximately 23 million acres.

Weyerhaeuser's combined 1996 timber inventory on U.S. and Canadian lands was approximately 26.6 billion cubic feet of which 75% was softwood. To increase the yield and quality of the timber supply from its U.S. timberlands, the company engages in the active suppression of nonmerchantable species, precommercial and commercial thinning, fertilization, and operational pruning.

PULP, PAPER, AND PACKAGING

In 1996, the Pulp, Paper, and Packaging (PPP) sector had sales of $4.6 billion on $6.7 billion in assets, representing approximately half of the company's asset base. While earnings soared in 1995 due to unusually strong pulp and paper prices, low PPP prices in the balance of the 1990's depressed PPP earnings relative to Timberlands and Wood Products (see Exhibit 1). PPP products include market pulp, newsprint, fine paper, containerboard packaging,

---

[1] In contrast to hardwoods which are deciduous (i.e. leaf bearing).

and paperboard. The PPP business uses wood chips and other residuals from the timber and wood products businesses, but still sources raw material from non-Weyerhaeuser suppliers. PPP also secures a growing percentage of its raw material needs from recycled fiber.

WEYERHAEUSER REAL ESTATE COMPANY

The final business, Weyerhaeuser Real Estate, is a remnant of the company's diversification efforts from the 1970s to the early 1980s. With the aim of countering the cyclical nature of the core businesses, Weyerhaeuser invested in ventures ranging from hydroponic farming to disposable diapers. Efforts to refocus the company begun in the late 1980s have led Weyerhaeuser to shed most of these operations.

Weyerhaeuser Real Estate develops single-family housing and residential lots, as well as master-planned communities. Operations are concentrated in selected metropolitan areas in Southern California, Nevada, Washington, Texas, Maryland, and Virginia. In 1996, the real estate business made $35 million in operating earnings on sales of $804 million. Over the past several years, a number of operating units within Weyerhaeuser Real Estate have been sold.

Evolution of Weyerhaeuser Forestry

Weyerhaeuser was one of the first U.S. forest products companies to move to the west coast from the midwestern lake states at the turn of the century. As the midwestern forests of Michigan, Wisconsin, and Minnesota were depleted, it seemed clear that America's westward expansion would require additional timber. The great northwestern forests in California, Oregon, and Washington clearly held the most promise. In 1900, Frederick Weyerhaeuser and a number of partners founded the Weyerhaeuser Timber Company with a purchase of 900,000 acres of forestland in Washington belonging to Northern Pacific Railroad.

Initially, Weyerhaeuser was not oriented toward sustained-yield timber growing practices. In the early years, timber was harvested to provide returns to the original investors and to pay for the survey of the land holdings. By the late teens and early 1920s, the company decided to invest in milling facilities rather than only selling logs on the open market. The 1920s and early 1930s witnessed a substantial increase in the harvest levels along with additional acquisition of lands and mills. During this period, regeneration of logged-over land was left to natural regeneration. However, the success of such regeneration, especially with the more desirable Douglas fir, was mixed at best.

TREE FARMING

By the late 1930s, the company was faced a decision: Should the land that had been cut be sold or could it be retained and continue to earn a profit?[2] Contrary to prevailing industry wisdom, Weyerhaeuser believed that timber could be grown like a crop on a sustainable and profitable basis. The decision was made to retain and reforest the land holdings.

In 1941, the company established the nation's first privately-owned tree farm to demonstrate that intensive management of previously harvested forestlands could be profitable. The Clemen's Tree Farm became a laboratory where the company undertook experiments in forest regeneration, aerial seeding, fire protection, and brush control techniques. This research indicated that an economically viable tree farming operation would require more than merely allowing secondary forests to regenerate where virgin forests had once stood.

---

[2] Much of the motivation for "cutting and running" had to do with the tax consequences of holding on to cut over land: The conventional wisdom suggested that it would take too long for the next generation of trees to mature.

The effort during the 1940s was therefore focused on how to better regenerate cut-over land to Douglas fir through seeding methods. At the same time, the company also worked to change tax policies that discouraged the long-term holding of productive forestlands, and helped to shape fire suppression policies that reduced the risk of long-term investment. By the end of the 1940s, Weyerhaeuser had begun the transition to a "tree growing" company in contrast to much of the rest of the industry, which was still largely focused on liquidating old-growth stands or cutting timber from public lands to feed downstream mill operations.

HIGH YIELD FORESTRY

By the early 1950s, it was evident that while seeding methods were superior to natural regeneration, a more systematic, research-based approach would be required if tree farming were to be profitable. Research efforts were concentrated first on how to make the dense, second-growth stands of Douglas fir established after logging the old-growth saleable in a shorter time. By the late 1950s, however, it became increasingly clear that planting seedlings following harvesting was a better and more reliable method of forest regeneration.

As a consequence, the company began, by the mid-1960s, to develop a new forestry model which would come to be known as High Yield Forestry (HYF). Research initially concentrated on management practices such as planting and stocking control to achieve the goal of rapid forest regeneration. Forest nurseries were established to produce the seedlings required for replanting. Biological and financial variables were linked to develop practices that provided a continuous cycle of planting, growth, harvesting, and replanting.

Improving a Good Idea

To that end, soil surveys were conducted across the company's entire land base. The resulting data was used to construct simulations of forest growth and yield. These simulations helped to improve the technical and scientific aspects of HYF, which include:

- planting techniques to optimize siting, spacing, and rooting;

- pest and noncommercial vegetation control to minimize competition;

- fertilizing regimes and thinning practices to maximize growth;

- road-building techniques to minimize erosion and compaction;

- harvest timing to maximize yield, while maintaining wildlife habitat;

- protecting soil productivity; and

- ensuring rapid reforestation.

Once established, forests managed under these practices grew at unprecedented rates, compared to unmanaged natural stands.

The company then began to work to decrease seedling mortality rates in the nurseries that up to that time had a dismal 50% survival rate. Additional applied sciences aided in the development of seed orchard techniques for efficient cone and seed production and collection, as well as nursery techniques for establishing reliable germination, culturing and fertilizing practices that led to healthy seedlings with strong root systems. Seedling mortality has dropped from 50% to less than 5% today.

Breeding Better Trees

In the early 1970s, the company began to make collections of healthy, vigorous trees from natural stand populations of both Douglas fir and loblolly pine. These "plus trees" formed the basis of the tree improvement program for each species. Through selective breeding, trees that were inferior for growth, form, and adaptability were truncated out of the population. Beginning in the 1980s, the tree-improvement program began to focus on breeding varieties for specific end-use characteristics such as increased volume (fast growth), strength (specific gravity), and stem quality (straightness, branching habit).

While HYF seeks to maximize growth of the primary commercial species at each site while suppressing other noncommercial species, the resulting forest is not a pure monoculture. Complete repression of undesired species is impossible because seeds are dispersed by wind, water, and animals. The result, while significantly different from a natural forest, is far from sterile either visually or ecologically.[3] Furthermore, HYF does not imply introducing the wrong species onto unsuitable sites. In addition to Douglas fir and loblolly pine, there are also active regeneration programs in hemlock, noble fir long-leaf pine, and some species of hardwood.

Today, a cycle of HYF begins when harvest plans are drawn up and information on physiographic and soil features of the next round of planting sites are analyzed to indicate which of the currently available genetic varieties is optimum. A region forester will place an order for seed of that variety that will be germinated and grown at the nurseries for 1-3 years before it is planted on those sites and managed according to the optimum silvicultural practices (planting density, thinning, pruning, etc.) which are designed to lead to a particular desired forest community and final harvest product.[4]

Large quantities of seedlings are sold to other forest landowners, but the most advanced seedlings are used exclusively on Weyerhaeuser lands. Within a year of harvesting any given region, those seedlings are planted by hand following predetermined spacing and patterns that will maximize the desired

qualities. Over the forest growth cycle, most sites are thinned to remove less desirable trees, which provides more growing space to ensure that the remaining trees are larger and straighter. Precommercial thinning allows the company to leave organic matter in the form of trunks and branches on the forest floor where their slow decay adds to the soil's organic matter and nutrient base, which promotes further growth. Commercial thinning initially yields trees large enough for use in the pulp mills, and in subsequent thinnings, for lumber. Trees may also be pruned once or more during their life. Lower branches are removed to reduce the presence of knots in future lumber products and increase the production of clear wood. Aerial fertilization may be used to promote growth spurts throughout the rotation.

Patience Is Key

The goal of HYF is to produce wood that is, ultimately, more valuable to the end user. After thirty years in the Southeast (loblolly pine) and forty-five years in the Pacific Northwest (Douglas fir), specific sites are harvested consistent with the needs of the company's downstream businesses. To ensure a sustainable flow of product, on average, only 2% of Weyerhaeuser's forest sites in the West and 3% in the South are harvested in a given year. None of the harvest is wasted. Crowns and branches are left on site to return nutrients to the soil through decomposition. Bark may be burned for energy or sold as mulch for landscaping. Depending upon the region where it is processed, the bulk of a log becomes either lumber and other solid wood products or ends up as chips for pulp production. The rest, comprised of sawdust, is used for particleboard, mulch, or burned for energy.

HYF has resulted in significant increases in forest productivity on Weyerhaeuser's lands since the 1960s. Yields per acre have doubled in the Northwest and have increased fourfold in the Southeast, compared to a natural forest, which makes the company's lands some of the most pro-

---

[3] For example, in coastal Washington, if 100% Douglas fir is planted, the forest matures into a mix of approximately 50% Douglas fir and 50% hemlock and other canopy species.

[4] The variation of 1-3 years in the nursery is due to site-specific issues such as elevation, frost susceptibility, animal damage, etc.

ductive in the world. However, HYF has also required great patience: Investments made in the 1960s will begin to yield economic returns early in the next century, as genetically improved plantations begin to reach maturity.

Since the decision to pursue tree farming in the late 1930s, the company's strategy has depended on a long-term outlook. Investment in lands that will not realize significant returns for thirty to fifty years requires planning and commitment. The company takes care not to overreact to the cyclical nature of the market. Instead, Weyerhaeuser believes that measured, deliberative actions that do not over react to short-term changes in the marketplace produce a more stable operating environment. As one manager stated, "...deals are available all the time, companies are always making mistakes." This long-term commitment has made it possible for the company to take the time and make the investments necessary to effectively implement HYF management practices.

WEYERHAEUSER FORESTRY

With its heritage as a pioneer in forest management, Weyerhaeuser developed a deep sense of commitment to environmental stewardship. However, during the 1970s and 1980s, the public's concern over forest practices escalated and regulatory pressures upon the industry increased. Objections to clearcutting and concern about dwindling old-growth forests resulted in increasingly strict regulations on the use of public and private lands for timber harvesting, particularly in the Pacific Northwest.

Since Weyerhaeuser owned much of its own timberland and did not source significant amounts of wood from public lands in the United States, the company focused its attention on its own specific issues and tended to disregard these concerns. Weyerhaeuser countered criticisms by reemphasizing their commitment to growing trees, and providing scientific data to justify their practices and positions. Increasingly, however, environmentalists were not persuaded by scientific and technical defenses of the company's practices. By the end of the 1980s, Weyerhaeuser's reputation began to suffer as it was judged by the industry's lowest common denominator.

Private Forest as Public Trust

Dogged by poor financial performance relative to competitors, Weyerhaeuser underwent a series of changes beginning in 1989 when incoming President and CEO John Creighton, Jr. started to refocus the organization on core forest products operations. In the early 1990s, the company set about regaining the public trust by seeking to understand the changing social forces that had led to the gap between public perception and Weyerhaeuser's practices. The company came to understand that the public viewed *all* forests as public trust resources, regardless of ownership. As one Weyerhaeuser manager noted, "We were not acting as stewards of the resource in the eyes of the people." Weyerhaeuser's attempts to justify its practices through mounds of data had failed to address fundamental public fears over the health and future of forests and ecosystems.

This understanding allowed Weyerhaeuser to move beyond the sustained yield practices of HYF into a broader definition of sustainability: HYF evolved into Weyerhaeuser Forestry as management moved to incorporate the concerns of external stakeholders into operational decisions. Through regional Forest Councils and town hall-style meetings in the communities directly affected by the company's operations, senior management began to enter into a dialogue which allowed them to better understand where public values dovetailed or diverged from those of the company.

The result was the creation of a stewardship statement and a set of resource strategies that provided a guideline for the company's more holistic model of sustainable forestry which now includes commitments to water quality, fish and wildlife habitat, soil productivity, biodiversity, and aesthetic, cultural and historical values (See Exhibits 2 and 3).

Weyerhaeuser Forestry has also challenged the company to assume a more proactive role in the protection of endangered species habitats, and has facilitated the development of new tools and data for better management of the natural environment. For example, the company's Habitat Conservation Plans (HCPs) reduce operational impacts on threatened and endangered species through multispecies management, while watershed analyses allow the company to operate more flexibly by managing forests as systems rather than addressing the scores of regulations that apply to forest lands one at a time.

Weyerhaeuser Forestry was put to the test in 1989 when the Spotted Owl was listed as a threatened species in the Pacific Northwest. The management plan proposed by the Federal Fish and Wildlife Service, would have defined a "no cut" zone 1.2 miles in radius around each identified Spotted Owl nest. Through its HCP approach, however, Weyerhaeuser was able to demonstrate a more effective conservation plan for the Owl which also allowed more flexible use of its forest lands, opening up timber assets that had previously been defined as off limits under the "owl circle" management plan.

With five steelhead trout populations recently listed as endangered or threatened by the National Marine Fisheries Service, Weyerhaeuser Forestry will again be put to the test: Proposed management strategies by regulatory agencies include leaving 300-foot wide riparian zones on each side of fish-bearing streams. The company's challenge once again is to define an environmentally appropriate and socially acceptable management plan while also retaining flexibility in the use of their lands: Reasonable set-asides of harvestable timber to achieve environmental goals can be offset by the higher yields associated Weyerhaeuser's forestry practices.

Weyerhaeuser Forestry Stewardship Statement

Our commitment:
- To continuously improve our performance as responsible stewards of the environmental quality and economic value of the forests we manage.
- To actively listen to and act upon public expectations.
- To communicate consistently to ensure understanding of our forest stewardship goals, practices, and accomplishments.

What our commitment means:
We will manage our forestlands for the production of wood. In addition, our goals are to protect, maintain, or enhance other important environmental values, such as:
- Water quality and fish habitat.
- Wildlife habitat.
- The productivity of the soil.
- Aesthetics.
- Plant and animal species diversity.
- Culturally or historically unique areas.

We will accomplish this by:
- Practicing sustainable forestry to meet increasing worldwide demand for wood and wood products.
- Performing to standards set for all forest operations.
- Basing our management processes and practices on scientific research and technology.
- Leading cooperative efforts with public agencies and other groups interested in forest resources to develop balanced, cost-effective forest practices and regulations based on sound scientific standards.
- Meeting specific resource goals set by our regional Forest Councils.

Exhibit 2

Weyerhaeuser Forestry Resource Strategies

Principles

While managing our forestlands to provide a competitive return to shareholders, we are committed to:
- Using scientifically based practices.
- Communicating with our stakeholders.
- Actively listening to their concerns.

Strategies

Forest Products

We manage our forestlands for the sustainable production of wood and other forest products by:
- Reforesting promptly after harvesting.
- Practicing intensive forestry to produce wood and wood products that meet our customers' needs.
- Maintaining healthy forests—minimizing loss due to fire, insects, and disease.
- Harvesting at sustainable rates over the long term.
- Minimizing waste in our harvesting.
- Encouraging the use of other products from the forest.

Water Quality

We protect water quality and fish habitat by:
- Developing and implementing processes to assess the impact of our forest practices on these resources, and adapting our practices accordingly.
- Practicing sound road construction and maintenance.
- Continuously improving management practices in streamside areas.

(continued)

Exhibit 3

ECONOMIC AND ENVIRONMENTAL SUSTAINABILITY

As indicated above, the company believes Weyerhaeuser Forestry can achieve both environmental and economic goals, the latter through higher yields that reduce the need to cut or manage additional forestland; the former through initiatives like HCPs and watershed analyses. Weyerhaeuser's high-yield plantation model also facilitates cleaner air and offsets the production of greenhouse gases through the higher oxygen production/carbon-dioxide fixing associated with forests comprised of young, fast-growing trees.

It may be that the company's high-yield model can help fulfill society's growing demand for wood products into the next century without shifting that demand onto the world's remaining old-growth and native forests. By harvesting more wood on less acreage, potentially more forestland can be reserved on public lands for other nonwood purposes, including the development of nontimber forest resources, wilderness, wildlife habitat, and recreation. Indeed, the Weyerhaeuser Forestry model raises interesting possibilities for the preservation of public lands, parks, and reserves around the world.

Market Analysis[5]

Population growth and economic development pushed worldwide consumption of forest products upward at a rate of 1.3% per year over the period 1983-1993. However, this seemingly modest rate of growth cloaks wide regional variations: Most developed countries had little or no growth in their

---

[5] This section is based upon data from: Lent, T., Propper de Callejon, D., Skelly, M., and Webster, C. 1997. *Sustainable Forestry within an Industry Context.* New York: Environmental Advantage.

consumption of forest products; the countries of the Pacific Rim, however, experienced huge growth in demand for forest products. The use of fuelwood for cooking and heating still accounts for about half of the world's annual roundwood harvest. The second leading category is lumber and sawnwood products, followed by pulp and paper products.

GLOBAL SUPPLY AND DEMAND

Global demand for forest products is predicted to increase into the future, consistent with population growth and rising standards of living, just as it has during this century. By 2010, total annual paper and paperboard demand is expected to rise from approximately 250 million tons in the mid 1990s to over 450 million tons. At the same time, global demand for solid wood products is expected to increase almost 60% from 630 million cubic meters annually in 1989 to over 1 billion cubic meters annually in 2010. Demand for softwood logs is expected to grow just under 20% while softwood panels are expected to grow approximately 13% by 2010. Global demand for hardwood logs and panels are expected to grow 11% and 40%, respectively during the same period.

For a variety of reasons, however, the supply of timber and fiber is likely to become tighter over the next twenty years. Some countries' forests have almost reached the point of commercial extinction (e.g. West Africa, Malaysia, and parts of Latin America). Other key factors putting downward pressure on supply include:

- stricter regulations on public forest lands in the United States and Canada;

- withdrawal of some forest lands from production due to environmental pressures (e.g. the Pacific Northwest);

- overharvesting or lack of reforestation on private, nonindustrial forest lands from which half of the U.S. harvest is sourced;

Weyerhaeuser Forestry Resource Strategies

(continued)

Wildlife Habitat
We provide habitat for wildlife by:
- Identifying and protecting unique sites.
- Implementing landscape planning for wildlife, using wildlife-habitat data gathered on our lands.
- Protecting threatened and endangered species on our lands.
- Cooperating with government agencies to determine how Weyerhaeuser forestlands can contribute to the conservation of threatened and endangered species.

Soil Productivity
We protect soil stability and ensure long-term soil productivity by:
- Using equipment and practices appropriate to the soil, topography and weather to minimize erosion and harmful soil disturbance.
- Using forestry practices and technology to retain organic matter and soil nutrients.

Cultural, Historical, and Aesthetic Values
We consider cultural, historical, and aesthetic values by identifying sensitive areas and adapting our practices accordingly.

Exhibit 3

- land conversion from forests to agricultural, industrial, or residential uses;

- lack of infrastructure to harvest timber cost-effectively, such as in the Russian Far East; and

- relatively small contributions made by plantations, despite their rapid growth in the Southern Hemisphere.

SOFTWOOD MARKET[6]

Softwood forests are found for the most part in the temperate and boreal regions of the Northern Hemisphere. The United States, Scandinavia, Russia and Canada account for two-thirds of the industrial softwood harvest. Softwood production overall is estimated to increase from 939 million cubic meters to a total of 1085 million cubic meters, for growth of about 15% over the next 25 years. Given that demand growth is conservatively estimated at 1.3% per year, the softwood market would appear to be on a collision course, with demand outstripping supply for the foreseeable future.

Indeed, a number of regional developments, many of them driven by environmental concerns, will hinder softwood supply expansion over the next twenty years.

1. While Russia holds half of the world's softwood forests (over 500 million hectares), production is stagnant and well below cutting potential. The country contains enough softwood timber to supply total world demand for 66 years at current levels, but an unstable political environment, poor infrastructure, and outdated harvesting methods keep it from becoming a top supplier to nearby Asian markets.

2. Douglas fir harvests in the U.S. Pacific Northwest are expected to remain well below the levels that existed during the 1980s due to a combination of tightening regulations, overharvesting during the late 1980s, and reductions in logging on public lands.

3. In the U.S. Southeast, production on private forest lands will increase marginally, but not enough to offset the drop in sales from public lands from the Pacific Northwest. Because small, private landowners tend not to manage their forests in a high-yield manner, quality and volume of timber from these sources will remain on balance below potential indefinitely.

4. In Canada, softwood harvests are expected to decline as provincial governments implement increasingly strict timber harvest controls in response to concerns about environmental issues.

5. Scandinavia's harvests are expected to increase incrementally even though forests are already approaching the limits of sustained yield management.

6. Production in Latin America and Asia will be concentrated on the growth of short-rotation hardwood species. Despite the rapid rise of softwood plantations in Chile, Brazil, New Zealand, and Australia, the annual cutting potential of these plantations is expected to rise only from 100 to 160 million cubic meters between 1996 and 2010, a large percentage increase, but relatively small contribution to the total. Furthermore, these fast-rotation plantations will, for the most part, produce radiata pine, which is structurally inferior to Douglas fir.

These trends indicate a market where softwood timber demand will outpace supply, creating a softwood timber deficit. As softwood timber products become more expensive due to restricted supply, substitute products such as steel, concrete, and engineered wood products, such as wood I-beams, laminated veneer lumber, and oriented strandboard, may grow in popularity.

---

[6] This section is also based upon the Environmental Advantage report cited previously and does not represent Weyerhaeuser projections.

Competitive Position

Weyerhaeuser's strategy is centered on the production of the highest value softwood timber in one of the most developed, environmentally-conscious regions in the world. The company has focused its principal business on softwood species with structural and performance characteristics that yield superior quality lumber and wood products. For Douglas fir, this means virtually 100% of the harvest goes directly to the sawmill; for loblolly pine, more than half the harvest goes to the sawmill or plywood plant. Even with this sawlog mentality, however, over half of the cubic volume of wood ends up as residual materials, since only the highest-grade portions of each tree can be used for grade lumber. Most of the remaining material becomes pulping fiber that can be used in Weyerhaeuser's pulp and paper mills.

HYF as Core Competence

Weyerhaeuser has spent almost fifty years developing a knowledge of its forestlands. This enables it to evaluate site conditions that influence breeding programs so that tree varieties are matched to sites, ensuring high levels of productivity. Stocking control in the form of spacing, thinning, and fertilizing, help to produce straight, fast-growing trees with a maximum amount of clear wood for added value. These traits fulfill customers' needs while sustainable yields guarantee a reliable supply of product, and an advanced infrastructure guarantees the product reaches the customer in a timely manner.

Forestry initiatives that yield consistent, dependable results garner strong customer relationships and allow the company to enjoy premium prices on certain product lines where demand often exceeds supply. "The Weyerhaeuser name gets us in a lot of doors," noted one manager. This is particularly important to customers in the company's most important export market of Japan where reputation and relationships are a prerequisite for continued success, and the company enjoys a large market share (10% of company sales are to Japan alone). Thus, a cycle exists where Weyerhaeuser Forestry helps to retain existing customers and attract new ones, which in turn legitimizes and drives the continuation of the forestry operations.

Because the company's asset base resides in North America, it is beholden to longer investment and growth cycles than competitors in the warmer climates of Latin America and the Pacific-Asian region. Furthermore, with thirty years of experience in softwood genetics, most other companies find it difficult to emulate Weyerhaeuser's productivity gains. Three decades of capability building in forestry have enabled the company to establish a substantial first mover advantage over its competitors, especially in the Pacific Northwest.[7]

"Wall of Wood"

Weyerhaeuser's early investment in the land base enabled it to build a portfolio comprised of some of the best forest sites in the world. The majority of the company's lands in the Northwest are more moderately sloped, lower-elevation lands almost impossible to acquire today. Indeed, through almost 100 years of carefully planned acquisitions, the company has been able to accumulate a superior portfolio of forest lands.[8]

---

[7] For example, Weyerhaeuser has now entered a second generation in its Douglas fir genetics program, and a third generation in its loblolly pine genetics program. The closest competitors appear to be a full generation behind in each of these species.

[8] Site quality and slope are particularly important in the Pacific Northwest where mountainous terrain is the norm. Moderately sloped lands increase harvesting ease, reduce risks of landslides and accidents, make pruning easier, etc. In the Southeast, much less emphasis has been placed upon forest land holdings, since site quality varies less significantly in the dominantly flat terrain.

Through superior site quality and forest management methods, Weyerhaeuser has boosted forest productivity. The company estimates that its lands currently generate twice the wood per acre in the Pacific Northwest and four times as much in the Southeast compared to unmanaged (natural) forest land. For example, U.S. net annual tree growth in the 1990s is about 54 cubic feet per acre, twice the level of the 1950s. Average growth on Weyerhaeuser land, however, is more than 108 cubic feet per acre, with the company's prime, intensively managed lands producing an annual tree growth of up to 240 cubic feet per acre.

Furthermore, by 2010-2020, as the genetically-improved plantations reach maturity, the company estimates there will be an *additional* increase in yield on its lands in both the Pacific Northwest and the U.S. Southeast. These yield improvements should translate into an increased wood harvest per acre of 50% in the Northwest and 100% in the Southeast. The result is a reliable, high quality, and growing volume of softwood timber (referred to internally as a "Wall of Wood"), which is expected to reach maximum yield in a time of constricting supply, beginning around the year 2005 (see Exhibit 4).

ENVIRONMENT AS COMPETITIVE ADVANTAGE

Through its initiatives to establish a dialogue with communities, indigenous peoples, and environmental groups, the company has developed skill at cultivating stakeholder relationships. The company's experience with town hall-style meetings to solicit stakeholder opinions has allowed it to develop an openness and willingness to accept outside views and an understanding of how best to engage in that kind of dialogue. Such constructive engagement has resulted in better experiences and relationships with both regulating agencies and the public through their open implementation of alternative regulatory initiatives like HCPs and Watershed Analyses.

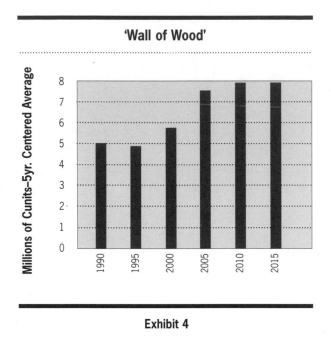

Exhibit 4

Weyerhaeuser's willingness to cultivate stakeholder relationships positions it well to capture the environmental high ground in the forest products industry. Indeed, environmentally-driven constraints on timber harvests, particularly on public lands, provide a competitive advantage. Given their substantial reliance on well-managed, company-owned lands, Weyerhaeuser is well positioned to provide a growing supply of high quality softwood in a world of tightening environmental regulation and constricting supply. In fact, if the competitive game shifts away from sustainable forestry and toward timber mining—if the Russian Far East were to be logged at an accelerating pace to generate hard currency—then the Weyerhaeuser Forestry model would lose much of its basis for competitive advantage. In a very real sense, the environment can be a competitive ally for Weyerhaeuser. It is possible that a restricted supply of quality wood along with increased demand for softwood could amplify this cashflow trend over time.

FINANCIAL PAYOFF

Each stage of forest operations has financial implications: Investment decisions must be discounted over thirty to fifty years placing great importance on the discount rate chosen. Because of HYF management, however, investments of $300 to $600 per acre can be recouped through doubled, tripled, or even quadrupled wood yields.

While Weyerhaeuser underperformed the industry during the 1980s in the equity markets, the company has performed significantly better over the past few years, due to refocusing efforts, reaching and maintaining its top quartile performance goal since 1993 (see Exhibit 5). Increased yields from Weyerhaeuser Forestry, at today's market prices would significantly improve the company's cash flow position over the next twenty-five years (see Exhibit 6).

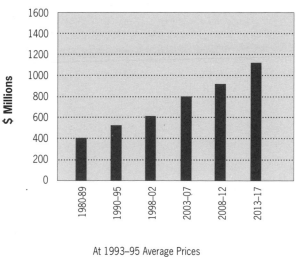

Pre-Tax Free Cash Flow from Timberlands

At 1993–95 Average Prices

Exhibit 6

Total Shareholer Return

1985-90 Percent Change		1990-95 Percent Change	
MacMillan Bloedel	153%	Willamette	210%
International Paper	146	Louisiana-Pacific	202
S&P 500	80	**Weyerhaeuser**	**136**
Temple-Inland	71	Georgia-Pacific	112
Willamette Ind.	63	S&P 500	111
Westvaco	62	Bowater	96
Georgia-Pacific	49	Champion Int.	72
Union Camp	46	Westvaco	70
Louisiana-Pacific	39	Potlatch	67
Weyerhaeuser	**28**	Temple-Inland	59
Champion Int.	22	International Paper	59
Potlatch	13	Union Camp	56
Boise Cascade	-9	Boise Cascade	45
Bowater	-18	Stone Container	32
Stone Container	-19	MacMillan Bloedel	-1

Exhibit 5

Future Challenges[9]

Weyerhaeuser appears to have successfully integrated environmental concerns into its core strategy and value proposition. Significantly, environmental concerns have been incorporated within a broader model of product quality and competitive advantage, offering the potential for long-term sustainability, both economically and environmentally. However, a number of strategic challenges face the company as it looks toward the next century.

---

[9] This section consists of the reflections of the case writers and does not represent Weyerhaeuser policy on these issues.

STRATEGIC QUESTIONS

Should the company seek certification for its company-managed timberlands or its products?

The concept of environmental certification and product labeling emerged during the late 1980s from the nongovernmental (NGO) sector to provide a driver for voluntary, market-based mechanisms to improve environmental performance. Since then, industry-specific initiatives have emerged with Scientific Certification Systems (SCS), Smart Wood, and the Forest Stewardship Council (FSC) programs leading the way in the forest products industry. These programs all set forth principles and criteria for site-specific forest management, and require third-party evaluation. A few niche players have used third-party certification to advantage. Virtually all of the large companies, however, have resisted the certification programs advocated by NGOs. Companies have preferred instead to identify outcome-based goals and design management systems that foster measurable progress toward those goals: The U.S. American Forest and Paper Association's (AFPA) Sustainable Forestry Initiative (SFI) represents such a "management system" approach, as does the ISO 14001 Environmental Management Standard.

So far, demand for certified wood products has been limited to niche markets in western Europe and North America: Worldwide, buyers have consistently shown that they will not pay a premium for environmental wood products. However, should a large industry player break from the pack and seek third-party certification, the competitive dynamics of the industry may change as other large competitors feel compelled to follow. Such a move might have special consequence in the environmentally-sensitive European market and in developing countries where certification could establish barriers to entry in an otherwise under-regulated environment.

Third-party certification might make competitive sense for a large player that was able to integrate sustainable forestry and environmental performance into its core business strategy—that is, a firm that could offer a quality product at competitive prices *plus* satisfy the environmental criteria. Weyerhaeuser may be positioned to gain such an advantage. It may therefore make sense for the company to consider third-party certification, at least for several of its forest sites, if not for its products in general.

How should the company handle chain-of-custody issues for the wood it buys from outside suppliers?

Despite high yields from its own well-managed forest lands, Weyerhaeuser is not self-sufficient in wood. Both the Wood Products and the Pulp, Paper, and Packaging businesses rely on non-Weyerhaeuser timber for their wood and fiber needs. Reliance on outside, usually small, private, sources of wood and fiber suggests that it may become increasingly important to manage the impacts of these suppliers' practices, particularly if third-party certification is deemed useful or necessary in the future. Indeed, working with small, private suppliers to improve their forest management practices could become a critical element in the company's management system should product certification ever be seriously considered as an option. Chain-of-custody tracking might also become important should environmental concerns force large consumer products companies such as Procter & Gamble, into requiring first-tier suppliers to provide such data. Moving the supplier base toward sustainable forestry will not happen overnight. There may be first-mover advantage for the large forest products player that can gain preferred access to the best, meaning sustainable, outside wood suppliers in specific geographic regions.

Is the company's current position in Canada consistent with the objectives of Weyerhaeuser Forestry?

With 20% of its assets residing north of the border, Weyerhaeuser is a substantial player in the Canadian forest products industry. In contrast to the United States, however, where only 39% of the forest lands are publicly owned, 94% of Canadian Industrial Forestlands are owned by the provincial governments. Crown ownership prevents private ownership of that country's forestlands and undermines the company's ability to apply Weyerhaeuser Forestry practices for the lands on which it operates. Given the opportunity to follow its best management practices, the company could derive substantially higher yields from less land in Canada. However, lack of land ownership makes it risky to establish high-yield plantations on the Crown concessions granted to the company, since there is no guarantee that the company will be able to maintain long-term control over these sites.

As a result, the company continues to log native forests on concessions covering more than 18 million acres granted to it by the Provincial governments of British Columbia, Alberta, and Saskatchewan. And while the company avoids the controversial "ancient forests" of western British Columbia, the inconsistency between its U.S. and Canadian forest operations is evident. As environmental pressure mounts on the Canadian and Provincial governments to alter their current concession systems, which favor job creation through harvesting at the expense of long-term management, Weyerhaeuser's position will be improved. If sustainable forestry is to be a centerpiece of the company's competitive strategy, it is in Weyerhaeuser's self-interest to push for policy changes consistent with the principles of Weyerhaeuser Forestry in Canada and gradually move away from its practice of harvesting native forest in favor of long-term plantation forestry.

What should Weyerhaeuser's posture be with regard to the Russian Far East?

In Weyerhaeuser's initial venture into Siberia it became a partner with a local company in an effort to gain a timber concession. Following an experiment involving the planting of 1.2 million seedlings, the company withdrew from the venture citing

infrastructure and corruption problems that made the practice of Weyerhaeuser Forestry impossible.

Some in the environmental community saw Weyerhaeuser's entry into Russia as a preliminary step intended only to gain preferred access to Siberia's vast old-growth softwood forests. As in Canada, if Weyerhaeuser is serious about centering its strategy on sustainable forestry, it is in the company's self-interest to collaborate with environmental groups, NGOs, and governments to preserve as much of the Russian Far East old growth as possible, while working to identify and secure access to the few highest potential sites for Weyerhaeuser Forestry.

Does the Weyerhaeuser Forestry model transfer easily as the company looks to expand its operations internationally?

As Weyerhaeuser currently produces exclusively in the United States and Canada, it has learned a great deal about producing timber under conditions of intensive regulation and environmental scrutiny.[10] This should serve the company well as it seeks to expand its forest and production base offshore, where the growth markets of the future reside. Both Weyerhaeuser's forestry competence and its skills in engaging stakeholders should be transferable internationally. However, extensions and adjustments for each new location will be required. Expansion into less-developed and under-regulated countries will require the company to expand its stakeholder involvement model to address more difficult tests than it has dealt with so far, including issues of indigenous peoples, poverty, rural migration, energy generation, and a host of others related to sustainable development.

Weyerhaeuser will have to apply its core capabilities in selective breeding, tree improvement and planta-

---

[10] Weyerhaeuser has recently reached agreement to purchase timberland and related assets in New Zealand.

tion management to species other than Douglas fir and loblolly pine. This should be possible to achieve so long as the company sticks with its focus on softwoods. Particularly attractive are overseas lands that have already been converted to plantations or marginal farming/grazing uses with the soil potential for the company to implement Weyerhaeuser Forestry practices. Opportunities in the temperate tropics of Argentina, Chile, Brazil, Australia, and New Zealand, where it is possible to grow softwoods on a fast-rotation basis, are especially desirable. However, should Weyerhaeuser decide to move into fast-rotation hardwoods, a great deal of new learning and perhaps finding a partner would be required to counter the already strong positions held by substantial competitors such as Aracruz, Mondi, and others.

The company appears to have ruled out investment opportunities that preclude full use of Weyerhaeuser Forestry. For example, they have steered clear of tropical rainforests due in part to the controversial nature of these ecosystems, but also because of their comparative lack of competence with tropical hardwood forests and soils. Similarly, the company has avoided entering politically unstable regions where long-term land ownership and stewardship become high-risk propositions. As a case in point, political instability caused the company to back away from its venture in the Russian Far East.

Does Weyerhaeuser's high-yield plantation forestry constitute a model of sustainable forestry?

Perhaps the biggest challenge facing the company in the coming years will be that of convincingly demonstrating that Weyerhaeuser Forestry constitutes a model of sustainable forestry. Given the lack of operational definitions concerning "sustainable development" and "sustainable forestry" it will be important that the company proactively make the case that its plantation-based model of high-yield,

environmentally-sensitive forest management constitutes a viable path toward sustainability.

Recognizing that many believe that plantation-based models are unsustainable by definition (since they imply alteration of the composition and character of native forests), Weyerhaeuser will need to engage a broad range of interests if it is to gain general acceptance for its methods. This will probably require the development of a strong argument as to how such intensive, plantation-based models fit into the big picture regarding the state of the world's native forests, biodiversity, and the need for sustainable development, especially in developing and emerging economies. A vision of how plantations, managed native forests, and preserved old-growth forests might sustainably coexist is probably essential if such intensive management practices are to gain broad acceptance.

Woodmaster Company Ltd.: The Trials of Sustainable Tropical Forestry

CASE STUDY

PREPARED BY:
ROBERT DAY
NIGEL SIZER
PETER ZOLLINGER

A Case Study from "The Business of Sustainable Forestry"
A Project of The Sustainable Forestry Working Group

The Sustainable Forestry Working Group

Individuals from the following institutions participated in the preparation of this report.

Environmental Advantage, Inc.

Forest Stewardship Council

The John D. and Catherine T. MacArthur Foundation

Management Institute for Environment and Business

Mater Engineering, Ltd.

Oregon State University
Colleges of Business and Forestry

Pennsylvania State University
School of Forest Resources

University of California at Berkeley
College of Natural Resources

University of Michigan
Corporate Environmental Management Program

Weyerhaeuser Company

The World Bank
Environment Department

World Resources Institute

Woodmaster Company Ltd.: The Trials of Sustainable Tropical Forestry

CASE STUDY

PREPARED BY:
ROBERT DAY
NIGEL SIZER
PETER ZOLLINGER

A Case Study from "The Business of Sustainable Forestry"
A Project of The Sustainable Forestry Working Group

Contents

This fictional case was written by Peter Zollinger, Nigel Sizer and Robert Day of the World Resources Institute based upon the experiences of actual companies. The case is intended to serve as a basis for discussion rather than to illustrate either effective or ineffective handling of an administrative situation.

Woodmaster Company Ltd.: The Trials of Sustainable Tropical Forestry

Executive Summary

Woodmaster Company Ltd. (WoodCo) is a plywood and veneer producer in a lesser-developed tropical country. Wood harvested from the company's 0.9 million hectares of natural forest in several concessions is converted to plywood and veneer at the company's manufacturing plant within the country. The operation has an annual capacity of 96,000 cubic meters of plywood, which is sold to customers in the local region, the United States, Asia, and Europe.

For the major investor, WoodCo represents an important strategic entry into the region. WoodCo, however, has presented numerous challenges to its investors. Financial performance has lagged expectations; the concession has not been as productive as expected; the plywood and veneer mill has not operated at capacity; costs have been high; work force problems have hampered productivity; and WoodCo has generated significant controversy in the environmental community. A turnaround management team was put in place five years after the inception of the venture. Since then it has vigorously engaged the business challenges and worked to demonstrate credible leadership in environmental and social stewardship.

WoodCo's management believes there are significant gaps between perception and reality surrounding their operation:

• WoodCo has found it difficult to achieve normal profits, and has invested millions in the development of the local economy. Meanwhile, many critics believe the company has an excessively favorable concession arrangement with the government that does not accrue fair rents to the country.

• WoodCo has found the local work force demanding and difficult to train and manage. But critics have alleged that the company has not been fair to workers.

• WoodCo is under pressure from stakeholders to implement sustainable forest management (SFM) practices, and have third party certification bodies evaluate its operations. The buyers of plywood, however, will not pay a premium for sustainably-produced product.

• WoodCo is striving to drive costs down, so that it can break even on its investment, while SFM appears to increase costs.

• WoodCo has found that information about its concessions and operating conditions in the country is vital for effective SFM, yet difficult to obtain.

Nowhere are forestry practices more controversial than in the tropics. Sustainable forestry is increasingly touted as a potential solution to the overharvesting of these valuable forests. While WoodCo's experience demonstrates the potential of sustainability as a business strategy for tropical forestry operations, it also underscores the difficulties of successfully implementing sustainable practices.

Investors' Strategy: Establishing a Regional Beachhead

WoodCo, as mentioned earlier, is a forest products manufacturer incorporated in a lesser-developed tropical country. The company is a wholly-owned venture of a foreign investor. Established in the early 1990s, WoodCo produces plywood and a limited amount of veneer primarily for export. The company holds forty-year rights to a total of 0.9 million hectares in several concessions of natural forest. WoodCo harvests its concessions to supply its mill within the country

The parent company is well suited for this venture. This company is a forest products company that has been involved in logging and forest products manufacturing for many years in its home country, and is now looking to expand overseas. The company brings to WoodCo extensive experience in tropical forest harvesting and plywood manufacturing.

Strategically, the WoodCo venture is important for the investor. It is looking to expand abroad, and so is trying to establish toeholds in the world's remaining tropical forest reserves. The company is therefore committed to responsible business practices in the WoodCo venture. The success of the WoodCo venture is not only important for this company's credibility, it may also be vital for the success of any future company ventures elsewhere.

As a consequence of a management change five years after the inception of the venture, two managers from the parent company have been called in as "trouble-shooters" to deal with WoodCo's initially poor performance.

SUSTAINABILITY DEBATE AND WOODCO: TIPTOEING THROUGH A MINEFIELD

As a harvesting and manufacturing operation in a tropical rain forest region, almost every aspect of the WoodCo operation has environmental and social implications. The investor, as a major tropical timber producer, has extensive experience with these issues. In its home country sustainable forestry is emphasized, as a result of international pressure, quality management systems, and the precipitous decline of the domestic forest base. Sustainable forestry was applied to WoodCo from the outset. In the absence of a universally accepted definition of SFM, the WoodCo venture operates on sustained yield principles as well as responsible management practices in dealings with the local communities.

Management has declared its intention to remain in operation for at least forty years. Of course, economic success is the venture's paramount goal.

WoodCo Country

This region holds one of the world's largest remaining reserves of virgin tropical forests. Political confusion in the WoodCo country since independence has left a legacy of a weak economy, with little domestic manufacturing. It has also led to a serious "brain drain" as many of the best educated and most affluent citizens left the country permanently. Those circumstances have left the country with a small, poorly skilled workforce, and relatively underdeveloped resources and infrastructure. In recent years the country has piled up a massive national debt, and the government is looking for ways to exploit its vast forest resources to bring in much needed cash. As of yet, logging and mining are the overwhelming industrial uses of the forests.

Most of the country's population lives near or around the capital. The remainder tends to cluster near the coast because the interior is undeveloped. A number of indigenous communities, however, live deep in the country's forest. For those indigenous communities who have been given official status, the local law mandates that concessions for mining and logging cannot overlap with their lands. Nevertheless, not all indigenous communities have been identified and granted this status, and the actual borders of the officially-recognized indigenous land often conflict with the local communities' actual area. Many of the communities also log within their own land.

This region's forests represent a substantial reserve of biodiversity. Over 8,000 species of plants are found in the WoodCo country alone, of which about half are found nowhere else. The current local government is interested in promoting sustainable development, but it must also use resources to bring in cash, so it has begun to allow logging concessions in extensive areas of the forest.

Forest Inventories

The investor originally undertook a feasibility study, which included some investigation of the forest's characteristics, to determine whether to pursue the WoodCo project. This study relied upon previous studies describing the country's forests.

When the venture began, management hired an independent consulting firm to conduct a more detailed inventory of the forest and to evaluate the sustainability of WoodCo's practices and operations in the forest. The company hoped that this independent analysis would help WoodCo communicate its commitment to responsible practices.

The consulting firm's studies yielded some surprises. Namely, that the original forest inventories were overly optimistic. The consulting firm estimated that the previous studies had overestimated timber volume in the concessions by 30%. They found that the forests of that part of the country tend to be relatively short in stature and that most of the more valuable species appear relatively infrequently compared to other regions of the country. The investors drew up their business plans based in part upon these erroneous estimates of the forest's quality. As a result, harvest volumes have been lower than expected, which has hindered WoodCo's performance.

Throughout its planning, WoodCo's management assumed an annual growth rate in the concession's forest of 1 cubic meter per hectare. They adopted this conservative assumption, based on their experiences elsewhere, because it would take so long to test any estimate reliably in the field. Plans called for a twenty-year logging rotation, which led management to conclude that it could harvest as much as 20 cubic meters per hectare through selective cutting. Management figured that this rate of harvest could supply the plywood mill with enough wood to operate at full capacity, while remaining sustainable. WoodCo's progress toward financial and socioecological sustainability, however, has been hampered by a variety of issues in its forestry operations.

Harvesting Issues

The company's logging practices are designed to minimize damage to the forest. Eight logging teams operate in the concession. Each is led by a foreman, who is responsible for ensuring that the team only harvests in its assigned plots and that residual damage to the forest from felling and extracting is minimized. Plot inspectors are responsible for determining which trees will be harvested and for planning proper consolidated skidding trails, which minimize damage to the forest. Only recently, however, were the plot inspectors reorganized into an inspection department, so they could cover every team. Fellers cut down the trees using chainsaws and determine the direction in which the tree will fall. Tractor drivers then maneuver the fallen log so that it is easier for the skidder drivers to grab the log and pull it to the nearest logging road.

In its most recent report the consulting firm points out that actual team practices do not fully achieve the company's stated practice of preventing residual damage to the forest. Workers often see the guidelines as hampering productivity (they are given pay incentives based upon productivity measures), may not know the reasons why the harvesting guidelines are in place, and may not even know what the guidelines are. While training was done primarily on the job during the first years of operation, efforts are underway to establish a more formal system. The upshot, however, is that the consulting firm estimates that WoodCo's felling operations damage 8 cubic meters per hectare, including poles and saplings.

However, WoodCo is extracting only 12 cubic meters per hectare, well below its initial estimates. The density of trees is lower than expected, and useable timber is even less available. The new management team is undertaking a detailed inventory to determine the location, density, and volume of commercial and potentially commercial species. Preliminary results, however, do not look promising. The southern concessions have not yet been logged at all by WoodCo's operations, and timber availability looks to be even worse in that area. Furthermore, 30% of the concessions have been excluded from operation because it is indigenous land, and another 10% will not be worth operating because the timber density is so low. The consulting firm concluded that the operation is damaging (8m3/ha) and extracting (12m3/ha) at a "sustainable" rate (20m3/ha, based upon the 1m3/yr/ha regrowth estimate). But these harvest rates cannot fully supply the plywood mill.

Road construction has also been an issue. Road building is a major concern in any forestry operation. Cutting paths through a virgin forest facilitates erosion, creates barriers to animal migration, and in the process, fells a large number of trees. In areas of poor enforcement and underdevelopment, roads facilitate even worse deforestation by enabling people to colonize previously impenetrable forest. WoodCo's operations require cutting a forty meter wide path through the forest for main roads and a twenty meter wide path for secondary roads. Although the actual road is not necessarily this wide, drainage ditches are built along the sides of the roads, and space must be allowed for logging equipment to travel between skid trails along the side of the road to prevent erosion of the road surface.

The consulting firm reports that WoodCo built 276 kilometers of roads in the concession during the first three years of operation, and some of those roads may have been built to places where there was nothing to harvest. The consulting firm also reports that many roads were built through areas with few useable trees to reach pockets of timber deeper in the forest. Recently, WoodCo was given authority to control traffic on the roads in the concessions, but the company still finds it difficult to deny routine access to outsiders.

Productivity has been hampered in other phases of the felling operation, as well. The road surface in the concessions is often clay, which during the rainy season can fail, preventing any passage until it

can be repaired during the dry season. This has been a major obstacle in making the operation more productive. Equipment is also often left running or is operated incorrectly, leading to high maintenance and repair-related costs.

Operation of the loading facility and the barges has also proved costlier than expected. At Log Base One, in one of the concessions, logs are brought in from the forest, separated by grade, and then loaded onto barges for transportation to the plywood mill. Each barge has a capacity of 800 cubic meters. The initial plan called for three barges, each with a tug, making four trips per month to the mill and back. So far, though, the barges make only two trips per month. However, three additional barges and two tugs were subsequently bought.

The productivity of the forestry workforce is a major issue. One foreign road-clearing crew is still being brought in at great expense because it is 280% more productive (16.8 km/mo vs. 6 km/mo) than local crews. In addition, three of the logging teams are made up of expatriates. Overall in the harvesting operation, costs are higher than expected even though productivity is lower than had been planned upon.

Community and Indigenous Issues

Complex and intractable social problems have also hindered the productivity of WoodCo's operations, including the lack of local government authority, a high level of alcohol use within the community, and the low skill level of the local workforce. WoodCo has also been forced to grapple with employee misunderstanding and dissatisfaction generated by its efforts to be what it considers a responsible company.

The logging operation is based out of the remote community of Log Base One. To make it habitable for employees, WoodCo had to revive the entire community. The company has so far provided the community with all of its basic services, including electricity, water, medical care, access to television, food supply, and limited air transportation. It has built a number of facilities including an office building, a guesthouse and canteen, staff houses, a hospital, a powerhouse, a repair shop for heavy duty vehicles, a log pond, and a loading port. While the overall investment in the camp's infrastructure is relatively modest, the amount of management time spent "running" this little community is more than WoodCo management expected it to be.

Three officially recognized indigenous communities, and at least one unofficial one, exist within the concession. The population in the officially recognized communities is around 560. Under the law, their land is excluded from WoodCo's concession rights. But the unrecognized lands are not officially excluded. Nevertheless, WoodCo has agreed that any questionable lands, whether or not they are officially recognized by the government, are to be set aside. WoodCo has hired an indigenous liaison officer to help address these and other cultural and economic issues. It is clear that WoodCo's intention is to be a positive factor in the local economy and culture. Nevertheless, due to the value-laden nature of the questions and issues faced, it would be inappropriate for this case to draw any conclusions regarding the overall effects of WoodCo's presence upon the indigenous communities in the region.

As a foreign-owned company, WoodCo will always be vulnerable to criticism, especially in the absence of clear and enforced government rules. WoodCo has tried to address the indigenous and local community issues as openly as possible. But in this undeveloped community, routine corporate operations can provoke misunderstandings and unintended consequences. The following two examples illustrate the types of issues that have confronted WoodCo.

- **Cash Salaries:** WoodCo pays its staff in cash. The local economy was nonexistent before WoodCo moved into the area, so at first there was no easy way to spend cash in Log Base One except on alcohol. Excess drinking led to low

productivity and morale among the local workforce, almost all of whom work in WoodCo's logging operation. The company decided to try to provide consumer goods, such as television sets and household goods, for the workers to purchase. The company then set up a satellite dish in the village so that the community could get a television signal. The workers began watching television all night long, and soon accidents and other losses in productivity rose. When WoodCo turned off the satellite antenna at eleven every night to impose a curfew and encourage workers to sleep, the workers went on strike for two weeks.

- **Taxes and Raises:** The company is legally required to withhold income tax from workers' pay. To the workers, however, unfamiliar with concepts of taxation and withholding, the company was simply keeping a substantial part of their paycheck. Since income taxes can be as high as 26%, the cash handed over to employees at payday was significantly less than they expected to see. After much dissention and unrest, WoodCo agreed to raise wages, so that workers took home the amount they had expected to see. But the pay issue continues to be a problem. In the past, the country has been plagued by runaway inflation, with workers' pay increasing proportionately. Recently, inflation abated, but workers are still demanding the kind of pay raises they received when inflation was high. WoodCo is now trying to educate workers to understand that low inflation means lower salary increases this year.

WoodCo management considers its efforts to support the local community as part of its commitment to a responsible operation. The company perceives that it undertakes costly investments and provides services which, in other circumstances, the government should provide or reimburse. Under the circumstances, WoodCo is being forced to consider the long-term effects of its actions. On the one hand, WoodCo's support for the community should pay off in increased workforce productivity long term—many of the activities are necessary for the logging camp to even function. On the other hand, each time the company agrees to another measure requested by the community, expectations rise for greater "gifts" in the future.

Plywood and Veneer Production

Although the new management team now appears to be turning things around, WoodCo's plywood operations initially suffered from inefficiency and low quality, which in turn affected both the financial and eco-social sustainability of the venture.

WoodCo's plywood and veneer mill operates in Ply Base Two, near a major city on the coast. For plywood, logs are first peeled into thin sheets. These sheets are trimmed and stacked together, then glued and pressed and dried. The resulting product is again trimmed and rated for quality. WoodCo is self-sufficient in providing its own fuel, in the form of wastewood, for drying the plywood.

WoodCo's own forests do not produce enough logs to keep the mill operating at capacity. Maximum capacity, so far, is 8,000 cubic meters of plywood a month, plus a limited amount of veneer. The volume of logs needed to achieve that is 18,000 to 20,000 cubic meters each month. As of last year, the Log Base One operation was supplying the plywood and veneer facility with an average of only 11,600 cubic meters each month. Since WoodCo cannot supply enough logs from its own concessions, it was until recently buying some 2,400 to 4,800 cubic meters each month from third parties. Neither the company nor the consulting firm monitors the responsibility of these suppliers. To whatever extent WoodCo manages its own concessions sustainably, it may be purchasing from less responsible suppliers. Nor has the company been able to build up a large enough log inventory to serve as an effective buffer against supply bottlenecks.

Initially, the plywood and veneer mill suffered from a low recovery rate (the amount of log volume actually in the finished product). A mill's recovery rate is important: a low rate means significantly higher operating costs, and requires more logs. The plywood manufacturing industry standard is around 50-55%, while some Indonesian plywood manufacturers, benefiting to an extent from large logs, have reported 65% recovery rates. At first, WoodCo's mill achieved only a 40-45% recovery rate. Through management's continuous efforts, recently this rate reached close to 50%.

Eventually, plans call for expanding the mill from its present four lines to ten lines. Preliminary results from the new inventory suggest, however, that WoodCo's forests could only support five lines' worth of production under current practices. Development of new species of logs is, therefore, a critical issue. The company is currently assessing the suitability of using various other species, and considering the addition of a steam chamber to "cook" logs, which would make it feasible to peel some currently unusable species.

Poor workforce productivity is also an issue. Two shifts of local workers (800 total) have been trained on the job for low-skilled labor, and thirty-five expatriate technicians and supervisors make up the rest of the workforce. Morale is low among the expatriate workforce, who consider this a hardship assignment. Furthermore, local entrepreneurs confirm that to educate new local staff and maintain performance and quality requires the permanent presence of a dedicated supervisor. To find, train, and monitor local staff is therefore a permanent challenge for WoodCo. Quality of the products has suffered at times due to low-quality logs combined with poor labor skills. Management attention to this issue, appears to have paid off: the company is now producing a normal percentage of export quality plywood.

WoodCo's Business Strategy: "Location, Location, Location" and Value-Added

WoodCo operates with a three-part business strategy:

1) Secure a huge, long-term, timber base for the production of value-added products. Locate in a country with favorable investment conditions.

2) Use low labor cost to produce labor-intensive value-added products.

3) Use the comparative advantage of relatively short transportation distances to lucrative markets to maximize margins.

Furthermore, from the beginning WoodCo emphasized communicating its commitment to SFM as defined by responsibly-sustained yield. If the forest had provided the expected volumes of the species targeted for plywood and veneer production throughout the entire concession, there would have been no apparent conflict between the needs of the company and this commitment. Now that it is apparent that the wood supply is insufficient, at first glance it might appear that the smart short-term financial decision for management would be to back away from that commitment. So far, however, WoodCo's management is determined to continue their sustainability efforts because they realize that sustainability will pay off in the long run, both within WoodCo's operations and for future expansion efforts.

Nevertheless, the company has already reduced the minimum tree width that it was willing to harvest. Management had wanted to leave some relatively mature seed trees in the forest to aid regeneration. They now must harvest some of these trees to increase their supply of logs. This type of incremental revision of logging guidelines may continue to be necessary.

AN UNFRIENDLY MARKET FOR PLYWOOD PRODUCERS

Product

Plywood is a structural panel product consisting of several thin peelings of wood, called "veneers" or "plies," glued together so that the wood grain of each ply is perpendicular to that of the adjacent plies. This allows for the maximum resistance to warping or splitting. Structural panels, which also include oriented strandboard (OSB) and medium density fiberboard (MDF), substitute for solid wood in a wide variety of applications from construction to furniture manufacturing, and other uses where weight and costs must be kept low. The best quality veneers can be sold separately for use on structural panels elsewhere or other applications.

Plywood can be made out of both hard- and softwood with no differences in performance. The number of plies varies, but is always an odd number for symmetry around a center ply. Ply quality is generally judged on an A, B, C scale according to its appearance (A being the highest-quality, with the best appearance—no knotholes or splits), although the quality scales vary some according to region and custom. The quality of the plywood itself is judged by the quality of the outside, or "face," plies. For instance, the highest quality plywood would be A-C-A, with top-quality face plies around an indeterminate number of inner plies.

High quality plywood is often sanded and is used where the panel will be seen, including such uses as furniture, cabinet doors, partitions, or shelving. Lower quality plywood is used for hidden, structural applications like subfloor and roof sheathing. Of course, use also varies with region and custom.

Tropical Plywood Market

For tropical plywood, the market demands a low price and secure delivery. Products are categorized by species, geographical origin, thickness, and quality grades, but within these categories product differentiation is not important. This means that WoodCo's competitive position in this market depends upon ensuring the quality of its plywood, keeping costs low, and maintaining a consistent supply.

The structural panel market overall is expected to grow dramatically as the emerging economies in Asia and Latin America continue to expand, requiring large supplies of inexpensive construction materials. Demand for structural panel products is also expected to rise in the United States (up 33% by 2010) and Europe as those economies continue to grow, albeit more slowly. Plywood substitute products, however, are expected to make major inroads into these markets. OSB and MDF are made from low-cost waste wood and have similar performance qualities, which make them more attractive than plywood. In the United States, in fact, plywood demand is projected to decline over the next fifteen years, even as overall structural panel demand increases. OSB mills are even starting to appear outside of the Organisation for Economic Co-Operation and Development (OECD) countries. Nevertheless, plywood's competitive position is expected to remain strong in the fastest growing economies.

Hardwood plywood is produced primarily in the tropics from wood harvested in rainforests. Historically, Indonesian producers have dominated the hardwood plywood markets. In recent years, however, Malaysian and Brazilian producers have gained significant marketshare. These established production regions command premiums that make it difficult for other regions to gain footholds in the market.

Advantages and Disadvantages of WoodCo's Region

The nature of the customer relationships, however, is an idiosyncrasy of the hardwood plywood market. Although plywood is typically a commodity with prices determined by global market trends, buyers tend to be very large and to develop long-term relationships with their suppliers. Often buyers will go to the same producers year after year. Therefore it is vital for plywood producers to establish—and even differentiate—themselves to attract the most reliable and stable customers.

The species of tree used is an important factor in the plywood market. Each species has its own unique strength, appearance and performance characteristics. It is difficult to introduce new species into the market because they must first be rigorously tested for each of these characteristics, and then buyers must be convinced of the species' quality. WoodCo is at a disadvantage because they are not working in an established area and they must use local species that are not favored by the market.

On the other hand, geographically, the country is much closer to Europe than some other hardwood plywood supplying countries. For sales into Europe, for example, WoodCo saves US$20/cubic meter in freight cost over Asian suppliers (US$44 total vs. US$64), although these numbers vary with the scale of the shipment. WoodCo, in its marketing literature, emphasizes reliability of shipment, the quality of its species, and the short shipping times required to these markets.

So far, WoodCo has successfully attracted customers from around the world, as shown in Table 1. Because importers of plywood tend to be large, WoodCo's total number of customers is only twelve.

WoodCo produces plywood of various thickness and quality grades, and a limited amount of veneer. The company uses three primary species with colors varying from red to white. The average selling price for the plywood in a recent downturn was between US$280 and US$320 per cubic meter, but this tends to vary widely.

X-Co's Customers by Country or Region

Country/Region	1996 % of sales	Target % of sales
U.S.	25	40
Latin America	10	10
Asia	15	20
Europe and others	50	30

Table 1

So far, markets have absorbed whatever WoodCo has produced without WoodCo making extraordinary marketing efforts. The market, however, is not yet paying any premium for sustainably-produced plywood. WoodCo's management is aware of this. Nevertheless, the new management is exploring options including green marketing and possibly a certification effort to establish that its products are sustainably produced. Even then, WoodCo would face a problem. If consumers in the United States, Europe, and possibly elsewhere begin demanding certified wood products, they would probably still be leery of WoodCo because of the stigma associated with tropical logging operations.

ELUSIVE PROFITS, UNCLEAR SUSTAINABILITY COSTS

Investment

Overall, the company planned to invest US$123 million through the first eight years of the project, the period for which it had been granted a tax holiday from the local government. This would include an expansion of the plywood mill to ten lines by 2000. The plywood mill is currently running four production lines, giving it a capacity of 96,000 cubic meters of plywood production per year. In addition, a limited amount of veneer is produced every year.

Lower than expected volumes have made it unclear whether major further investments are warranted. Table 2 describes the difference between the initial expectations and what has actually occurred over the first five years of the project's operation. Initial investment in WoodCo was roughly US$43 million (US$14 million equity and US$29 million debt). By the end of the period total investment had reached US$80 million (US$38 million equity and US$36 million long term debt, less accrued operating loss-

es and currency exchanges). Short-term debt appears to consist primarily of payables.

Although the company is interested in finding a local partner for the operation, WoodCo does not need additional investment from outside interests. The shareholder is financially strong and already has established bank relationships. The relatively low long-term debt to total capitalization ratio is indicative of the perceived riskiness of the project; WoodCo management report that they could expect better leverage for similar projects back in their home countries, with long-term debt to total capitalization ratios near .70.

Production and Pricing

As mentioned above, production has lagged behind expectations. Table 3 provides log and plywood production for the most recent three-year period. Current plywood production capacity is approximately 96,000 cubic meters, which means that in the last year total plywood production was only about 80% of capacity. As Table 3 indicates, the trend is positive, and recent news suggests that the company continues to improve its production to capacity ratio.

Project Expectations vs. Reality

Initial Expectations		Reality	
Payback period (initial investment) = 5 years		Payback period of total investment = 10-12 years	
Scale of plywood production: 2-yr: 4 lines, 3-yr: 6 lines, 9-yr: 10 lines		Scale of plywood production: mid-1990's: 4 lines (max. 8,000 m3/month)	
Total investment (8 yrs): US$123 million		Actual investment (5 yrs): US$ 80 million	
Phase I:		**Phase I:**	
Plywood factory	23.6		32.0
Barges & Tugs	3.0		8.0
Logging	15.0		22.0
Administration	3.4		18.0
Total Phase I:	**45.0 million**	**Total:**	**80.0 million US$**

Table 2

Production, 3-yr. Period

	Year x	Year x+1	Year x+2
Log extraction (m³)	30,700	133,600	154,000
Log exports (m³)	4,000	17,800	10,000
Plywood production (m³)	7,200	42,800	77,000
Plywood exports (m³)	4,700	37,200	72,900

Table 3

Estimated Revenues, 3-yr Period

	Year x	Year x+1	Year x+2
Veneer exports (m³)	1,000	4,400	2,500
Plywood exports (m³)	4,700	37,200	72,900
Value of veneer exports (US$)	396,800	1,759,200	835,680
Value of plywood exports (US$)	42,012,400	12,948,640	24,552,160
Total value of exports (US$)	2,409,200	14,707,840	25,387,840

Table 4

As mentioned above, several market factors mean that WoodCo realizes lower prices than competitors in the major tropical plywood producing countries of Brazil, Malaysia, and Indonesia. The cyclical plywood market has also been in a downturn recently, further depressing prices.

Costs

The company does not release specific cost data, but cost control is a priority for the new management team. The Year x+2 cost per cubic meter of logs delivered to the mill, for instance, was not only substantially higher than WoodCo's own budget estimate, but also was much higher than the cost of logs purchased from third parties.

WoodCo's high costs are the result, in part, of high fixed costs. Almost all of WoodCo's costs are fixed, which means that the unit cost can only be brought down by increasing production or lowering overhead. The strong presence of unions in the local workforce means that most employees receive fortnightly salaries with no or only minor variable elements based upon performance. Capital costs are also fixed. Fuel is theoretically variable, but becomes somewhat fixed if machines tend to run for a certain number of hours without regard to actual production—making empty, but scheduled barge trips, for instance. Under these circumstances enhancing production would significantly lower costs.

Some of the causes for the productivity difficulties are out of WoodCo's control, such as bad weather that prevents log transport, or the low availability of the desirable tree species. Nevertheless, variable costs are also higher than the optimum. Repairs and maintenance, as well as general overhead in both the logging and plywood operations are of special concern to management. At any given time up to 40% of the equipment may be idle, compared to rates as low as 5% in foreign operations. The new management team has concentrated on bringing these costs down, reportedly with some success.

Overall Financial Performance

So far, indications are that WoodCo is at about the break-even point financially, with revenues roughly equivalent to costs. Table 4 provides a chart of the revenues from the same three-year period as used above.

WoodCo management report an accrued operating loss of around US$12 million, and now estimate a payback period of ten to twelve years on the total investment. Nevertheless, reports indicate that the financial performance is improving, bringing the company closer to financial sustainability.

Conclusion—More Than a Simple Business Management Challenge

It is possible to enhance WoodCo's business performance, and the changes needed may go hand in hand with a strong management commitment to sustainability. The company has not yet been able to extract the sustainable yield rates that it had targeted from the concessions. The harvest, however, could be increased by using lesser-known species. This practice would not necessarily be better for the forest itself. But it would require fewer purchases from third parties and would give WoodCo greater control over the overall environmental impact of the operation. Management could minimize waste throughout the entire process, which would not only reduce the impact on the environment, but also lower costs. In this sense, WoodCo's

sustainable forestry management efforts are analogous to quality management requirements. Sustainable forestry management requirements are likely to bring about improved management systems, higher quality, cost savings, productivity gains and a better market position, thereby allowing the company to adhere to its commitment of sustained yield production.

At this point, WoodCo's management must decide how best to deal with its current difficulties. While the company makes its strategic decision, the local government needs to make every effort to strengthen its institutions, monitor closely the impacts of WoodCo's operation on forests and people, and draw the right conclusions for future forest use. Important lessons for the future of sustainable tropical forestry, both for this country and elsewhere, may be revealed by WoodCo's actions in the near future, as the company changes its operations or possibly even abandons the venture. One lesson is already clear: tropical forest management is difficult and complex, and not an assured success.

Copies of individual cases studies, or a bound set of all the case studies listed below, are available for purchase from the distributor, Island Press, which in 1998 plans to publish a book-length study based on this material entitled *The Business of Sustainable Forestry.*

For purchasing information contact:

Island Press
Phone 800.828.1302
Fax 707.983.6414

The working group has an Internet web-site at http://www.sustainforests.org

The Cases

Overall Market Analyses:

OVERVIEW OF SUSTAINABLE FORESTRY
A conceptual and illustrative framework for sustainable forestry.

SUSTAINABLE FORESTRY WITHIN AN INDUSTRY CONTEXT
Defines the relationship between sustainable forestry and the entire forestry industry.

MARKETING PRODUCTS FROM SUSTAINABLE FORESTS: AN EMERGING OPPORTUNITY
The current demand for sustainable forest products and the likely demand over the next two to five years.

A REVIEW OF EMERGING TECHNOLOGIES
New technologies which influence investment decisions in sustainable forest management.

Business Case Studies on Companies or Landowners:

**ARACRUZ CELULOSE S.A. AND
  RIOCELL S.A., BRAZIL**
COLLINS PINE COMPANY, U.S.
COLONIAL CRAFT, U.S.
**J SAINSBURY PLC AND
  THE HOME DEPOT, U.K./U.S.**
MENOMINEE TRIBAL ENTERPRISES, U.S.
PARSONS PINE PRODUCTS, U.S.
PORTICO S.A., COSTA RICA
PRECIOUS WOODS, LTD., BRAZIL
STORA, SWEDEN
VERNON FORESTRY, B.C., CANADA
WEYERHAEUSER COMPANY, U.S.

Case Studies on Small Private Landowners:

(7 representative U.S. properties)

BRENT PROPERTY
CARY PROPERTY
FREDRICK PROPERTY
FREEMAN PROPERTY
LYONS PROPERTY
TRAPPIST ABBEY
VAN NATTA TREE FARM